Naturalizing Logico-Mathematical Knowledge

Sorin Bangu has assembled an important collection of papers on the project of naturalizing mathematics and logic. It will be a must read both for those who favour that project and also for those who disfavour it.
—Jeff Buechner Rutgers University—Newark, USA

The book is meant to be a part of the larger contemporary philosophical project of articulating a naturalized epistemology for mathematics and logic. It addresses the key question that motivates most of the work in this field: what is philosophically relevant about the nature of logico-mathematical knowledge in the recent research in psychology and cognitive science? The question about this distinctive kind of knowledge is rooted in Plato's dialogues, and virtually all major philosophers have expressed interest in it. The essays in this collection tackle this important philosophical query from the perspective of the modern sciences of cognition—namely, cognitive psychology and neuroscience. *Naturalizing Logico-Mathematical Knowledge* contributes to consolidating a new emerging direction in the philosophy of mathematics, which, while keeping the traditional concerns of this sub-discipline in sight, aims to engage with them in a scientifically informed manner. A subsequent aim is to signal the philosophers' willingness to enter into a fruitful dialogue with the community of cognitive scientists and psychologists by examining their methods and interpretive strategies.

Sorin Bangu is Professor in the Department of Philosophy, University of Bergen, Norway. He works in philosophy of mathematics and philosophy of physics, having long-standing interests in epistemology and the history of analytic philosophy. He is the author of *The Applicability of Mathematics: Indispensability and Ontology* (2012).

Routledge Studies in the Philosophy of Mathematics
and Physics
Edited by Elaine Landry
University of California, Davis, USA
and
Dean Rickles
University of Sydney, Australia

Naturalizing Logico-Mathematical Knowledge

Approaches From Philosophy,
Psychology and Cognitive Science

Edited by Sorin Bangu

Routledge
Taylor & Francis Group
New York London

First published 2018
by Routledge
52 Vanderbilt Avenue, New York, NY 10017

and by Routledge
2 Park Square, Milton Park, Abingdon, Oxon OX14 4RN

First issued in paperback 2020

Routledge is an imprint of the Taylor & Francis Group, an informa business

Library of Congress Cataloging-in-Publication Data
A catalog record for this book has been requested

ISBN 13: 978-0-367-66711-5 (pbk)
ISBN 13: 978-1-138-24410-8 (hbk)

Typeset in Sabon
by Apex CoVantage, LLC

Contents

Acknowledgments

Questions related to the cognitive basis of logico-mathematical knowledge have been with me for quite a while. I first felt their major importance and difficulty while taking a graduate seminar on Frege with Professor Alasdair Urquhart at the University of Toronto in 2004—an amazing intellectual experience for which I am forever grateful. The present volume originates in a conference I organized in November 2015 at the University of Bergen, in the Philosophy Department. I would like to offer thanks to Professor Reidar Lie, the Head of the Department, for supporting the idea; to Professors Andrea Bender and Sieghard Beller, from the Department of Psychosocial Science at the University of Bergen, for getting involved; and to all the participants for an excellent meeting. I am also thankful to those directly contributing to this project: the authors themselves; the two editors of the series for Routledge, Professors Dean Rickles and Elaine Landry; the publishers, Andrew Weckenmann and Alexandra Simmons; and the two anonymous reviewers. I am indebted to my colleagues in the Philosophy Department for many illuminating conversations over the years: Richard Sørli, Kevin Cahill, Alois Pichler, Simo Säätelä, Ole Hjortland, Pål Antonsen, Ole-Martin Skilleås, Harald Johannessen, Mette Hansen, and Espen Gamlund, as well as to our graduate students Mark Young, Ragnhild Jordahl, and Sindre Søderstrøm. Finally, I thank Professor Karen Wynn and Taylor and Francis for the permission to reprint, as Chapter 6, her paper "Origins of Numerical Knowledge" from the journal *Mathematical Cognition*, 1, 35–60.

I dedicate the volume to my parents: my mother, Liliana, and my late father, Constantin.

Bergen, Norway
December 2017

1 Introduction

A Naturalist Landscape

Sorin Bangu

I asked a question about a human being.

L. Wittgenstein, in conversation with A. Turing[1]

1. Themes and Motifs

The truths of mathematics and logic are special in several well-known respects: they are seemingly impossible to challenge on empirical grounds—hence they are traditionally called 'a priori'; there is also a sense in which they are considered to be 'necessary'. Yet, while stressing their specialness, we should not lose sight of the obvious fact that these propositions are, first and foremost, *beliefs* that we, human beings, often assert. As such, important questions about them arise immediately—e.g., how did we *acquire* them? (Or are they, or some of them, *innate*? If so, what does this mean?); What *actually* deters us from challenging them? What makes a proof of such a proposition *convincing*? What should we do when no proof is available? Or, what does it mean for such a proposition to be *self-evident*? And so on and so forth.

For all their naturalness, these kinds of queries were dismissed by Gotlob Frege (1848–1925), the most important logician since Aristotle. He argued that they are *completely* misguided, since what drives them—an interest in the psychological underpinnings of logico-mathematical thinking—is prone to engender confusion: one should not focus on how humans operate within the logico-mathematical realm, but rather on how they *ought* to do it. As part of this crusade to uphold (this kind of) normativity, Frege insisted that the only efforts worth undertaking consist in extracting, from the morass of the ordinary ways of speaking, the network of objective relations holding—whether or not individual people realize it—between the contents encapsulated into the logico-mathematical assertions. This line of thinking,

---

1 Recounted in Wittgenstein's *Lectures on the Foundations of Mathematics* (1976, 32).

unsurprisingly dubbed 'anti-psychologism', expelled a whole family of questions from the agenda of the philosophers of logic and mathematics.[2]

The sharp separation of the 'logical' from the 'psychological' became enormously influential in analytic philosophy; it still remains so, although it has constantly been challenged in various ways.[3] However, despite the name of this orientation, the intention behind it was not to dismiss psychology per se as an empirical science aiming to reveal, among other things, contingent truths about how people actually (learn to) reason, calculate, construct, or become convinced by proofs. The intrinsic legitimacy of this kind of research was not contested, only its relevance—for the normative questions about how we ought to reason. Thus, perhaps not even aware of the Fregean attitude, entire branches of psychology and cognitive science have developed and thrived for more than a century now,[4] investigating precisely the kinds of questions Frege took to be immaterial for genuinely understanding what mathematics and logic are about.

With rare but notable exceptions, the mainstream work in the epistemology of logic and mathematics has until recently barely intersected the trajectories taken by the flourishing cognitive sciences.[5] Yet this is not the case with epistemology in general, and this discrepancy is not that surprising given that mathematics and logic are traditionally believed to be the most resistant to naturalization. It will soon be almost half a century since W. v.

2 Frege's most important ally was Edmund Husserl (1859–1938). The literature on (anti-) psychologism is enormous, but see Kusch (1995) for a comprehensive discussion of Frege and Husserl's positions in an historical and sociological perspective. For Frege's epistemology, especially his conception of normativity, see Kitcher (1979), Conant (1991), Cohen (1998), Burge (2005), and Travis (2008); some of the essays in Ricketts and Potter (2010); and Maddy (2014). In my opinion, the most serious problems for Fregeanism arise upon reflection on rule-following (see Bangu (2016)) and on Frege's understanding of privacy (as in the 'private language argument'; see Baker and Hacker (1989)).

3 In philosophy of science, for instance, anti-psychologism arguably appears in the form of the distinction between the context of discovery and the context of justification; as soon as the 1960s, Kuhn and Quine, among others, began to confront this position.

4 Collections such as Butterworth and Cipolotti (1996), Campbell (2005), and Cohen Kadosh and Dowker (2012) offer overviews of the results obtained in the case of mathematics. For a focus on logic, see, e.g., Johnson-Laird and Byrne (1991), Johnson-Laird (2006), and Stenning and van Lambalgen (2008).

5 A footnote is perhaps not the place to attempt to document this claim, but let me quickly note that before 2000, for instance, one could mention, from a short list, Maddy (1980), Kitcher (1984), and Giaquinto (1992, 1996). Lately, the situation has improved; in addition to them, and to the authors contributing to this very volume, other philosophers tackled the issue, sometimes in papers co-authored with psychologists (e.g., Cappelletti and Giardino (2007) and Longo and Viarouge (2010)). Jenkins (2008) occasionally discusses cogntive matters, especially Butterworth (1999). Very recent work includes Dutilh Novaes (2013) and Pantsar (2014), as well as Pantsar and Quinon's (ms) discussion of Sarnecka and Carey (2008). For logic, see, e.g., the collaboration between Adler and Rips (2008) and Dutilh Novaes (2012).

O. Quine, in his famous programmatic "Epistemology Naturalized" (1969) asked philosophers to recognize that the traditional Cartesian "quest for certainty" is "a lost cause" and thus urged epistemologists "to settle for psychology". Consequently, "Epistemology, or something like it, simply falls into place as a chapter of psychology and hence of natural science" (1969, 82).[6] Such provocative statements may have been useful to reorient philosophical agendas 50 years ago, but nowadays, very few philosophers take them literally. It is quite clear that this radical 'replacement' naturalism, as Kornblith (1985, 3) calls it, is not the best option for a naturalistically bent philosopher, especially one of logic and mathematics (and perhaps not even for the scientists themselves). A better alternative seems to be a moderate view, sometimes called 'cooperative' naturalism, which, as the name indicates, encourages the use of the findings of the sciences of cognition in solving philosophical problems.[7] Yet what I take to be an even better approach is to understand 'cooperation' in a more extensive fashion, as promoting interactions that go in both directions; it is a reasonable thought that the scientists too may profit from philosophical reflection. Thus, fostering such a dialogue is the primary aim of the present project. The way to achieve it here is by displaying, for the benefit of both the philosophical and scientific audiences, a sketch of the landscape of the current research gathered under the heading 'naturalized epistemology *of mathematics and logic*'.[8]

Before I briefly present the contents of the chapters, it may be useful to set the reader's expectations right. Perhaps the first point to make is that, although traditionally it was the *concept* of knowledge that took pride of place in the writings dealing with the epistemology of these two disciplines, in what follows, this centrality is challenged. In a naturalist spirit, many contributors here can be described as shifting their attention to the very *phenomenon* of knowledge[9]—that is, the remarkable *natural fact* that human beings, of all ages and cultures, are able to navigate successfully within the

6 In terms of priorities, it is, however, A. Goldman's (1967) paper "A Causal Theory of Knowing", which, in my opinion, serves as a first illustration of a naturalist approach. Yet, importantly, in that paper, he explicitly sets to deal only with empirical knowledge—while, also importantly, Quine does not limit the scope of his naturalism. Naturalist ideas were, of course, expressed much earlier by Hume and Reid, as Rysiew's (2017) very useful overview notes.
7 As Feldman (2012) notes, citing S. Haack's characterization of it in Haack (1993, 118). Feldman also mentions several other prominent voices, more friendly to these ideas—e.g., Goldman (1992, 1999), Stich and Nisbett (1980, 118), and Harman (1986, 7).
8 I do not have my own definition of naturalism; from among the many varieties available (see Papineau (2016) and Paseau (2005)), my sympathies go toward the 'second-philosophical' version outlined in Maddy (2007).
9 The distinction is Kornblith's (see, e.g., 1995, 2002), although not drawn especially for the logico-mathematical domain; Goldman (2005) expresses some discontent with this particular way to understand the naturalist project. See also Goldman (1986).

realm of abstraction. In this type of analysis, it is not so much the symbolism itself that is being investigated, nor how a generic mind relates to abstraction, but rather the way in which the (presumably) abstract content is assimilated and manipulated by *concrete* epistemic agents in *local* contexts. Indeed, at least when it comes specifically to mathematics, there is no better way to summarize the issues investigated here than by citing the felicitous title of Warren McCulloch's (1961) paper, "What is a number, that a man may know it, and a man, that he may know a number?" Thus, it is causal stories, sensory perception, material signs and intuition, testimony, learning, neural activity, and other notions of the same ilk that now hold center stage in most of the chapters.

Consequently, the elements of the logico-mathematical practice under examination here no longer retain the purity and perfection traditionally associated with these two fields. Few, if any, of the perennial (and perennially frustrating) *in principle* questions are asked or debated. As expected, of major interest here is to probe to what extent a robust sense of normativity can be disentangled from an enormously complicated network of causal connections involving *nonidealized* epistemic agents ratiocinating in, and about, a material world. A central question is not only *how* but also *whether* normativity is possible *in practice*, or despite all the imperfections, approximations, and errors people are so prone to.[10] Both the friends and the foes of these naturalistic approaches will recognize the pivotal issue as being the following: does revealing the cognitive basis of mathematics and logic affect (threatens? supports?) the putative *objectivity* of mathematical and logical knowledge?

Another aspect worth pointing out is that the collection has not been conceived to promote a specific philosophical position, hiding, so to speak, behind the avowed naturalist attitude 'let us first *look and see*—what is the evidence'.[11] Thus, both the empiricistically inclined philosophers/scientists and their opponents are, I believe, represented; there are chapters inclining toward what is traditionally labeled as mathematical 'realism', while others display a preference for different metaphysical camps. There is also variety in terms of the methodological assumptions and conclusions among

10 Here I talk about normativity in general terms, glossing over the various subtleties identified in the literature. I regard it as an umbrella-notion, covering large issues such as the justification of logico-mathematical beliefs, the kind of warrant they may have—'internal' or 'external', the reliability of mathematical proofs, and so on.

11 This alludes to (what I take to be) the naturalistic spirit of Wittgenstein's advice to '*look and see*' in his *Philosophical Investigations* Section 66 (emphasis in original). For the therapeutical Wittgenstein, however, there is basically nothing left to do after one is able to attain this übersicht and thus 'cure' oneself. The next thing he says is 'don't think, but look' (one must admit, this is quite an advice to receive from a philosopher!) Yet the present project is undertaken on the premise that this is meant to be only (piercing) sarcasm: there may actually be a lot to think about after one is done 'looking'.

the scientifically oriented contributions. Moreover, it is my hope that the collection *as a whole* manages to avoid being biased in either of the two usual ways. It was *not* meant to provide empirical evidence that certain philosophical theories are true (or false), nor was it meant to provide reasons of a philosophical-conceptual nature that certain research programs in psychology and cognitive science are misguided. Importantly, however, note that acknowledging this is consistent with some individual chapters having such goals—although in most cases, a firm dichotomy empirical/conceptual is implicitly questioned. After all, not only should philosophers *look* at the empirical evidence first but also, as the scientists are often aware, what *counts* as evidence (i.e., which findings they are *justified* to present as evidence) may be influenced by deep commitments of a philosophical-conceptual nature.

To sum up, a reader motivated primarily by philosophical interests is invited to reflect on a (meta-)question, which I take to be both fundamental and insufficiently explored: what, if anything, is relevant about the nature of logico-mathematical knowledge in the recent research in psychology and cognitive science? Naturally, a corresponding question can be formulated for the more scientifically inclined reader: what, if anything, offers valuable insight into the nature of logico-mathematical knowledge in the philosophical work in this field? Although I regard these two questions to be equally urgent, and in fact entangled, the potential reader should be advised that the majority of the contributions here are philosophically oriented, as the table of contents and the brief presentations of the chapters that follows show. Moreover, most of the work deals with mathematics (and of a rather elementary kind), so logic per se receives less coverage than would be ideal.

2. Overview of the Volume

As noted, the material collected here is meant to join, in a balanced manner, the larger debate around the contemporary philosophical project of naturalizing logico-mathematical knowledge. The question about this distinctive kind of knowledge has its roots in Plato's dialogues, and virtually all major philosophers afterward have expressed interest in it. It remains alive today and, in light of the wealth of evidence collected by the cognitive sciences so far (and unavailable to these illustrious predecessors[12]) invites

---

12 As Stanislas Dehaene, a major contemporary figure in the field of mathematical cognition, remarks, "I often wonder how the great philosophers of the past would have welcomed the recent data from neuroscience and cognitive psychology. What dialogues would the images of positron emission tomography have inspired in platonists? What drastic revisions would the experiments on neonate arithmetic have imposed on the English empiricist philosophers? How would Diderot have received the neuropsychological data that demonstrates the extreme fragmentation of knowledge in the human brain? What penetrating insights would Descartes have had if he had been fed with the rigorous data of contemporary neuroscience instead of the flights of fancy of his time?" (1997, 231)

new, interdisciplinary approaches. Thus the shared goal of the chapters is to tackle this venerable query by taking into account the perspectives provided by the modern sciences of cognition (cognitive psychology, linguistics, neuroscience, etc.)

Here is a brief presentation of each of the next 14 chapters.[13] The volume opens up with Penelope Maddy's contribution, titled "Psychology and the A Priori Sciences". The a priori sciences she deals with are, in order, logic and arithmetic, and the 'psychology' includes experimental, especially developmental, psychology, neurophysiology, and vision science. Maddy investigates the role these empirical theories can play in the philosophies of those sciences, or, more precisely, the role she thinks they should play. She draws on a number of psychological studies that are most likely known to the readers of this volume. The next chapter, "Reasoning, Rules, and Representation" is co-authored by Paul D. Robinson and Richard Samuels. Their starting point is a trend observed in recent years in the philosophical theories of reasoning in general and logical inference in particular—namely, the impact that a regress argument had on them. The aim of this argument is to challenge a conception of reasoning adopted by most psychologists and cognitive scientists. In this chapter, they discuss this view—the *intentional rule-following account*—and begin by emphasizing its virtues. Then they reconstruct the Regress argument in detail and, essentially, show that it is unsound. Specifically, they point out that in cognitive science, many mainstream accounts of psychological processes actually have the resources to address the (putative) regress. In Chapter 4, titled "Numerical Cognition and Mathematical Knowledge: The Plural Property View", Byeong-uk Yi begins by noting that it is difficult to account for how we humans can know even very basic mathematical truths—a challenge first raised, as we saw, by Plato and whose urgency was more recently stressed by Paul Benacerraf (1973). The chapter aims to outline an account of this kind of knowledge that can meet this famous challenge. Yi elaborates two views: (i) natural numbers are numerical properties (the *property view*) and (ii) humans have empirical, even perceptual, access to numerical attributes (the *empirical access thesis*). As Yi remarks, this approach has been taken before by Maddy (1980, 1990) and Kim (1981) (in an attempt to deal with the Benacerraf problem). But Yi argues that a truly viable version of this idea needs an important modification. While Maddy and Kim hold that natural numbers are *properties of sets*, Yi bases his approach here on his well-known view that they are *plural properties*, that is, properties that relate in a sense to *many* things *as such* (this is the *plural property view*). In addition to outlining this position, the chapter examines recent psychological studies of numerical

---

13 The presentations of the chapters is in essence the authors' themselves; the text that follows
   incorporates and paraphrases the abstracts of the chapters. The abstract of Karen Wynn's
   chapter is part of Wynn (1995).

cognition, which, in his view, support the empirical access thesis. In the next chapter, Mark Fedyk deals with the relations between "Intuitions, Naturalism, and Benacerraf's Problem". While few would deny a connection between our knowledge of logical truths and mathematics and intuition, many contemporary naturalists adopt a skeptical attitude toward the ability of intuitions to actually generate substantial amounts of (this kind of) knowledge. Fedyk's approach, however, departs from this tendency. He sketches a naturalistic theory of intuitions, which is then used to respond to Benacerraf's problem. If Fedyk is right, then a naturalist is actually in the position to account for how intuition and mathematical knowledge are connected. However, a somewhat surprising conclusion also follows from his argument—namely, that mathematical knowledge may consist, at least partially, in *approximate* truths.

Chapter 6 is Karen Wynn's paper "Origins of Numerical Knowledge". It was published in 1995, and it is the only chapter not written especially for this collection. However, I was keen on including it here since I believe that the philosophical significance of this kind of work has not received enough philosophical attention.[14] (Another important, though quite different, work in the naturalist orientation, Lakoff and Núñez's *Where Mathematics Come From* (2000) is, it seems to me, in pretty much the same situation.)[15] Wynn's chapter documents the evidence she found for the age-old and very controversial hypothesis that some of our mathematical abilities may be innate, or, as she puts it, that infants possess "a system of numerical knowledge". Thus one may now be entitled to believe that Plato's and the rationalists' nativist speculations have received a scientifically respectable form after all. The system she identifies consists of a mechanism (the 'accumulator') for determining and representing small numbers of entities; it also accounts for the procedures for operating over these representations in order to extract information about the numerical relationships between them. Wynn presents a model for this mechanism and investigates its relation to the acquisition of further numerical knowledge. In Chapter 7, Kristy vanMarle asks the fundamental question "what happens when a child learns to count?" The work deals, from a current scientific perspective, with the development of the number concept. (The chapter was commissioned with the intention to supplement, and to update,

---

14 This (anti-Piagetian; see Piaget (1965)) line of thinking builds up on earlier work by— e.g., Gelman and Gallistel (1978), Starkey and Cooper (1980), and Baillargeon, Spelke, & Wasserman (1985), among others. Perhaps the very first philosophical reaction to Wynn's results was Giaquinto (1992); see also Giaquinto (2001). Maddy (2007) contains more recent discussions. For nativism in general, see Cowie (1999) and Carruthers, Laurence, and Stich (2005–9), a three-volume collection in which some contributions deal with logico-mathematical cognition. My view on the possible philosophical relevance of the Wynn-type of results is in Bangu (2012).

15 See, however, Parsons and Brown (2004) and Brown (2012).

the Wynn-type of work.) The author examines whether we can claim that learning how to count amounts to nothing less than a 'conceptual revolution'. One important motivation for the investigation here is an old question about the match of two kinds of principles: are the principles that define the formal system of the positive integers the same as the principles that guide our cognitive intuitions about numbers and counting? Moreover, as vanMarle points out, in developmental psychology, it is generally accepted that learning to count involves the formation of an understanding of the "counting principles" (e.g., one to one correspondence, stable order, cardinality). The debate continues, however: are these principles innate to children's cognitive systems, discovered in the world, or constructed through experience and logical induction. VanMarle draws on recent findings from a large longitudinal study in which she and her collaborators examined the growth and change of different number skills, formal and informal, over two years of preschool. The findings identify and emphasize the relevance of two core capacities for the acquisition of children's first symbolic mathematical knowledge—the counting routine. The chapter also discusses previous proposals in light of these findings, considering whether they support the notion of conceptual change; it also raises the question about the consistency of these data with the hypothesis of conceptual elaboration. Chapter 8, by Max Jones, is suggestively titled "Seeing Numbers as Affordances". Jones begins by pointing out an important tension: recent research into the nature of mathematical cognition clearly suggests that humans (and other animals) can somehow perceive numbers, on one hand; on the other, this is in conflict with both how mathematical entities are traditionally conceived and how we traditionally understand perception. Yet Jones notes that, as intimated by Philip Kitcher a while ago, there are perhaps ways to harmonize the two sides somehow. One such way is *the ecological theory of perception*, which can account for our perceptual access to numerical content. According to this theory, we perceive numbers by perceiving *affordances* for enumerative activity. However, Jones also believes that Kitcher's account needs updating to accommodate recent empirical evidence. What is needed is a more fine-grained notion of *enumerative activity* in terms of sequential spatial attention. Jones also reflects on the major metaphysical implications of this ecological approach to perception. He notes that this type of account potentially affects the traditional debates in the philosophy of mathematics. Furthermore, this demonstrates the need for naturalistically inclined philosophers of mathematics to look beyond the study of mathematical cognition and to become familiar with the developments in the sciences of the mind.

In "Testimony and Children's Acquisition of Number Concepts" (Chapter 9), Helen De Cruz takes up an enduring puzzle in philosophy and developmental psychology: how young children acquire number concepts. Her particular interest is the concept of natural number. First, she points out that most solutions to this problem conceptualize young learners as "lone mathematicians", who individually reconstruct the successor function and other sophisticated mathematical ideas. Yet, in this chapter, she argues this view needs to be changed: there is a crucial role of testimony in children's acquisition of number concepts. This role is manifested in two ways, not only in the transfer of propositional knowledge (e.g., the cardinality concept) but also in knowledge-how (e.g., the counting routine). Jean-Charles Pelland asks, in Chapter 10, "which came first, the number or the numeral?" He begins with a critical examination of some recently proposed accounts of the development of numerical cognition (Dehaene, 1997/2011; Carey, 2009; De Cruz, 2008). Then he argues that they all share a critical flaw: they rely on the presence of *external* numerical symbols in the environment. His claim is that this overall strategy—i.e., to explain how number concepts develop by appealing to external symbols—is akin to putting the cart before the horse. Such symbols, Pelland points out, cannot appear in the environment without there being number concepts in someone's head beforehand. But can't adopting the well-known *extended mind* perspective help here? Pelland agrees that this is indeed a tempting way to bridge the gap between the output of our innate cognitive machinery and the content of advanced number concepts. Yet, he argues, such an externalist framework simply cannot yield a complete description of the origin of numerical cognition. The most serious problem is that they cannot explain where our first number concepts came from! The relation between numbers and their concrete representation is also the main theme in Dirk Schlimm's contribution titled "Numbers Through Numerals: The Constitutive Role of External Representations" (chapter 11). Schlimm begins by noting that our epistemic access to mathematical objects, such as numbers, is "mediated" through our external representations of them—and numerals are such representations. Nevertheless, he points out that systems of number words and of numerals should not be treated alike since they have crucial structural differences. A key idea here is that one has to understand how the external representation works in order to form an advanced conception of numbers. Schlimm examines the relation between external representations and mathematical cognition by drawing on experimental results from cognitive science and mathematics education. The thesis the chapter proposes is that external representations play a *constitutive* role in the development of an advanced conception of numbers.

Beginning with Chapter 12 by Karim Zahidi and Erik Myin, we move on to other facets of naturalism. They engage in the project of "making sense of numbers without a number sense". As some previous chapters in the volume illustrate, there is a lot of sympathy among both cognitive scientists and philosophers for the idea that the number-sensitive behavior of young infants is best explained by attributing to them (innate) capacities for conceptual knowledge of numbers (and numerical operations). Furthermore, such knowledge is deemed necessary as a basis to acquire more sophisticated arithmetical knowledge later on. In this chapter, however, the authors will take issue with these ideas. They argue that, although one cannot deny some initial plausibility to the ascription of innate conceptual knowledge, appealing to it in fact *fails* to account for the observed behavior. In particular, they maintain that the key assumption—i.e., that the behavior is to be explained as a product of knowledge deployment—suffers from two major flaws: not only is it unwarranted, but also postulating the required knowledge raises new, unsolved problems. They make a case for the claim that these problems can be avoided if one adopts a *socio-historical view* of the ontogenesis of arithmetical knowledge. On this account, knowledge acquisition requires a community of already accomplished knowers. To substantiate their claims, they sketch an account inspired by Wittgensteinian models of language acquisition: the central idea is that a child can acquire numerical knowledge on the basis of universally shared non-epistemic behaviors, with the support of a community of already accomplished reckoners. In Chapter 13, Josephine Relaford-Doyle and Rafael Núñez aim to go "beyond Peano (by) looking into the unnaturalness of natural numbers". They note a widespread assumption made by philosophers, mathematicians, and developmental psychologists—namely, that the earliest number concepts we develop are concepts of *natural number*. Furthermore, they also point out another common supposition: that the conceptualizations we humans develop are consistent with Peano-like axioms. Against this background, the key question they ask is whether these claims hold up to scrutiny, and they present empirical evidence that they actually do not. Moreover, this evidence suggests that even highly educated adults routinely conceptualize the natural numbers in ways that are radically different from, or even at odds with, Peano-like characterizations. Among the positive claims they make is the one that the collection of *counting numbers* differs in important ways from the formal mathematical set of *natural numbers*. An important upshot of this analysis is that natural number concepts may *not* arise naturally from our counting experience. The next chapter, "Beauty and Truth in Mathematics: Evidence From Cognitive Psychology", by Rolf Reber, also takes up important philosophical issues from a scientific perspective. He first notes the existence of many reports about the role of beauty for finding truth in mathematics. For a long time, philosophical theorizing has dealt with the epistemic justification of aesthetic factors. Yet, as Reber notes, empirical evidence, and a theory about the underlying mental mechanisms, are largely missing. He discusses a particular theory, the *processing fluency theory* of aesthetic pleasure, which claims that the ease with which information can be processed is the common mechanism underlying perceived beauty and judged truth. The chapter summarizes some of the existent philosophical discussion on this topic; after this, it introduces the assumptions of the processing fluency theory before reviewing the empirical evidence for this account. Three major questions arise. (i) Is beauty truth indeed? (ii) Is there any relationship between the perceived aesthetic quality of a mathematical solution and its truth? And (iii), perhaps the most difficult issue, is processing fluency the common mechanism? Earlier analyses investigated the role of fluency in subjectively judged and objective truth, and they dealt with the evidence for the claim that familiarity with, and the coherence of, information results in fluency. These investigations also outlined the conditions under which fluency is epistemically justified—that is, related to actual truth. Likewise, according to Reber, it seems plausible that the familiarity of mathematicians with their field, and the coherence of mathematical rules, results in fluency. Hence a positive relationship between beauty and truth is plausible.

Chapter 15 consists of Fabio Sterpetti's contribution, "Mathematical Knowledge, the Analytic Method, and Naturalism". It fittingly closes the volume, as it deals with a key question for any naturalist-friendly philosophy—namely, how should we conceive of the very method of mathematics *if* we take a naturalist stance? Sterpetti indicates the origin of the problem: mathematical knowledge has long been considered to be the very paradigm of certain knowledge, and, as he also points out, this is so because mathematics is taken to be based on the *axiomatic method*. A source of additional problems is that natural science is thoroughly mathematized and thus accounting for the role of science and its relation to mathematics are crucial tasks in outlining any version of the naturalist perspective. Sterpetti examines, and expresses doubts about, the idea to naturalize the traditional view of mathematics by relying on evolutionism. The chapter makes the case for the claim that the naturalization project for mathematics can only succeed if we take the method of mathematics to be the *analytic method* rather than the axiomatic method. One interesting upshot is that we must then conceive of mathematical knowledge as *plausible* knowledge.[16]

References

Adler, J., & Rips, L. (2008). *Reasoning: Studies of human inference and its foundations*. Cambridge: Cambridge University Press.

Baillargeon, R., Spelke, E., & Wasserman, S. (1985). Object permanence in five-month-old infants. *Cognition, 20*, 191–208.

Baker, G. P., & Hacker, P. M. S. (1989). Frege's anti-psychologism. In M. A. Notturno (Ed.), *Perspectives on psychologism* (pp. 74–127). Leiden: Brill.

Bangu, S. (2012). Wynn's experiments and later Wittgenstein's philosophy of mathematics. *Iyyun: The Jerusalem Philosophical Quarterly, 61*, 219–241.

Bangu, S. (2016). Later Wittgenstein on the logicist definition of number. In S. Costreie (Ed.), *Early analytic philosophy: New perspectives on the tradition* (pp. 233–257). Western Ontario Series in Philosophy of Science. Dordrecht: Springer.

Benacerraf, P. (1973). Mathematical truth. *Journal of Philosophy, 70*, 661–679.

Brown, J. R. (2012). *Platonism, naturalism and mathematical knowledge*. Abingdon, UK: Routledge.

---

16 I thank Mark Fedyk for conversations on these issues and for punctual suggestions on these introductory remarks.

Burge, T. (2005). *Truth, thought, reason: Essays on Frege*. Oxford: Oxford University Press.

Butterworth, B. (1999). *The mathematical brain*. London: Macmillan.

Butterworth, B., & Cipolotti, L. (Eds.). (1996). *Mathematical cognition* (Vol. 1). Hove, UK: Psychology Press.

Campbell, J. T. (Ed.). (2005). *Handbook of mathematical cognition*. Hove, UK: Psychology Press.

Cappelletti, M., & Giardino, V. (2007). The cognitive basis of mathematical knowledge. In M. Leng, A. Paseau, & M. Potter (Eds.), *Mathematical knowledge*. Cambridge: Cambridge University Press.

Carey, S. (2009). *The origin of concepts*. New York: Oxford University Press.

Carruthers, P., Laurence, S., & Stich, S. (Eds.). (2005–9). *The innate mind*. (Vols. 1–3). Oxford: Oxford University Press.

Cohen, J. (1998). Frege and psychologism. *Philosophical Papers, 27*(1), 45–67.

Cohen Kadosh, R., & Dowker, A. (2012). *Oxford handbook of mathematical cognition*. Oxford: Oxford University Press.

Conant, J. (1991). The search for logically alien thought: Descartes, Kant, Frege, and the Tractatus. *Philosophical Topics, 20*(1), 115–180.

Cowie, F. (1999). *What's within? Nativism reconsidered*. New York: Oxford University Press.

De Cruz, H. (2008). An extended mind perspective on natural number representation. *Philosophical Psychology, 21*(4), 475–490.

Dehaene, S. (1997). *The number sense: How the mind creates mathematics* (2nd ed., 2011). Oxford: Oxford University Press.

Dutilh Novaes, C. (2012). *Formal languages in logic: A philosophical and cognitive analysis*. Cambridge: Cambridge University Press.

Dutilh Novaes, C. (2013). Mathematical reasoning and external symbolic systems. *Logique & Analyse, 56*(21), 45–65.

Feldman, R. (2012). Naturalized epistemology. In E. N. Zalta (Ed.), *The Stanford encyclopedia of philosophy* (Summer 2012 ed.). Retrieved from https://plato.stanford.edu/archives/sum2012/entries/epistemology-naturalized/

Gelman, R., & Gallistel, C. R. (1978). *The child's understanding of number*. Cambridge, MA: Harvard University Press.

Giaquinto, M. (1992). Infant arithmetic: Wynn's hypothesis should not be dismissed. *Mind and Language, 7*(4), 364–366.

Giaquinto, M. (1996). Concepts and Calculations. In B. Butterworth & L. Cipolotti (Eds.), *Mathematical Cognition* (Vol. 1, pp. 61–83). Hove, UK: Psychology Press.

Giaquinto, M. (2001). Knowing numbers. *Journal of Philosophy, 98*(1), 5–18.

Goldman, A. (1967). A causal theory of knowing. *Journal of Philosophy, 64*, 357–372.

Goldman, A. (1986). *Epistemology and cognition*. Cambridge, MA: Harvard University Press.

Goldman, A. (1992). *Liaisons: Philosophy meets the cognitive and social sciences*. Cambridge: MIT Press.

Goldman, A. (1999). *Knowledge in a social world*. Oxford: Clarendon Press.

Goldman, A. (2005). Kornblith's naturalistic epistemology. *Philosophy and Phenomenological Research, 71*(2), 403–410.

Haack, S. (1993). *Evidence and inquiry: Towards reconstruction in epistemology*. Oxford: Blackwell.

Harman, G. (1986). *Change in view: Principles of Reasoning*. Cambridge, MA: MIT Press.

Jenkins, C. S. (2008). *Grounding concepts: An empirical basis for arithmetical knowledge*. Cambridge: Cambridge University Press.

Johnson-Laird, P. (2006). *How we reason*. Oxford: Oxford University Press.

Johnson-Laird, P., & Byrne, R. (1991). *Deduction*. Hillsdale: Lawrence Erlbaum Associates, Inc.

Kim, J. (1981). The role of perception in a priori knowledge: Some remarks. *Philosophical Studies, 40,* 339–354.

Kitcher, P. (1979). Frege's epistemology. *Philosophical Review, 86,* 235–262.

Kitcher, P. (1984). *The nature of mathematical knowledge*. Oxford: Oxford University Press.

Kornblith, H. (1985). Introduction: What is naturalistic epistemology? In H. Kornblith (Ed.), *Naturalizing epistemology*. Cambridge, MA: MIT Press.

Kornblith, H. (1995). Naturalistic epistemology and its critics. *Philosophical Topics, 23*(1), 237–255.

Kornblith, H. (2002). *Knowledge and its place in nature*. New York: Oxford University Press.

Kusch, M. (1995). *Psychologism: A case study in the sociology of philosophical knowledge*. Abingdon, UK: Routledge.

Lakoff, G., & Núñez, R. (2000). *Where mathematics comes from: How the embodied mind brings mathematics into being*. New York: Basic Books.

Longo, G., & Viarouge, A. (2010). Mathematical intuition and the cognitive roots of mathematical concepts. *Topoi*, special issue on *Mathematical Knowledge: Intuition, Visualization, and Understanding,* 29(1), 15–27 (Eds. L. Horsten and I. Starikova).

Maddy, P. (1980). Perception and mathematical intuition. *Philosophical Review, 89*(2), 163–196.

Maddy, P. (1990). *Realism in mathematics*. Oxford: Oxford University Press.

Maddy, P. (2007). *Second philosophy: A naturalistic method*. Oxford: Oxford University Press.

Maddy, P. (2014). *The logical must. Wittgenstein on logic*. Oxford: Oxford University Press.

McCulloch, W. S. (1961). What is a number that a man may know it, and a man, that he may know a number? The Ninth Alfred Korzybski Memorial Lecture, in *General Semantics Bulletin*, No. 26–27:7–18. Lakeville, CT: Institute of General Semantics.

Pantsar, M. (2014). An empirically feasible approach to the epistemology of arithmetic. *Synthese, 191*(17), 4201–4229.

Pantsar, M., & Quinon, P. *Bootstrapping of the natural number concept: Regularity, progression and beat induction*. Manuscript, available at http://www.paulaquinon.com/publications/ Accessed 22 Nov. 2017.

Papineau, D. (2016). Naturalism. In E. N. Zalta (Ed.), *The Stanford encyclopedia of philosophy* (Winter 2016 ed.). Retrieved from https://plato.stanford.edu/archives/win2016/entries/naturalism/

Parsons, G., & Brown, J. R. (2004). Platonism, metaphor, and mathematics. *Dialogue, 43*(1), 47–66.

Paseau, A. (2005). Naturalism in mathematics and the authority of philosophy. *British Journal for the Philosophy of Science, 56,* 399–418.

Piaget, J. (1965). *The child's conception of number*. New York: W. Norton Company & Inc.

Quine, W. V. O. (1969). Epistemology naturalized. In *Ontological relativity and other essays* (pp. 69–90). New York: Columbia Press.

Ricketts, T., & Potter, M. (Eds.). (2010). *Cambridge companion to Frege*. Cambridge: Cambridge University Press.

Rysiew, P. (2017). Naturalism in epistemology. In E. N. Zalta (Ed.), *The Stanford encyclopedia of philosophy* (Spring 2017 ed.). Retrieved from https://plato.stanford.edu/archives/spr2017/entries/epistemology-naturalized/

Sarnecka, B. W., & Carey, S. (2008). How counting represents number: What children must learn and when they learn it. *Cognition, 108*(3), 662–674.

Starkey, P., & Cooper, R. (1980). Perception of numbers by human infants. *Science, 210*, 1033–1035.

Stenning, K., & van Lambalgen, M. (2008). *Human reasoning and cognitive science.* Cambridge, MA: MIT Press.

Stich, S., & Nisbett, R. (1980). Justification and the psychology of human reasoning. *Philosophy of Science, 47*, 188–202.

Travis, C. (2008). Psychologism. In E. Lepore & B. C. Smith (Eds.), *The Oxford handbook of philosophy of language.* Oxford: Oxford University Press.

Wittgenstein, L. (1953). *Philosophical investigations* (G. E. M. Anscombe & R. Rhees, Eds., G. E. M. Anscombe, Trans.). Oxford: Blackwell.

Wittgenstein, L. (1976). *Lectures on the foundations of mathematics* (C. Diamond, Ed.). Ithaca: Cornell University Press.

Wynn, K. (1995). Origins of numerical knowledge. *Mathematical Cognition, 1*, 35–60.

2   Psychology and the A Priori Sciences

Penelope Maddy

The 'a priori sciences'[1] to be considered here are logic and arithmetic;[2] the 'psychology' includes experimental, especially developmental psychology, neurophysiology, and vision science. My goal is to examine the role these empirical theories can play in the philosophies of those sciences, or more precisely, the role I think they should play. Most of the psychological studies referred to here will be familiar to readers of this volume, though perhaps not the use to which I hope to put them.

1.   Logic

Common sense tells us that much of the world comes packaged into middle-sized objects—stones, coins, snails, apples, trees, bodies of cats, apes, human beings—and not without reason; these items are what we see and touch, encounter, and engage with, in everyday life. Of course, common sense doesn't always hold up under scrutiny, but meticulous science confirms that each of these is a rough collection of molecules held together by various forces, resisting penetration due to other forces, moving as a bounded unit on a continuous spatiotemporal path.[3] Scientifically refined common sense also reveals that these objects have properties and stand in relations: stones come in a variety of sizes and shapes, apples in a various

---

1 I use the term 'a priori science' as the customary label, not to endorse the view that these disciplines are in fact a priori in some sense or other.
2 With a nod toward set theory in footnote 47.
3 Some philosophers question this simple view on the grounds that the commonsense table is intuitively 'solid', while the scientific object is largely empty space, so the two cannot be the same. Even assuming that common sense does picture things as continuous matter (over and above being impenetrable), it seems more natural to say science has taught us that the objects of common sense are different than we first imagined, not that they don't exist. (For a bit more on such thinking from Eddington, Sellars, and Ladyman and Ross, see Maddy 2014c, 95–97, 99, footnote 9.) Other philosophers go further, rejecting everything in science and common sense on radical skeptical grounds, but this challenge, too, I set aside for present purposes (for more, see Maddy, 2017).

colors, domestic cats are generally smaller than adult humans.[4] This simple structuring validates a certain amount of rudimentary logic: if the apple is either red or green and it's not red, then it must be green. This might all seem so obvious, so unavoidable, as to be true no matter what, true in 'all possible worlds', but in fact it breaks down at the quantum level: the particles don't behave as bounded units on continuous paths; the sense in which they enjoy properties (such as position and momentum) is problematic; some simple logical laws (e.g., the distributive law) appear to fail.[5] The inferences of this rudimentary logic are reliable as long as the requisite structure is in place, but not otherwise.

Psychology comes into this story as an investigation of how we come to those early commonsense beliefs about objects and their features. The groundbreaking developmental work of the 1980s and '90s[6] showed that young infants track cohesive, bounded, solid[7] individuals, despite occlusions, using spatiotemporal criteria such as contiguity, common fate, and continuous motion; though they're aware of object's other properties, they typically don't use regularities of shape, color, texture or motion, or features or kinds to determine object boundaries or identity.[8] Studies of neonates and nonhuman animals suggest a distant evolutionary origin.

There's some disagreement over the precise interpretation of these experimental results. The contents of 'the object concept' vary slightly from writer to writer;[9] disagreements arise over whether the abilities catalogued are

---

4 Dependencies between one situation and another are also important—the coin is on the floor because the cat shoved it off the table—as are universal properties, but I leave these aside for simplicity in this quick sketch of Maddy, 2007, III.4 (also (2014c)).

5 In another skeptical move, it's sometimes suggested that science can't serve to ratify the objects of common sense, because any science that begins with those objects will inevitably end up ratifying them. In fact, a science (ours) that begins with them has ended up without them in the quantum world.

6 For a summary with references, see Maddy (2007, 245–258). Carey (2009) (Chapters 2 and 3) is a much-discussed survey and philosophical elaboration by one of the leading researchers in the area. This work swept away the earlier seminal theories of Piaget (featured in Maddy (1990, 54–5)), according to which the ability to represent objects comes later in development. Carey (2009, 46–55) gives a fascinating reanalysis of Piaget's evidence.

7 That is, impenetrable (a feature of both of Eddington's tables in footnote 3).

8 Animate/inanimate, human/nonhuman appear to be exceptions. See Carey (2009, 263–284), especially pp. 276–277, for more on this point.

9 In one of the more dramatic examples, Burge departs from many psychologists (from Piaget on) in holding that the 'constitutive conditions' for representing bodies as such don't include tracking through occlusion: "A capacity to perceptually track a body as a three-dimensionally bounded and cohesive volume shape while it remains in view . . . suffices" (Burge, 2010, 460). In contrast, Hatfield (2009b) requires tracking through occlusion, but holds that the developmental evidence doesn't conclusively show infants are representing objects as "individual material objects (not as mere local collections of properties) . . . that . . . occupy . . . distinct . . . space-time worms . . . throughout their existence" Hatfield (2009, 241).

purely perceptual or somehow conceptual,[10] and so on.[11] Fortunately, these niceties needn't trouble us here because any of these options will be enough to serve as the building blocks, the 'objects with properties', in the rudimentary logical structure described earlier.

But 'object with properties' aren't all there is to that rudimentary structure: a stone has a size and a shape, an apple can be red or green, a coin can fail to be a quarter. More developmental work of the '80s and '90s shows that young infants classify cats and dogs so as to exclude birds, and even cats so as to exclude superficially similar dogs. They're also sensitive to correlations of features: infants[12] aware of three possibilities for each of the features A, B, C, D, and E (that is, A1, A2, and A3, and so on), habituated to items with correlations between these features— for example, items with $(A1 \wedge B1 \wedge C1) \wedge (D1 \vee D2) \wedge E1 \vee E2)$ and items with $(A2 \wedge B2 \wedge C2) \wedge (D1 \vee D2) \wedge (E1 \vee E2)$—find a new correlated combination (such as A2-B2-C2-D1-E2) familiar but an uncorrelated combination (such as A1-B2-C1-D1-E2) just as novel as one that's entirely new (such as A3-B3-C3-D3-E3)! In addition to these conjunctions, infants also appear sensitive to disjunctions—habituated to cats or horses, they find a dog novel—and to relations—for example, 'above', 'below', 'between'. Results like these strongly suggest that we humans are sensitive to rudimentary logical structures from an early age.[13]

Still, as is well-known, it's entirely possible to respond to a feature of the world without representing it: the frog's visual system might allow it to detect (then catch and eat) flies without representing them as flies. On this point, I'm less confident than Tyler Burge that 'representation' is a psychological natural kind[14] and even more doubtful that its contours can be

---

10 For example, Burge (2010, 438–450), in disagreement with Spelke (1988).

11 It could be that some of these disagreements run deeper than the sort of thing scouted in footnote 9. Sticking with Burge and Hatfield as our examples, notice that Burge takes the goal of the project to be determining what's 'constitutive' of objecthood—"Our question concerns necessary minimal constitutive conditions for having the capacity to attribute the kind *body* in perception" (Burge, 2010, 465)—where this presupposes a fact of the matter to be discovered (perhaps by rational intuition, perhaps with a hint of essentialism in the appeal to 'natural kinds'). In contrast, Hatfield (2009b, 241) only claims, "We as adult perceivers typically see (things) as individual objects" with the features listed in footnote 9 and that the developmental evidence doesn't establish that infants do this too. He describes this situation by referring to 'the adult concept', but there's no indication of an underlying Burge-like metaphysics; he could just be using the phrase to highlight the possibility of a significant cognitive shift.

12 This pattern and the next emerge in 10-month-olds, still pre-linguistic.

13 For a summary, with references, see Maddy (2007, 258–262).

14 See Burge (2010, 291): "Psychological explanations have a distinct explanatory paradigm. Psychology . . . discovers its own kinds. One of them is the kind representation".

discovered, as he suggests, by uniquely philosophical means.[15] In contrast, Gary Hatfield (1988) undertakes a more modest task, firmly grounded in contemporary vision science.[16] An ongoing debate pits those who believe that the visual system employs symbolic representations in an internal symbolic 'language'—that rules are encoded and applied, hypotheses formed and tested (in the tradition of Helmholtz)—and those who insist that the visual system is not representational, that it's simply tuned to register directly the rich and complex information available in the ever-changing array of ambient light (Gibson and his followers). Both sides acknowledge that processing takes place between the retina and the visual experience. The debate between them hinges on the question, is this processing purely physiological or does it break down into psychologically significant components and, in particular, into components with a characteristically representational role?[17] Hatfield threads the needle between the two schools of thought, arguing that there are representational components, but that they needn't involve a symbolic system.

To see how this goes, consider using a slide rule to multiply n times m: locate n on the A scale; slide 1 on the B scale beneath n on the A scale; find the number on the A scale that's above m on the B scale. The procedure works because the scales are laid out logarithmically and $n \times m = \ln^{-1}(\ln(n) + \ln(m))$. That same equation could be programmed into a digital computer and multiplication carried out in that way, in which case, the logarithmic algorithm itself would be encoded, represented, in the computer's program, but this isn't true for the slide rule: there the algorithm is effectively followed, but it isn't literally represented.[18] The lengths on the slide rule represent numbers because of what the device is designed to do (multiply, among other things) and how it was designed to do it (relying on, but not representing, the properties of logarithms). The computer is also designed to multiply (among other things), but it's designed to do so quite differently, by applying explicit rules in an internal symbolic system. So what a device does or doesn't do represent depends on how it does what it's designed to do.

15 See Burge (2010, xviii): "Philosophy has . . . a set of methodological and conceptual tools that position it uniquely to make important contributions to understanding the world. . . . Many of its topics remain of broadest human concern. Where, constitutively, representational mind begins is such a topic".

16 Obviously, this contrast (elaborated in footnotes 14 and 15) is reminiscent of the one in footnotes 9 and 11.

17 See Ullman (1980, 374) and Hatfield (1988), Section 1.

18 As Burge (2010, 504) points out, an odometer's computation of the distance traveled depends on the circumference of the tires (it records a tick for each rotation), but the circumference is nowhere represented. Hatfield (1988, 75) makes a similar point about a 'tension adder': n and m are represented by small weights placed on a pan and their sum registered by a pointer on the front of the device, but no algorithm is encoded.

If we now replace the slide rule designer or the computer programmer with the evolutionary pressures on our species,[19] then the representational status of some element of the visual system can be assessed in the same way: it depends on the function of the visual system in the evolved human organism, the function of that element within the visual system, and the method it uses to perform that function.[20] Hatfield (1988, 63–65) gives the example of seeing a circle at a slant rather than an ellipse. This function could be achieved by registering the retinal ellipse, registering slant information from shading, and computationally combining these two, or it might be achieved by a single registration of shading across the retinal ellipse. Obviously, it's an empirical matter which of these algorithms is actually implemented; it can be investigated by psychological experiments with carefully timed disruptions or by physiological investigation of the neuroanatomy. In these ways, we could determine whether or not, say, the projective retinal shape by itself is represented. But either way, it's not at all obvious that this sort of representation would involve a symbolic system.[21]

So, to return to our theme, we know that infants respond to conjunctions, disjunctions, negations, and so on, but do they actually represent them as such? As Burge notes (2010, 406), the fact that we infer in accord with a logical rule doesn't imply that the rule is somehow encoded in our psychology, presumably in some language of thought. Our concern here, though, isn't with inference, but with simple logical structuring, and (following Hatfield) the representation needn't be symbolic. The question is whether the infant represents the stone as small and round, the apple as red or green, the coin as not a quarter. Assuming that sensory sensitivity to these worldly features is adaptive,[22] a Hatfield-style answer to this question hinges on how that sensory sensitivity is achieved: does the scientific story of that ability break down into psychologically significant parts, into representational components, such as the separate representations of projective shape and

---

19 In practice, determining what aspects of the visual system are adaptations and which are spandrels is a very difficult undertaking. See, e.g., Warren (2012) and Anderson (2015).

20 Burge (2010), Chapter 8, soundly rejects accounts of representation based in biological function for reasons I don't fully understand and won't attempt to explicate.

21 Does this mean that the frog is representing flies? Opponents of biological function views suggest that evolutionary considerations aren't enough to show that the frog is representing flies as opposed to moving black dots, or even flies as opposed to nothing at all given that a frog's detector will occasionally go off on its own. Regarding the first point as subject to further investigation, granting it now only for the sake of argument, Hatfield contends that, nevertheless, biological function "can serve as the basis for ascribing to states of the frog's visual system the content *target fly/moving dot*, or some such coarse-grained content", that "[a]mong the functions of the frog's visual system is to represent small moving things as being there when they are, and not to represent them as being there when they aren't" (Hatfield, 1991, 122–123). In other words, the biological function account has room to regard the frog's detector as having misfired when it goes off on its own.

22 See footnote 19.

of shading information in the multistage algorithm for seeing the circle at a slant? For our case, given that the infant can represent stones, smallness, and roundness, is her representation of a small, round stone related to other representations in a way that merits describing it as a conjunctive representation? This needn't involve encoding in some language of thought any more than the circle-at-a-slant case does, but it is a straightforward empirical question for experimental psychology and neuroscience.

If a definitive answer to this question is known, it isn't known to me, but the study of visual working memory offers a hint of how a small part of it might go. Evidence suggests that we're able to store information about a limited number of objects (around four) and their features over short periods of time. This raises the question of how several features of one object are bound together: what distinguishes a scene with a vertical red bar and a horizontal green bar from one with a vertical green bar and a horizontal red one? One proposal is synchronized neural firing: a particular neuron fires repeatedly to encode a single feature; when the repeated firings of the neurons for two separate features are synchronized, they form a unit: 'cell assembly'.[23] In Hatfield's terms, the initial firings represent red, green, horizontal, and vertical bars; when the 'vertical' and 'red' neurons fire in unison, the resulting assembly represents a vertical red bar. The position I'm proposing, on pure speculation, requires that this isn't exceptional, that rudimentary logical structuring is widely represented, one way or another.

If all this is granted, what role is psychology playing in this philosophy of logic? The ground of logical truth, what makes it true (where it is true), is the objective logical structuring in the world, so there's no trace of psychologism. Psychology's role, then, might be thought to be epistemological. For example, a sufficiently externalist epistemologist, one who thinks the evolutionary pressures responsible for our logical cognition produce a reliable process,[24] might conclude that we know (at least some of) the world's logical structure a priori. I prefer to leave the policing of 'know' and 'a priori' to the specialists and to say only this much: we come to believe what we do about the logical structures in the world on the basis of primitive cognitive mechanisms, many of which we share with other animals, but our evidence for the correctness of those beliefs comes from common sense and its subsequent (partial) ratification by scientific means.

So far, this is a fairly slight philosophical impact for psychology, but I think there's an important moral concerning our philosophical preconceptions

---

23 See Vogel, Woodman, and Luck (2001) for discussion and references. Also Olson and Jiang (2002).

24 This needn't be a fallacious argument of the form 'this evolved, therefore it's reliable'. Instead, it might run roughly along the lines traced here: first science establishes that much of the world is logically structured; then psychology defends an evolutionary story of how we come to detect and represent that structure.

about logical truth. Because our logical beliefs rest on such primitive cognitive mechanisms, it's hard for us to see how things could be otherwise, how a world failing to instantiate those rudimentary logical forms is even possible. When quantum mechanics shows us not only that a world can fail to do this but also that our very own micro-world so fails, often the result is that we find quantum mechanics deeply problematic, not that we take logic itself to be contingent.[25] It seems to me that the psychology here is showing us *why* we're so easily inclined to believe that logical truth is necessary, a priori, certain—a stubborn preconception that vastly distorts our theorizing about it.[26] It's hard to imagine a more valuable lesson for the philosophy of logic!

2. Arithmetic

Obviously, any patch of the world with logical structuring into objects with properties, standing in relations, will also have number properties: so many objects, so many with this particular feature, so many standing in this relation to this particular individual, and so on. When it comes to our cognitive access to those number properties, though, it's well-known that the first four or so have special status: infants' expectations about how many objects will appear behind a screen after individual objects have been added or removed are accurate up to three; adults can hold three to four objects in working memory[27] and track three to four objects through complex motions, but these abilities break down quickly for higher quantities. Nonhuman animals share these abilities and limitations, indicating another primitive cognitive system.[28]

The mechanism underlying these abilities—the object-tracking or parallel individuation system—apparently includes so-called object files of mid-level vision,[29] which follow objects spatiotemporally and encode features as they go ("it's a

25 I suspect many of us have heard our fellow philosophers assert with great confidence that quantum mechanics must be false on a priori grounds.
26 See Maddy (2014a), Chapter 6, for a comparison of this conclusion with the late Wittgenstein's take on logic.
27 Feigenson (2011) describes how visual working memory can encode more than four slots worth of information by 'chunking', as when we remember a phone number by dividing it into three blocks of digits, or one of her infant subjects remembers two cats and two cars, but not four individual cats. (See also Carey (2009, 149–150).) The 'chunk' is often referred to as a 'set', exhibiting the higher ranks that differentiate sets from mere aggregates. I once appealed to analogous considerations (e.g., in Maddy (1990, 165), but for what it's worth, I'm no longer convinced anything essentially 'higher order' is involved in such cases. Seeing two cats and two cars could just be a particular way of seeing the cats and the cars, not a way of seeing something else (a set of cats, a set of cars).
28 See Maddy (2007, 319–326), for more on the story in this and the following three paragraphs, with references. See Carey (2009), Chapter 8, for her elaboration.
29 Kahneman et al. (1992). See Maddy (2007, 255–257, 319–320) for a brief discussion with references.

bird, it's a plane, it's Superman"), and visual working memory, which keeps visual information accessible over short periods.[30] The two are closely intertwined, with some evidence of complementary emphasis on tracking over motion and retention of object properties, respectively.[31] Though the infant expectation experiments are often described in arithmetic terms—$1 + 1 = 2$, $3 - 2 = 1$—it's widely agreed that these representations are not truly numerical: not '3', but the simply logical 'a thing, another thing, and yet another thing', most likely the opening of three successive object files.[32] In cases of 'subitizing'—immediate recognition (without counting) of up to three or four objects—perhaps visual working memory is engaged, but again, mostly likely through the opening of three distinct information slots[33] rather than an explicit numerical representation.

Yet another primitive system we share with other animals is sensitive to approximate quantities: it can distinguish one dot from three more easily than two dots from three (the 'distance effect'); it can distinguish two dots from three, but not eight dots from nine (the 'magnitude effect').[34] The mechanism for this is so far unknown (at least to me), but neurological studies on monkeys suggest a two-step process that begins with a group of neurons that encode locations of objects, ignoring other features, and then feeds into an array of neurons whose responses are bell-shaped curves, each peaking at a certain number.[35] This model would explain the distance effect—the ranges of firing for 'one-neuron' and 'three-neurons' overlap less than those for 'two-neurons' and 'three-neurons'—and the magnitude effect—the bell curves for large numbers are broader. In any case, this is clearly a more quantitative system than the object tracker, but it can't truly be said to represent cardinality. Burge (2010, p. 482) suggests a return to the ancient notion of 'pure magnitude', neither continuous nor discrete, but nevertheless stands in ratios. However that may be, what matters for our purposes is that features of the world's logical structure are being represented, albeit only approximately.[36]

---

30 See, e.g., the references in footnote 23.
31 See Hollingworth and Rasmussen (2010). The two are often lumped together without comment, or even identified.
32 Burge points out, "There need be no use of conjunction or negation in the perceptual representation (as in: this is a body and this is a body and this is not that)" (Burge (2010, 486). He's right: it's unlikely that anything like this is encoded in a language of thought. But in Hatfield's terms, the opening of three successive object files could represent the corresponding logical feature of the scene.
33 See Chesney and Haladjian (2011) for evidence that subitizing and object tracking rely on a shared visual mechanism.
34 See, e.g., Carey (2009, 118–137) and Dehaene (2011).
35 See Dehaene (2011, 247–254) and Nieder (2011).
36 Oddly enough, on small numbers, where the two systems overlap, the infant's object-tracking system appears to override the approximate system. For example, they prefer a box where three treats have been placed to a box where one or two treats have been placed, but when the numbers are two and five, beyond the object tracker's capacity, they perform at chance—despite the fact that the ratio is big enough for the approximate system to detect easily. See Carey (2009, 84–85, 139–141, 153–155).

So far, we're in step with the nonhuman animals, still far short of human arithmetic. The leading theory is that what sets us apart is the child's ability to combine the proto-numerical fruits of the object-tracking system and the approximate system via her command of the counting sequence.[37] 'One, two, three . . .' is first learned as a verbal nonsense scheme—such as 'eeny, meeny, miney, moe . . .'—and the act of reciting it while pointing to each of a group of objects in turn is just play, of no numerical significance. Young children do realize that the use of the word 'one' correlates with the presence of a single object, with a single opened object file or a single item in visual working memory, but the sense of larger number words comes only gradually, between two and a half and three and a half: first 'two' is associated with the presence of an object and another; a few months later, 'three', and maybe even 'four', gains meaning from the object-tracking system.[38] This far the nonhuman animals can follow, but what happens next is uniquely human: apparently the child notices that an extra object in the scene corresponds to the next number in the counting sequence, and suddenly, the true meaning of counting becomes clear: the last number recited in the procedure is the number of objects in the scene.

It's sometimes assumed that this is the end of the story of how humans come to a full understanding of arithmetic, but it isn't, for at least two reasons. First, consider a child who knows how to count and knows there are 'just as many' of these as those when the same number word results from counting these as those. That is, she knows that if she counts *n* children and *n* cookies, she'll be able to give each child exactly one cookie with no cookies left over. Richard Heck makes the case that a child can know all this without having the notion of a one-to-one correspondence, which is, after all, 'very sophisticated' Heck (2000, 170). Of course, when she counts, she forms what we understand to be a one-to-one correspondence, but she needn't understand it as such; she's just implementing the counting procedure.[39] So this is one respect in which the child still hasn't grasped a notion some consider essential to the concept of 'cardinal number'.

Another tempting assumption is that a child who understands that one more object corresponds to the next number word must also understand

---

37 Here again, Piaget was in disagreement. See Dehaene (2011, 30–36) for an amusing account of how the empirical results were misinterpreted.

38 Some hold that the object-tracking system isn't involved, that the underlying mechanism here is the approximate number system (ANS) (see, e.g., Piazza (2011)), which is most precise for small numbers. Dehaene (2011, 256–259), who once entrusted small numbers to the ANS (what he calls 'the number sense'), explains what changed his mind.

39 Heck also notes that the child can understand 'just as many' without understanding counting: there are just as many cookies as children if she can make sure everyone has exactly one cookie with none left over. He then shows how the Peano axioms can be derived with 'just as many' in place of Frege's 'one-to-one correspondence'.

that there's no largest number.[40] The only empirical study touching on this question that I know of, Harnett and Gelman (1998), actually aims to show that it's relatively easy for children to learn that the number sequence has no end—easy compared to learning fractions!—so its design includes more coaching than would be ideal for present purposes. Still, children in kindergarten and first grade[41] did quite poorly on questions such as "is there a biggest number of all numbers?" and "is there a last number?" They did somewhat better but still far from perfectly on leading questions such as "if we count and count and count, will we ever get to the end of the numbers?" and "can we always add one more, or is there a number so big we'd have to stop?", despite having been primed with exercises in counting larger and larger numbers.[42]

Explaining their answers, the six-year-olds might suggest that we have to stop counting "'cause you need to eat breakfast and dinner" or "because we need sleep", or that we couldn't then start up again where we left off because "you forget where you stopped". There's even a hint at mortality: if we try to add one more after counting to a very big number, "I guess you'll be old, very old". Though answers like these were classified as 'unacceptable', there is a straightforward sense in which the children have it right: there *are* practical limitations on how far we're inclined to count, and even physical limits on how far we could count.[43] The young children aren't wrong exactly; they're just failing to grasp the spirit of the question. What's being asked is whether there's any limit to how far we could count, *in principle*.

In contrast to the kindergarteners and first graders, the second graders[44] in this study generally answered the questions as they were intended: there is no largest number, period. Closer analysis of the experimental results led Harnett and Gelman to the observation that the children in a position "to benefit from a conversation that offers cues" (p. 361) were those who could count beyond 100:

> Once children master the sequence from 1 to 20 and the list of decade words, they have most but not all of the vocabulary they need to apply the recursive procedures by which larger and larger numbers are generated. As they count beyond 100, they come to learn that not only the digits, but also the decade terms, are recycled over and over. [Younger]

---

40 For a bit more on the line of thought in the remainder of this section, see Maddy (2014b).

41 Averaging just under 6 and 7 years old, respectively.

42 One group of subjects in one of the studies was questioned about the largest number, etc., before the counting exercises. Their performance was even worse than the group who did the counting exercises first.

43 Russell once remarked that running through an infinite decimal expansion is "*medically impossible*" Russell (1935/6, 143).

44 Averaging just under 8 years old.

[c]hildren are still at work memorizing the teens and decade terms and are less able to appreciate that the count sequences is systematic.
(Harnett & Gelman, 1998, 361)[45]

This suggests, as the psychologist Paul Bloom proposes, that

the generative nature of human numerical cognition develops only as a result of children acquiring the linguistic counting system of their culture. Many, but not all, human groups have invented a way of using language to talk about number, through use of a recursive symbolic grammar.
(Bloom, 2000, 236)

This would mean that children's belief in the infinity of the numbers derives from their belief in the infinity of numerical expressions, not vice versa:

[I]t is not that somehow children know that there is an infinity of numbers and infer that you can always produce a larger number word. Instead, they learn that one can always produce a larger number word and infer that there must therefore be an infinity of numbers.
(Ibid., 238)

In this way, our question—how do we come to believe there's no largest number?—is pushed back one step to how do we come to believe that there's no largest numerical expression?

Harnett and Gelman's studies show that it's quite easy for children to come to this view once they've appreciated the intricacies of the systematic generation of numerical expressions. What's striking is that they don't seem bothered by concerns about the practical or physical limitations on, for example, the length of those numerical expressions or the breathe needed to utter them or the need to stop for lunch—all that apparently matters is grasping the recursive character of the rules of formation. Why is the intended 'in principle' reading of the question more natural here when it's posed for numerical expressions than it was when posed for the numbers themselves? To engage once again in rank speculation, I suggest that this traces to the recursive element of the innate linguistic faculty, whatever it is in our genetically endowed cognitive machinery that underlies our ability to understand and produce indefinitely varied and complex linguistic items:

All approaches agree that a core property of [the linguistic faculty] is recursion . . . [The linguistic faculty] takes a finite set of elements

---

45 Though Harnett and Gelman speak of "recursive procedures by which larger and larger numbers are generated"; obviously, they're talking about linguistic procedures that generate numerical expressions. (Understanding that adding one results in a larger number was another predictor for successful response to the cues.)

and yields a potentially infinite array of discreet expressions. This capacity . . . yields a discrete infinity (a property that also character-izes the natural numbers).

(Hauser, Chomsky, & Fitch, 2002, 1571)

The suggestion is that this linguistic capacity is what produces our intuitive grasp of the 'in principle' question.

Assuming this sketch of the psychology is roughly right—a big assump-tion, subject to empirical test—the consequences for the philosophy of arithmetic are fundamental. Simple arithmetical claims such as $2 + 2 = 4$ and $12 < 191$ are ordinary facts about worldly logical structures (where they're present), but the subject matter of mathematical arithmetic—the standard model, what we now think of as an omega-sequence—doesn't depend on any contingent features of the actual world, which may or may not be finite. Insofar as arithmetic is 'about' anything, it's about an intuitive picture of a recursive sequence of potentially infinite extent—an intuitive picture we humans share thanks to the evolved linguistic faculty common to our species.

Now, we all tend to believe that the structure of the standard model of arithmetic, that simple omega-sequence, is coherent, unique, and deter-minate. But if it's really just a matter of an intuitive picture, what reason do we have to believe these things? As Wittgenstein once asked, "What if the picture began to flicker in the far distance?" (*RFM*, V.10). Our innate cognitive structuring may well give rise to these firm convictions, but if the story told here is correct, our capacity for mathematical arithmetic could be a mere spandrel, generated just by the way we evolved toward lan-guage, and even if it is an adaptation in itself, that's no guarantee of reli-ability.[46] Under the circumstances, we reflective beings should want more support for our faith in the cogency of an omega-sequence than just our brute inclination to believe it. I think there are facts we can appeal to, but they're hardly conclusive: our biological similarity as humans is reason to think your intuitive picture is more or less the same as mine; the apparent coherence of the picture, plus long experience of the species with math-ematical arithmetic, provides some evidence for its consistency; the lack of any important independent statement comparable to the Continuum Hypothesis (CH) suggests it may be fully determinate.[47] But our sense that arithmetic is more secure than that may be an illusion—another valuable lesson from psychology!

46 See the fallacy described in footnote 24.

47 There's an analogous question for set theory, where the relevant intuitive picture—the it-erative hierarchy—seems to rest on three elements: recursion (presumably based in the same cognitive faculty as the standard model of arithmetic); the combinatorial notion of an arbitrary subset, not beholden to any rule, definition, or construction (perhaps related to Heck's 'very sophisticated' one-to-one correspondence?); and Cantor's gutsy bet on the

3. Conclusions

Though psychologists sometimes take their work to support a brand of anti-realism about mathematics—Stanislas Dehaene's influential *Number Sense*, for example, bears the subtitle *How the Mind Creates Mathematics*—in fact, their skepticism doesn't extend to the contingent logical/numerical structure I've been attributing to the world or our cognitive access to it:

> [A]rithmetic . . . draw[s] upon a store of fundamental knowledge accumulated over millions of years of evolution *in a physical world which, at the scale we live it, is . . . numerically structured.*
>
> (Dehaene in Dehaene and Brannon (2011, 187),
> emphasis added)

This type of straightforward realism breaks down, I've suggested, with the potential infinite, the standard model of arithmetic, where attention to the psychological facts reveals that our cognitive architecture does, in a sense, 'create' the subject matter under investigation. In addition to this positive semantic or metaphysical conclusion, empirical work in psychology also uncovers the less-than-firm underpinnings of some of our firmest philosophical preconceptions: that logic is necessary and that arithmetic is obviously cogent (coherent, unique, determinate). This valuable therapeutic helps free the philosophies of these subjects from traditional baggage and sets them on a more vital course. In these ways, psychological inquiry stands to play a central and highly beneficial role in our philosophizing about the a priori disciplines.[48]

References

Anderson, B. (2015). Can computational goals inform theories of vision? *Topics in Cognitive Science*, 7, 274–286.

Bloom, P. (2000). *How children learn the meanings of words*. Cambridge, MA: MIT Press.

Burge, T. (2010). *Origins of objectivity*. Oxford: Oxford University Press.

Carey, S. (2009). *The origin of concepts*. New York: Oxford University Press.

---

completed infinite (see Maddy (1988, I.5)). This picture isn't definitive of the field in the way the standard model is for arithmetic: it wasn't present when set theory was founded by Cantor and others, and it could be altered or replaced in the future (e.g., by the multiverse conception, though for now I'm skeptical about that (see Maddy (2017a, III)). In any case, given the added vagaries of the two additional elements, any case for cogency is correspondingly weaker: determinacy is undercut by independent statements such as the CH, and our biological similarity gives less support for uniqueness. Perhaps the apparent coherence of the conception delivers some evidence of consistency, but considerably less than in the case of arithmetic. Still, this would be a form of so-called intrinsic support distinct from the merely instrumental role described in Maddy (2011).

48 Thanks to Gary Hatfield, Ethan Galebach, Reto Gubelmann, and Sorin Bangu for helpful comments on earlier drafts.

28   *Penelope Maddy*

Chesney, D., & Haladjian, H. (2011). Evidence for a shared mechanism used in multiple object tracking and subitizing. *Attention, Perception, and Psychophysics, 73*, 2457–2480.

Dehaene, S. (2011). *The number sense: How the mind creates mathematics* (revised and expanded ed.). New York: Oxford University Press. (First edition, 1997).

Dehaene, S., & Brannon, E. (Eds.). (2011). *Space, time and number in the brain: Searching for the foundations of mathematical thought*. Amsterdam: Elsevier.

Feigenson, L. (2011). Objects, sets, and ensembles. In S. Dehaene & E. Brannon (Eds.), *Space, time and number in the brain: Searching for the foundations of mathematical thought* (pp. 13–22). Amsterdam: Elsevier.

Harnett, P., & Gelman, R. (1998). Early understandings of numbers: Paths or barriers to the construction of new understandings? *Learning and Instruction, 8*, 341–374.

Hatfield, G. (1988). Representation and content in some (actual) theories of perception. In *Perception and cognition: Essays in the philosophy of psychology* (pp. 50–87). Oxford: Oxford University Press.

Hatfield, G. (1991). Representation in perception and cognition: Task analysis, psychological functions, and rule instantiation. In *Perception and cognition: Essays in the philosophy of psychology* (pp. 88–123). Oxford: Oxford University Press.

Hatfield, G. (2009a). *Perception and cognition: Essays in the philosophy of psychology*. Oxford: Oxford University Press.

Hatfield, G. (2009b). Getting objects for free (or not): The philosophy and psychology of object perception. In *Perception and cognition: Essays in the philosophy of psychology* (pp. 212–255). Oxford: Oxford University Press.

Hauser, M., Chomsky, N., & Fitch, T. (2002). The faculty of language: What is it, who has it, and how did it evolve? *Science, 298*, 1569–1579.

Heck, R. (2000). Cardinality, counting, and equinumerosity. Reprinted in his *Frege's Theorem* (pp. 156–179). Oxford: Oxford University Press, 2011.

Hollingworth, A., & Rasmussen, I. (2010). Binding objects to locations: The relationship between object files and visual working memory. *Journal of Experimental Psychology: Human Perception and Performance, 36*, 543–564.

Kahneman, D., Treisman, A., and Gibbs, B. (1992). The reviewing of object files: Object-specific integration of information. *Cognitive Psychology, 24*, 175–219.

Maddy, P. (1988). Believing the axioms. *Journal of Symbolic Logic, 53*, 481–511, 736–764.

Maddy, P. (1990). *Realism in mathematics*. Oxford: Oxford University Press.

Maddy, P. (2007). *Second philosophy*. Oxford: Oxford University Press.

Maddy, P. (2011). *Defending the Axioms*. Oxford: Oxford University Press.

Maddy, P. (2014a). *The logical must: Wittgenstein on logic*. New York: Oxford University Press.

Maddy, P. (2014b). A second philosophy of arithmetic. *Review of Symbolic Logic, 7*, 222–249.

Maddy, P. (2014c). A second philosophy of logic. In P. Rush (Ed.), *The metaphysics of logic* (pp. 93–108). Cambridge: Cambridge University Press.

Maddy, P. (2017). *What do philosophers do? Skepticism and the practice of philosophy*. New York: Oxford University Press.

Maddy, P. (2017a). Set-theoretic foundations. In A. Caicedo, J. Cummings, P. Koellner, and P. Larson, (Eds.), *Foundations of Mathematics* (pp. 289–322). Providence, RI: American Mathematical Society, 2017.

Nieder, A. (2011). The neural code for number. In S. Dehaene and E. Brannon (Eds.), *Space, time and number in the brain: Searching for the foundations of mathematical thought* (pp. 103–118). Amsterdam: Elsevier.

Olson, I., & Jiang, Y. (2002). Is visual short-term memory object based? Rejection of the 'strong-object' hypothesis. *Perception and Psychophysics, 64*, 1055–1067.

Piazza, M. (2011). Neurocognitive start-up tools for symbolic number representation. In Dehaene and Brannon (Eds.), *Space, time and number in the brain: Searching for the foundations of mathematical thought* (pp. 67–285). Amsterdam: Elsevier.

Russell, B. (1935/36). The limits of empiricism. *Proceedings of the Aristotelian Society, 36*, 131–150.

Spelke, E. (1988). Where perceiving ends and thinking begins: The apprehension of objects in infancy. In A. Yonas (Ed.), *Perceptual development in infancy* (pp. 197–234). Hillsdale, NJ: Lawrence Erlbaum.

Ullman, S. (1980). Against direct perception. *Behavioral and Brain Sciences, 3*, 373–381.

Vogel, E., Woodman, G., & Luck, S. (2001). Storage of features, conjunction, and objects in visual working memory. *Journal of Experimental Psychology: Human Perception and Performance, 27*, 92–114.

Warren, W. (2012). Does this computational theory solve the right problem? Marr, Gibson, and the goal of vision. *Perception, 41*, 1053–1060.

Wittgenstein, L. (1937/44). *Remarks on the foundations of mathematics* (revised ed., G. Anscombe, Trans, G. von Wright, R. Rhees, & G. Anscombe, Eds.). Cambridge, MA: MIT Press, 1978.

3 Reasoning, Rules, and Representation

Paul D. Robinson and Richard Samuels

Introduction

Regress arguments have had a long and influential history within the philosophy of mind and the cognitive sciences. They are especially commonplace as a bulwark against representational or *intentional* theories of psychological capacities. For instance, arguments of this sort played a prominent role in debates concerning the theory of transformational grammar (Chomsky, 1969a; 1969b; Harman, 1967, 1969), the language of thought hypothesis (Fodor, 1975, 1987; Laurence & Margolis, 1997), the massive modularity hypothesis (Fodor, 2000; Collins, 2005), and intentional accounts of intelligent activity quite broadly (Ryle, 1949; Fodor, 1968; Dennett, 1978). Typically, the regress is presented as one horn of a dilemma:

> To explain the manifestation of some kind of capacity, *C*, the theorist postulates an (intentional) psychological process of kind *P*. But, the critic suggests, the successful operation of any *P* process itself depends upon some prior manifestation of *C*. Thus: Either it is necessary to postulate a second psychological process of kind *P*, and so on, *ad infinitum*, or alternatively, one must grant that *C* can be explained without positing *P*.

Thus the proposed intentional theory is either broken-backed or redundant. Or so proponents of regress arguments would have us believe.

Recently, a similar argument—which we simply call the *Regress*—has surfaced in philosophical debate regarding the nature of reasoning. Participants in this debate are not concerned with everything that gets called 'reasoning'. Rather, they focus on a relatively circumscribed range of reasoning-like phenomena—which they call *active reasoning* or *inference*—[1]a kind of

---

1 Three comments regarding terminology. First, as is common in the present context (but see Broome, 2013, 292), we use 'active reasoning' and 'inference' interchangeably. Second, although it is slightly infelicitous to use 'inference' in this restricted sense, it should be read as such unless explicitly modified—e.g., as in *sub-personal* inference. Third, as is typical, we take it to be true, more-or-less by definition, that a process or activity is active only if it is person-level. As such, we count no sub-personal processes as active.

person-level, conscious, voluntary activity, which at least in paradigmatic instances results in the fixation of belief. It is widely assumed that active reasoning in this sense is fairly pervasive among human beings; that it can involve attitudes with markedly differing contents; that simple, consciously made, deductive inferences are a prototypical case; and that errors in active reasoning are both possible and, indeed, fairly commonplace. For philosophers interested in active reasoning, then, the core explanatory challenge is to provide an illuminating account of the nature of this psychological capacity.

Within this context, the presumed significance of the Regress is that it (allegedly) undermines a family of highly influential accounts of inference—what might be called *intentional rule-following* (or IRF) theories. To a first approximation, such theories make a pair of commitments. First, they suppose that inference essentially involves following rules concerning the premises from which one reasons:

> (Rule-Following View): All active reasoning involves rule-following operations.

In addition, they impose the following necessary condition on rule-following:

> (Intentional View): All rule-following involves intentional states, which *represent* the rules being followed.

In brief, the Regress purports to show that if such accounts were correct, *any* instance of active reasoning—no matter how apparently simple—would be a *supertask* involving an infinite number of rule-following operations. In which case, contrary to fact, it would be impossible for finite creatures such as us to actively reason.

If the Regress were sound, it would have serious implications for philosophical debate regarding the nature of inference. What may be less obvious is that it would also have significant consequences for scientific theories of reasoning, and cognition more broadly. Within the psychology of reasoning, quite generally, and the psychology of *deductive* reasoning in particular, it is commonplace to suppose that reasoning relies on mentally represented rules.[2] This commitment is perhaps most apparent in *mental logic* accounts, where it is explicitly hypothesized that there are "deduction rules that construct mental proofs in the system's working memory" (Rips, 1994, 104). But the commitment is also apparent among dual-process theorists who routinely suppose

---

2 It is worth noting, in this regard, that philosophers writing on inference have tended to focus on cases in which we reason in accordance with logical rules, such as modus ponens. Further, Boghossian (2008, 499) claims that whereas denying the Rule-Following View of reasoning in general seems false, denying the Rule-Following View of deductive reasoning in particular seems 'unintelligible'.

that System 2 processes involve intentional states that represent rules (e.g., Sloman, 1996). Moreover, we suspect—though won't argue here—that even those who explicitly reject the mental logic approach *also* presuppose the existence of intentional states that represent rules. For example, *mental models* accounts seem to presuppose the existence of such states, albeit where the presumed rules are for the manipulation and inspection of iconic models denoting possibilities, as opposed to the construction of mental proofs through chaining linguistic entities such as sentences (e.g., Johnson-Laird, 2008).

Of course, such accounts of reasoning are contentious and may turn out to be false. But on the face of it, this should be an empirical issue, addressed by empirical means. If the Regress is sound, however, such theories should be rejected a priori. Further, as we will show, since the Regress does not turn essentially on assumptions about the nature of *active* reasoning per se, the argument, if sound, would apply to a far broader class of phenomena. Specifically, as we will see, it would apply, with minimal modification, to processes that are unconscious and sub-personal, and, hence, not active. If sound, then, the Regress would have ramifications for a wide array of theories in many regions of cognitive science, including theories of perception.

Fortunately, the Regress is not sound. Formulations of the argument are invariably underspecified, and once presented in suitably perspicuous fashion, it becomes clear that the Regress relies on assumptions no sensible version of IRF should endorse. The primary burden of this chapter is to show why this is so.

Here's how we proceed. In Section 1, we explain the IRF account of reasoning in more detail, and set out some of its *prima facie* virtues. In Section 2, we aim to explain the general structure of the Regress and provide the most charitable formulation of the argument that we can. In Section 3, we discuss a standard—and we think correct—response to this original regress: to posit sub-personal inferential processes. We show that this response provides a plausible way to block the original regress. But following suggestions from Boghossian and others, we also a) show how to develop a Revenge Regress, which targets IRFs about sub-personal processes, and b) explain how to use this result to develop a Strengthened Regress, which fills the gap in the original argument. Finally, in Section 4 we explain why the Strengthened Regress is still subject to a serious objection, and in Section 5 we address two responses to this objection.

1. The Virtues of Intentional Rule-Following Accounts of Inference

The IRF is not so much a single account of inference as a family of proposals that share a common commitment to the Rule-Following View of inference and to the Intentional View of rule-following. In our view, such proposals merit serious consideration because they possess a host of explanatory virtues. We are especially sympathetic to variants of IRF that incorporate some form of computationalism about mental processes—a class that includes the sort of 'classicism' advocated by Fodor and Pylyshyn (1988), versions of connectionism (e.g., Smolensky, 1988), and some recent Bayesian approaches to cognitive modeling (e.g., Perfors, Tenenbaum, & Regier, 2011). These frameworks are among the most plausible extant approaches to the study of higher cognition in general, and reasoning in particular.

Although this is not the place to discuss the virtues of IRFs in detail, a brief reminder should make clear that much is at stake if the Regress is sound. First, consider some of the prima facie explanatory virtues that accrue merely as a result of adopting the Rule-Following View (cf. Boghossian, 2014, 4, 12):

- *A theory of reasoning should discriminate reasoning from mere causation by belief (and other intentional states).* Not all instances of beliefs causing other beliefs are inferences. Notoriously, there are "deviant" causal chains involving beliefs that are obviously non-inferential.[3] The rule-following account helps to explain the difference. Very roughly, in the case of inference, the influence of belief is wholly mediated by rule-following operations and in the other cases, not.

- *Since not all reasoning is good reasoning, we should prefer, on grounds of generality, an account that covers both the good and the bad.* Rule-following accounts can capture this desideratum. On such views, one can reason badly, either by following a bad rule or by making mistakes in one's attempt to follow good rules. In contrast, good reasoning only occurs when one correctly follows a good rule.

- *A theory of reasoning ought to explain the sorts of generality that are exhibited by inference.* For example, it is widely recognized by philosophers and psychologists that we are capable of reasoning about an exceedingly broad array of topics—roughly, any topic for which we possess concepts. Moreover, our inferences often exhibit similar patterns or 'logical forms' across these various topical domains. Rule-following accounts provide promising explanations of such phenomena. Specifically, if some inferential rules are akin to logical rules in being largely 'content independent' or 'formal', then we have a partial explanation of why we are able to reason about so many different subject matters. Further, if we suppose that humans follow these rules in lots of different contexts, we will have an explanation of why inferences in different domains exhibit similar forms.

- *A theory of active reasoning should both subsume and explain the difference between deductive and inductive reasoning.* Once again, the rule-following picture offers a natural account. When reasoning deductively, the relevant rule-following operations involve deductive rules, and when one reasons inductively, the relevant rules are inductive ones.

---

3 Example: Suppose John believes that he's late for class and that this realization makes him sweat. If on the basis of this experience he came to believe that he was sweating, we could have a case of causation by belief, but not inference.

No doubt there are other issues that the Rule-Following View might help address, but let's turn to the Intentional View. As we see it, there are two deep and closely related explanatory motivations for this view. The first is what we call the *Guidance Problem*. The aforementioned explanatory virtues of the Rule-Following View all turn on the assumption that rules can in some sense *guide* our cognitive activities. But how is this possible? After all, a rule *qua* rule is 'just an abstract object' and so presumably incapable of exerting any causal influence (Boghossian, 2014, 13).

Here's where the Intentional View enters the picture. Although rules as such cannot guide cognition, intentional states that encode or represent such rules can. For in addition to their representational properties, intentional states have other properties that are causally relevant—various physical and structural properties, for example. On the Intentional View, then, rules guide behavior in an attenuated sense: they are the contents of intentional states—rule-representations—that are causally implicated in reasoning.[4]

A second and related virtue of the Intentional View is that it helps resolve a very old problem for rule-following accounts of cognition. In brief, such accounts presuppose a distinction between following a rule and mere *accordance* with a rule (see e.g., Hahn & Chater, 1998, 203f.). Without such a distinction, rule-following *per se* will be of little use in explaining what is distinctive of reasoning. For it will turn out that all processes describable by a rule—that is, all processes that display regularity in their behavior—are rule-following processes. In which case, it will be no more true of reasoning that it involves rule-following than it will be of, say, the planets that they 'follow' a rule when conforming to Kepler's laws of planetary motion (Fodor, 1975).

Again, we think that the Intentional View provides a credible approach to this problem. According to this approach,

> what distinguishes what organisms do from what the planets do is that a *representation of the rules they follow* constitutes one of the causal determinants of their behavior.
>
> (Fodor, 1975, 74)

---

4 Although it does not require endorsing the Intentional View in its full generality, it is also worth noting that the idea that rules are encoded by intentional states helps explain what is otherwise a puzzling fact about human beings—namely, we are capable of learning rules that influence our behavior on the basis of "one-shot" instruction or linguistic communication. For example, if, at passport control, the guard tells me, "Stand behind the yellow line, until you are called", I stand behind the yellow line and wait to be called! On the basis of one exposure to instruction, my behavior is modified so that I follow the rule. This is readily explained on the assumption that, on the basis of linguistic processing, I come to possess one or more intentional states that represent the content of the guard's utterance—i.e., the rule.

In contrast, where the planets are concerned,

> at no point in a causal account of their turnings does one advert to a structure which encodes Kepler's laws and causes them to turn. The planets *might* have worked that way, but the astronomers assure us that they do not.
>
> (ibid.)

In summary, if the Intentional View is correct, we have a prima facie plausible way both to resolve the Guidance Problem and to draw the rule-following/rule-accordance distinction. Moreover, since a solution to these problems is a prerequisite for the Rule-Following View to have any explanatory value, it is exceedingly attractive to combine the Intentional and Rule-Following views in the manner proposed by IRFs.

Of course, the aforementioned is defeasible, and matters would be quite different if there were powerful independent reasons to reject IRFs. With this in mind, we turn to the Regress.

2.   The Regress

Although a number of theorists have invoked variants of the Regress, Boghossian's discussion strikes us as the most perspicuous, to date, and for this reason, we focus primarily on it here. In Section 2.1, we lay out Boghossian's general strategy. In Section 2.2, we explain how he aims to establish a crucial premise of the argument, what we call the Rule Application Condition. Then in Section 2.3, we sketch the Regress itself, and in Section 2.4, we provide a more regimented formulation of the argument.

2.1   The General Strategy

It is important to distinguish the Regress, advocated by Boghossian and others, from a range of superficially similar worries. The relevant regress is *not* an epistemic one. It is not, for example, a regress with respect to justification, or reasons for belief. Nor is it a regress concerning the determination of meanings or contents of the sort associated with Kripke's Wittgenstein (Kripke, 1982). Finally, it is not a definitional regress wherein the definiendum—'inference'—is to be defined in terms of 'rule-following', which in turn is to be defined in terms of 'inference', and so on. Rather, the problem allegedly posed by the Regress is a regress of mental *operations*. The worry, in brief, is that IRFs place active reasoning beyond the grasp of finite creatures by turning every instance of inference into a *supertask*: an infinite sequence of rule-following operations to be performed in a finite period of time.

Here is the general strategy: to generate the desired regress, Boghossian seeks to show that the IRF entails the following interlocking pair of conditions:

Condition 1: Each inference I_i requires a rule-following operation R_i.
Condition 2: Each rule-following operation R_j requires some further inference I_j.

Given these conditions, we can generate a regress of mental operations by cycling between them:

Suppose I draw inference I_1;
by Condition 1: I perform a rule-following operation R_1;
by Condition 2: I draw an inference I_2.
by Condition 1: I perform a rule-following operation R_2;
by Condition 2: I draw an inference I_3.
And so on . . .

A regress of mental operations ensues. In which case, if the conjunction of the Rule-Following View and Intentional View entail these conditions, then IRFs turn all inferences into supertasks.

2.2.   Establishing Condition 2

How does Boghossian seek to establish that IRFs are committed to Conditions 1 and 2? Since the Rule-Following View *asserts* that all inference involves rule-following operations, Condition 1 is easily secured. Condition 1 is just a rendering of the Rule-Following View. In contrast, neither the Rule-Following View nor the Intentional View asserts Condition 2. Boghossian's main argumentative burden, then, is to show that they entail it.

How is this to be done? Rather than focusing on rules of inference, Boghossian initially discusses a simple *decision* rule with the aim of drawing out some general morals regarding what, on the Intentional View, would be required for active, person-level rule-following:

Suppose I receive an email and that I answer it immediately. When would we say that this behavior was a case of following:

(Email Rule) Answer any email that calls for an answer immediately upon receipt!

as opposed to just being something that I happened to do that was in conformity with that rule?

Clearly, the answer is that it would be correct to say that I was following the Email Rule in replying to the email, rather than just conforming to it, when it is because of the Email Rule that I reply immediately.

(2014, 13)

Of course, this immediately raises an instance of the Guidance Problem: what is it to follow this rule, as opposed to merely *conforming* to it? Since the E-mail Rule 'qua rule, is just an abstract object' it cannot *directly* guide behavior. Instead, if the Intentional View is correct,

> my behavior is to be explained via some state of mine that represents or encodes that rule.
>
> (Ibid.)

Merely positing such an intentional state does not, however, fully explain this particular instance of rule-following activity. There also needs to be a process in which this rule-representation might figure so as to guide my behavior. And, according to Boghossian, it is plausible that this process conforms to the following pattern:

> I have grasped the rule, and so am aware of its requirements. It calls on me to answer any email that I receive immediately. I am aware of having received an email and so recognize that the antecedent of the rule has been satisfied. I know that the rule requires me to answer any email immediately and so conclude that I shall answer this one immediately.
>
> (Ibid.)

Of course, this is only one specific instance of rule-following activity. Nevertheless, Boghossian takes it to illustrate what, on the Intentional View, active rule-following *in general* would require:

> On this Intentional construal of rule-following, then, my actively applying a rule can only be understood as a matter of my grasping what the rule requires, forming a view to the effect that its trigger conditions are satisfied, and drawing the conclusion that I must now perform the act required by its consequent.
>
> (Ibid.)

Notice—and this is the crucial point—that this appears tantamount to claiming that "on the Intentional view of rule-following, rule-following requires inference" (ibid.). More precisely, Boghossian appears to be insisting that, on the Intentional View,

> *Rule Application Condition:* For a person-level rule-following process to utilize a rule-representation, it must contain an inferential subprocess—an inference from the rule to what the rule calls for under the circumstances.

And, of course, if this is true, then so too is Condition 2. That is, the Rule Application Condition entails that each rule-following operation R_j requires some further inference I_j.

2.3.   *The Regress Within Reach*

If the aforementioned is correct, then an *intentional* rule-following account of inference is committed to both Conditions 1 and 2. As Boghossian puts it,

> On the one hand, we have the Intentional View of rule-following, according to which applying a rule always involves inference. On the other hand, we have the Rule-Following picture of inference according to which inference is always a form of rule-following.
>
> (2014, 14)

Further, if this is so, then it would seem that any instance of person-level rule-following must involve an infinite series of further rule-following operations. If, for example, I actively follow the E-mail Rule, then I must draw an inference in order to follow it, and since, by assumption, this involves rule-following, I must draw another inference, which requires another instance of rule-following, and so on *ad infinitum*. Thus Boghossian concludes,

> These two views . . . can't be true together. Combining the two views would lead us to conclude that following any rule requires embarking upon a vicious infinite regress in which we succeed in following no rule.
>
> (2014, 14)

Boghossian is not alone in drawing this pessimistic conclusion. For instance, for the same reason, Wright claims that if rule-following requires a state that carries a content that licenses the inferential transition, then it is "uncertain that any coherent—regress-free—model can be given of what inferring actually is" and hence,

> we must drop the idea that inference is, everywhere and essentially, a kind of rule-following. That, in outline, is the solution to the problem of the Regress.
>
> (2014, 32f.)

Similarly, Broome claims that if rule-following requires an explicit representation of a rule, then

> you would have to determine whether each particular case of potential reasoning falls under the rule. Doing so would require reasoning, which would again require following a rule. There would be a circle.
>
> (2014, 632)

In short, some very influential philosophers maintain that the Regress undermines IRF.

2.4. The Regress Regimented

With the earlier exegetical work complete, we are now in a position to set out the Regress in full dress. Our aim is to capture the details and spirit of Boghossian's discussion as charitably as possible, though without logical lacunae. What follows is our best effort.

The Regress proceeds from the assumption that IRF is true, to the untenable conclusion that active reasoning—or inference—is impossible for finite creatures. And since the Regress targets IRFs about *active* reasoning, it is natural to formulate the Rule-Following View and Intentional View in person-level terms. That is,

(1) Any process of inference is a kind of person-level rule-following.
(2) Any process of person-level rule-following utilizes a person-level rule-representation.

Here (1) and (2) clearly characterize a version of IRF about inference, or active reasoning. But as we saw earlier, without additional premises, they do not suffice to generate a regress. Rather, one must further maintain a version of what we earlier called the Rule Application Condition:

(3) For a person-level, rule-following process to utilize a personal-level rule-representation, it must contain an inferential sub-process—a person-level inference from the rule to what the rule calls for under the circumstances.

Here 'sub-process' is to be understood as referring to a *proper* part of the person-level, rule-following process.[5] It follows from (1), (2), and (3) that

(4) Any inferential process involves an inferential sub-process.

Since any such inferential sub-process is itself an inference, it will also involve an inferential sub-process, and so on, *ad infinitum*. Hence by iteration on (4) we have

(5) Any inference requires infinitely many inferential sub-processes.

But given that the performance of infinitely many inferences cannot be carried out in finite time, it follows from (5) that

(6) Inference is impossible for finite beings like us.

---

5 This is required to block an interpretation of (3) on which the rule-following process is *identified with* the inference from the rule to what the rule calls for. For if such an identification is made, (3) will not generate a regress of operations.

Yet active reasoning *is* possible for creatures such as us. At any rate, this is what Boghossian, Broome, Wright, and almost *everyone*—including the present authors—suppose.[6] In which case, on the assumption that the IRF is true, we appear to have reason to reject the IRF.

3.   Sub-personal Processes, Revenge, and the Strengthened Regress

3.1   *Getting Sub-personal*

How should proponents of IRFs respond to the Regress? If premises (1)–(3) are true, then regress ensues. But it is plausible to reject premise (2) in favor of a weaker requirement. Specifically, proponents of IRFs may allow that person-level rule-following *sometimes* involves person-level rule-representations, while insisting that *sub-personal* rule-representations may also play the requisite role. The resulting variant of the Rule-Following View can be formulated as follows:

(2**)   Any process of person-level rule-following utilizes a rule-representation that is either personal or sub-personal.

This modification evades the Regress. Moreover, it does so in an independently motivated and independently plausible fashion. Although the personal/sub-personal distinction is a notoriously vexed one (see Drayson, 2012), for present purposes, the crucial requirement is—as Boghossian recognizes—that person-level processes are "processes of which we are, in some appropriate sense, aware" (2008, 483). In contrast, sub-personal states are "not consciously accessible to the thinker" (2014, 15). Yet if this is how we are to draw the personal/sub-personal distinction, it should be clear that any plausible IRF will need to insist that person-level rule-following quite typically involves *sub-personal* rule-representations. This is because, as Boghossian, Broome, and many others recognize, active reasoners very typically *lack* conscious awareness of following a rule. In which case, proponents of IRFs have exceedingly good reason to insist that rule-representing states are often sub-personal. Moreover, this has nothing to do with regress worries per se. Rather, it is mandated by the antecedent assumption that active reasoning is a commonplace cognitive activity, along with the overwhelmingly plausible empirical claim that active reasoners very frequently lack conscious awareness of any relevant rule or rule-representing state. More generally, the point is that once a theory posits representational states to explain a cognitive phenomenon, the hypothesized states must be sub-personal if the agent lacks conscious awareness of them. In this regard, proponents of IRFs are

6 Presumably, some eliminativists, behaviorists, and the like would deny this.

in the same predicament as psycholinguistics who posit sub-personal representations of syntactic rules, or vision scientists who posit sub-personal representations of edges. And, in our view, this is not bad company to keep.

3.2. *The Revenge Regress*

We have argued that weakening (2) in the proposed manner both evades the earlier regress and is independently plausible. Nonetheless, the proponent of IRF is not out of the woods yet. For, as critics note, a closely analogous sub-personal regress can be generated. Thus Boghossian maintains,

> In the present context, going sub-personal presumably means identifying rule- acceptance . . . not with some person-level state, such as an intention, but with some sub-personal state . . . Let us say that [such a state] is some sub-personal intentional [i.e., representational] state in which the rule's requirements are explicitly encoded. Then, once again, it would appear that some inference (now sub-personal) will be required to figure out what the rule calls for under the circumstances. And at this point the regress will recur.
>
> (2008, 498)

The core insight of the earlier passage is that merely extending the Intentional View to cover sub-personal rule-following does little to alter the overall structure of the IRF. In which case, one might think that if utilizing a person-level rule-representation requires an inferential sub-process, then utilizing a sub-personal rule-representation will also require an inferential sub-process—albeit a *sub-personal* one. And if this is so, then we can generate a *Revenge Regress* that mirrors the original:

(1*)   Any sub-personal inference is a kind of sub-personal rule-following.
(2*)   Any process of sub-personal rule-following utilizes a sub-personal rule-representation.
(3*)   For a sub-personal rule-following process to utilize a rule-representation, it must contain an inferential sub-process—a *sub-personal* inference from the rule to what the rule calls for under the circumstances.

From (1*– 3*) it follows that

(4*)   Any sub-personal inference involves a sub-personal inferential sub-process.

And since any such inferential sub-process is itself a sub-personal inference, by iteration on (4*) we may infer the following:

(5*)   Any sub-personal inference requires infinitely many sub-personal inferential sub-processes.

Finally, given that the performance of infinitely many sub-personal inferences cannot be carried out in finite time, it follows from (5*) that

(6*) Sub-personal inference is impossible for finite sub-personal systems like ours.

3.3   *Regress Strengthened*

No doubt, this conclusion will be welcome to those already suspicious of intentional explanations of the sort found in cognitive science. For the proponent of IRF, however, the Revenge Regress is exceedingly unfortunate. Supposedly, by allowing for sub-personal rule-representation, IRFs have an independently plausible way to escape the original Regress. But if the Revenge Regress is sound, the escape route is blocked, and the IRF is left without a way to account for active reasoning.

In our experience, the significance of the Revenge Regress is not always clearly appreciated. One problem is that it targets a different phenomenon from the earlier Regress—i.e., *sub-personal* inference. Why, then, should it be relevant to theories of *person-level* inference? Another problem is that proponents of the Regress never spell out in detail how the Revenge Regress interacts with the original one in order to strengthen the case against IRFs further. In view of this, it would be helpful to fill the gap by showing how to combine the Revenge Regress with the original argument in order to develop a *Strengthened Regress*. Again, here is our best effort.

First, assume the Rule-Following View:

(1) Any process of inference is a kind of person-level rule-following.

Next, in view of the response to the original Regress, reject (2) and replace it with

(2**) Any process of person-level rule-following utilizes a rule-representation that is either personal or sub-personal.

Now we require two variants of the Rule Application Condition. The first we retain from the original argument:

(3) For a person-level, rule-following process to utilize a personal-level rule-representation, it must contain an inferential sub-process—a person-level inference from the rule to what the rule calls for under the circumstances.

However, the replacement of (2) by (2**) requires that we supplement it with another variant of the Rule Application Condition:

(3**) For a personal-level, rule-following process to utilize a *sub-personal* rule-representation, it must contain an inferential sub-process—a *sub-personal* inference from the rule to what the rule calls for under the circumstances.

The crucial difference between (3) and (3**) is, of course, that the former specifies what is involved in using person-level rule-representations, whereas the latter specifies what is involved when actively reasoning with sub-personal rules. These premises commit the intentionalist not to (4) but to

(4**) Any person-level inference involves either a person-level or sub-personal inferential sub-process.

Moreover, the Revenge Regress still commits the intentionalist to

(4*) Any sub-personal inference involves a sub-personal inferential sub-process.

Suppose we try to carry out a process of active reasoning. By (4**), it involves a sub-process of either active reasoning or sub-personal inference. If it involves the former, then (4**) will also apply to that sub-process. Thus if successive iterations were always to lead to a further sub-process of active reasoning, then they would generate the original regress. But if at any stage active reasoning involves a sub-personal inference, then by iteration on (4*), the Revenge Regress is generated. So we have shown not (5) but rather

(5**) Any inference requires infinitely many person-level or sub-personal inferential sub-processes.

Hence for the by now familiar reason,

(6) Inference is impossible for finite beings like us.

QED.

4.   Rejecting the Strengthened Regress

Although the Regress is widely supposed to show that IRFs are untenable, we maintain that such a view is unwarranted. Even in its strengthened form, the Regress is unsound.

Our first pass response to the Strengthened Regress is to reject (4*)—the claim that any sub-personal inference involves additional inferential sub-processes. We take it to be obvious that any remotely plausible theory of sub-personal inference—intentionalist, or otherwise—must reject this commitment, since it is viciously regressive all by itself. But, of course, (4*) is a consequence of premises (1*)–(3*) of the Revenge Regress. So if we are to reject (4*), it must be because one of those premises is false. Further, since our response to the original Regress was to advocate an IRF for sub-personal inference, we are committed to rejecting (3*) since (1*) and (2*)

simply describe the Rule-Following and Intentional Views as they apply to sub-personal inference. Our challenge, then, is to argue that it is legitimate to reject (3*).

How is this to be done? Premise (3*) is a sub-personal version of the Rule Application Condition. It maintains that sub-personal rule-following of the kind envisaged by the Intentional View requires an inferential sub-process from the rule to what the rule calls for under the circumstances. The obvious way to justify the rejection of (3*), then, is to explain how a sub-personal process might utilize a rule-representation without thereby containing an inferential sub-process. We think that this challenge can be met and indeed that the right response is an exceedingly familiar one.

4.1. The 'Primitivist' Strategy

Causal-explanatory regress is among the most commonplace theoretical challenges to intentional theories in cognitive science. It should come as no surprise, then, that cognitive scientists have a routine strategy for quashing such worries. In our view, this strategy works extremely generally, including for IRFs about sub-personal inference.

Obviously, the intentionalist about rule-following processes cannot maintain—on pain of regress—that rule-guided psychological processes *always* involve further rule-guided psychological processes. Yet the intentionalist need not make this commitment. Instead, they can—and often do—posit a level of *primitive* processing mechanisms. Such processors may take rule-representations as inputs. In which case, the primitive processes they subserve will, in a sense, be rule-guided—though only in the thin sense that a rule-representation is causally implicated in the process because it is an *input* to the processor. In contrast to non-primitive processes, however, primitive ones are *not* rule-guided in a richer sense. That is, they are not rule-guided in the sense that they involve *further* rule-guided or inferential sub-processes. Thus, if non-primitive processes—such as those involved in active reasoning—ultimately decompose into primitive ones, then we have a general view of psychological processes on which rule application regresses cannot occur.

4.2. Primitive Processes and Reflexes

The earlier primitivist proposal is, of course, exceedingly well-known (Block, 1995; Dennett, 1978; Fodor, 1968, 1987; Fodor & Pylyshyn, 1988; Pylyshyn, 1980). But to see how it helps address the Regress, it is useful to clarify the notion of primitive processes. We think that this is usefully done by comparing them with prototypical (monosynaptic) reflexes.[7] Primitive

---

7 Or, at any rate, a caricature of reflexes.

processes are closely analogous to prototypical reflexes in two crucial respects and, importantly, disanalogous in another. A first point of similarity is their *automaticity*. Given the relevant input conditions, a reflex generates a fixed behavioral output. Knock a knee, and it flexes. Analogously, provide input to a primitive processor, and it too flexes automatically—though not to lift a knee, but to output a representation.

A second similarity is that, in contrast to non-primitive intentional processes, the input-output relations of both reflexes and primitive processes are not inferentially mediated. Given a blow to the knee, it flexes, and as far as we know, no intervening stage of the process involves an inference or rule-following operation. The same holds for primitive psychological processes. They too have no intermediate stages that involve further inference or rule-following.

Yet there is, of course, an important difference between prototypical monosynaptic reflexes, and primitive psychological processes. In the case of reflexes, such as the patellar or corneal reflexes, the input is not a representation. Crudely put, it is a mere physical magnitude—a stimuli. In contrast, for primitive processes to play their assigned role within an intentional psychology, it is necessary that their inputs are representational. Indeed, in the cases of interest here, it is necessary that they represent *rules*. Primitive processors of the relevant sort, then, must be automatic, non-inferentially mediated, rule-applicators. By virtue of being rule-applicators, they can underwrite an intentional account of sub-personal inference, and by virtue of being automatic and non-inferentially mediated, they evade the concern that the application of any rule requires a further inferential step. They thus provide an alternative model of sub-personal rule application, which permits the proponent of IRF to reject (3*).

4.3.   *Primitive Processes and Stored Program Computers*

The aforementioned might well sound rather mysterious were we to lack any model of how sub-personal inference could "bottom out" in processes that are reflex-like and yet rule-guided in the thin sense outlined earlier. But we *do* possess a model of such processes. For what we are describing is closely akin to a core aspect of standard, stored program computers. Such computers take programs (rules) as input, and many of their sub-processes involve rule-governed sub-processes (inferences). But computers are organized in such a manner that sooner or later all this rule-governed activity decomposes in a set of reflex-like operations, which do not rely on any further inferential activity. Indeed, their possession of this characteristic is among the central reasons that the concept of a stored program computer became so important to cognitive science. For it provides a model of how a system can be rule-guided without thereby succumbing to regress problems (Dennett, 1982; Fodor, 1975, 2000). The notion of a primitive process is simply a generalization of this aspect of stored program computers, formulated in a manner that remains neutral regarding the precise nature of the processors and operations involved in human cognition.

4.4.   *Primitive Processes and Cognitive Architecture*

It is worth stressing that the aforementioned is little more than standard background theory in much of cognitive science. This is because a hypothesized set of primitive processes and operations is a core facet of what, by deliberate analogy with computer science, is ordinarily called *cognitive architecture*.

As Zenon Pylyshyn noted a very long time ago, a cognitive architecture consists, at least in part, of "those functions or basic operations of mental processing that are themselves not given a process explanation" (1980, 126). That is, they are psychological functions and operations that are not to be explained in terms of other *psychological* processes—specifically, processes that deploy rules and representations (ibid.). In this respect, Pylyshyn continues, they are quite unlike "cognitive functions in general . . . [which] are . . . explainable . . . in terms of rules and representations" (ibid.). Instead, primitive processes and operations are "appealed to in characterizing cognition" and "are themselves explainable biologically rather than in terms of rules and representations" (ibid.)

By broad consent, it is an empirical matter *which* specific cognitive processes and operations are primitive. That there *are* such processes and operations is, however, widely—and we think correctly—assumed to be a presupposition of any sensible intentional psychological science and for the very same reason that primitive operations are a prerequisite for any sensible version of IRF. Without such operations, regress ensues. Again, as Pylyshyn observed long ago, the positing of primitive processes and operations avoids "a regress of levels of interpreters, with each one interpreting the rules of the higher level and each in turn following its own rules of interpretation" (ibid.).

5.   Counterarguments

Positing primitive sub-personal processes allows proponents of IRFs to reject the Rule Application Condition—(3*)—and thereby neutralize the Regress, even in its strengthened form. Yet, as already noted, this regress-blocking strategy is an exceedingly familiar one from cognitive science. So it is somewhat surprising that it receives so little attention in the literature on active reasoning.

Why might this be? One obvious possibility is that the cost of primitivism is in some way too high—that it staves off the Regress, only to raise other no less serious problems for IRFs. In this section, we conclude by briefly considering two possible problems of this sort, which are hinted at in the literature on active reasoning.

5.1. *Positing Primitive Processes Addresses the Strengthened Regress Only at the Expense of Succumbing to Well-Known Kripkensteinian Rule-Following Problems*

Primitive rule-applicators take rule-representations as inputs. But one might find it deeply puzzling how such inputs could have rules as their *contents*. How, for example, might an input determinately represent modus ponens as opposed to some other rule? The obvious suggestion is that it represents the rule in virtue of the effects it has on the processor itself—that it induces modus ponens–like behavior in the processor. Yet this suggestion appears to raise Kripke's familiar Wittgensteinian worries about rule-following. Out of the frying pan and into the fire.

We are tempted to give the earlier worry short shrift. The Regress, as understood by its advocates (and by us), is entirely independent of Kripke's problem. Our aim here has been to address the Regress. If Kripke's problem remains unaddressed, so be it. That's a problem for another day.

Of course, if primitivism generated special Kripkensteinian worries for IRFs, then this quick-fire response would ring hollow. But we deny that it has such consequences. First, Kripke's problem is orthogonal to the issue of whether one adopts the primitivist proposal. If Kripke's problem is a serious one for IRF, then it applies equally to primitive processes *and* non-primitive ones. Kripke's problem concerns the possibility of internalizing a determinate rule, given that it is supposed to cover a potential infinity of cases. Assuming the Intentional View of rule-following, this reduces to the issue of what it is for an intentional state to determinately *represent* a specific, infinitary rule. Further, if IRF is correct, then both primitive and non-primitive processes rely on rule-representations of very much the same sort. In which case, it is hard to see why Kripkean concerns would not arise equally for both sorts of processes. In short, positing primitive processes should make no difference to whether or not Kripke's problem undermines IRF.

Second, we deny that Kripke's problem is especially troublesome for IRFs as such, whether or not they endorse primitivism. To be clear, IRFs are theories about a class of psychological processes—i.e., inferential ones. In contrast, as Boghossian notes,

> Kripke's problem arises against the backdrop of a naturalistic outlook relative to which it is difficult to see how there could be determinate facts about which infinitary rule I have internalized.
>
> (2014, 13)

For proponents of IRF, to "internalize" a rule is to represent it. In which case, Kripke's problem clearly arises for IRFs only when one further demands a *naturalistic* account of rule-representation. In contrast, the problem has no traction if one "waives naturalistic constraints"—e.g., by allowing for primitive facts regarding the content of rule-representing states.

Speaking personally, we are not much inclined toward this sort of non-naturalism. But that's beside the point. Our point is that Kripke's problem is not a problem for primitivism as such, or even IRFs as such. Indeed, it is not a problem about psychological processes at all. Rather, it is a problem for naturalistic theories of *content*. Moreover, it is one that arises for them entirely independently of issues to do with rule-*following*. If one accepts that we can so much as *think* about determinate, infinitary rules—e.g., modus ponens—then the problem arises for naturalistic theories of content.

5.2. *Although Positing Primitive Processes May Save an IRF for Sub-personal Inference, It Does So Only at the Expense of Rendering IRF Untenable for Active Reasoning*

At one stage, Boghossian considers a proposal about sub-personal inference, which may appear to resemble primitivism to a considerable degree. It goes like this:

> I consider [the premises] (1) and (2). I do so with the aim of figuring out what follows from these propositions, what proposition they support. A sub-personal mechanism within me "recognizes" the premises to have a certain logical form. This activates some sub-personal state that encodes the MP rule which then puts into place various automatic, sub-personal processes that issue in my believing [the conclusion] (3).
>
> (2014, 15)

Setting aside the challenge posed by the Revenge Regress, which he maintains is "importantly correct", Boghossian is prepared to imagine that some reasoning works like this. However, he continues,

> That is not the sort of reasoning that this paper is about—rather, it is about person-level reasoning, reasoning as a mental action that a person performs, in which he is either aware, or can become aware, of why he is moving from some beliefs to others.
>   No such process of reasoning can be captured by a picture in which (a) reasoning is a matter of following rules with respect to the contents of our attitudes and (b) our following rules with respect to the contents of our attitudes is a matter of automatic, subconscious, sub-personal processes moving us from certain premises to certain conclusions.

If this is so, then it may seem that, by introducing such sub-personal processes, we fail to account for the sort of active reasoning we sought to understand in the first place. And since this sort of sub-personal process looks much like primitive rule application, it may further appear that endorsing primitivism thereby undermines the prospect of an IRF about active reasoning.

Appearances, at least in this instance, are misleading. As we noted in Section 3, it is widely if not universally supposed that active reasoners often, though not invariably, lack conscious awareness of the rules they are following (cf. Boghossian, 2014, 12).[8] In which case, it ought to be common ground that there are two different kinds of active reasoning:

AR_1: Active reasoning for which there is conscious awareness of the premises and the conclusion, but not the rule.

AR_2: Active reasoning for which there is conscious awareness of the premises, the conclusion, *and* the rule.

If the representation of the rule is sub-personal, we have AR_1. If the representation of the rule is person-level, we have AR_2—the sort of reasoning that Boghossian says he seeks to explain.

In our view, a theory of reasoning ought to capture *both* these kinds of active reasoning. And, as far as we can tell, our response to the Regress in no way prevents us from doing so. If we insisted that *all* rule application was sub-personal, then we could not. But we make no such commitment. Indeed, we allow for at least four different sorts of rule-governed process.

- We allow for AR_2 because we accept that, in some cases, people are conscious of the rules they follow as well as their premise and conclusion attitudes.
- We allow for AR_1 because we posit rule-representations that are sub-personal and, hence, not the subject of conscious awareness.
- We allow for non-primitive, sub-personal inferences where one has no conscious awareness of premises, conclusion, or rule.
- Finally, we allow for primitive sub-personal rule application processes, which involve no inferential sub-processes but are rule-governed, in a thin sense, by virtue of taking rule-representations as input.

The main point of positing this hierarchy of processes was to allow for dependencies that block the Regress in a plausible fashion. Thus, for example, in some cases, AR_2 may rely on AR_1 so that a consciously accessible rule is applied via an inferential sub-process whose rule is not, itself, consciously accessible. Further, we allow that such AR_1 processes may rely on non-primitive, sub-personal inferences. And, of course, we *insist* that all such cascades must at some point rely on primitive rule application processes. Far from failing to accommodate AR_2, we maintain that we accommodate it, and many other sorts of inference beside.

---

8 Furthermore, that children can engage in active reasoning is taken to be an important reason to avoid conceptual oversophication in their accounts (see, e.g., Broome, 2013, 229, 236; Boghossian, 2014, 6).

6. Conclusion

We started by touting the prima facie explanatory virtues of IRFs. We then argued that, by positing a cascade of different sorts of rule application processes, IRFs can accommodate active reasoning in a non-regressive fashion. Specifically, we argued that these resources allow proponents of IRFs to address the Regress, even in its strengthened form. We concluded by suggesting that our solution does not generate any obviously untenable consequences for IRFs. In view of this, and contrary to the opinion of many, we conclude that the Regress fails to undermine intentional rule-following accounts of reasoning.

References

Block, N. (1995). The mind as the software of the brain. In E. Smith & D. Osherson (Eds.), *An invitation to cognitive science, volume 3: Thinking* (pp. 377–425). Cambridge, MA: MIT Press.

Boghossian, P. (2008). Epistemic rules. *Journal of Philosophy, 105*, 472–500.

Boghossian, P. (2014). What is inference? *Philosophical Studies, 169*, 1–18.

Broome, J. (2013). *Rationality through reasoning*. Oxford: Wiley-Blackwell.

Broome, J. (2014). Normativity in reasoning. *Pacific Philosophical Quarterly, 95*, 622–633.

Chomsky, N. (1969a). Linguistics and philosophy. In S. Hook (Ed.), *Language and philosophy: A symposium* (pp. 51–94). New York: New York University Press.

Chomsky, N. (1969b). Comments on Harman's reply. In S. Hook (Ed.), *Language and philosophy: A symposium* (pp. 152–159). New York: New York University Press.

Collins, J. (2005). On the input problem for massive modularity. *Minds and Machines, 15*, 1–22.

Dennett, D. (1978). *Brainstorms: Philosophical essays on mind and psychology*. Montgomery, VT: Bradford Books.

Dennett, D. (1982). Styles of mental representation. *Proceedings of the Aristotelian Society, 83*, 213–226.

Drayson, Z. (2012). The uses and abuses of the personal/subpersonal distinction. *Philosophical Perspectives, 26*, 1–18.

Fodor, J. (1968). The appeal to tacit knowledge in psychological explanation. *The Journal of Philosophy, 65*, 627–640.

Fodor, J. (1975). *The language of thought*. Cambridge, MA: Harvard University Press.

Fodor, J. (1987). *Psychosemantics: The problem of meaning in the philosophy of mind*. Cambridge, MA: MIT Press.

Fodor, J. (2000). *The mind doesn't work that way: The scope and limits of computational psychology*. Cambridge, MA: MIT Press.

Fodor, J., & Pylyshyn, Z. (1988). Connectionism and cognitive architecture: A critical analysis. *Cognition, 28*, 3–71.

Hahn, U., & Chater, N. (1998). Similarity and rules: Distinct? Exhaustive? Empirically distinguishable? *Cognition, 65*, 197–230.

Harman, G. (1967). Psychological aspects of the theory of syntax. *The Journal of Philosophy, 64*, 75–87.

Harman, G. (1969). Linguistic competence and empiricism. In S. Hook (Ed.), *Language and philosophy: A symposium* (pp. 143–151). New York: New York University Press.

Johnson-Laird, P. (2008). *How we reason.* Oxford: Oxford University Press.

Kripke, S. (1982). *Wittgenstein on rules and private language: An elementary exposition.* Cambridge, MA: Harvard University Press.

Laurence, S., & Margolis, E. (1997). Regress arguments against the language of thought. *Analysis, 57,* 60–66.

Perfors, A., Tenenbaum, J., & Regier, T. (2011). The learnability of abstract syntactic principles. *Cognition, 118,* 306–338.

Pylyshyn, Z. (1980). Computation and cognition: Issues in the foundations of cognitive science. *Behavioral and Brain Sciences, 3,* 111–132.

Rips, L. (1994). *The psychology of proof: Deductive reasoning in human thinking.* Cambridge, MA: MIT Press.

Ryle, G. (1949). *The concept of mind.* London: Hutchinson.

Sloman, S. (1996). The empirical case for two systems of reasoning. *Psychological Bulletin, 119,* 3–22.

Smolensky, P. (1988). On the proper treatment of connectionism. *Behavioral and Brain Sciences, 11,* 1–23.

Wright, C. (2014). Comment on Paul Boghossian, 'What is inference?'. *Philosophical Studies, 169,* 27–37.

4 Numerical Cognition and Mathematical Knowledge

The Plural Property View

Byeong-uk Yi

It does not even pertain to the forms to be known if they are such as we say they must be.

—Plato (*Parm.*/1996)

What is a number, that a man may know it, and a man, that he may know a number?

—McCulloch (1960)

1. Introduction

We humans know a great many mathematical truths. But it is difficult to understand and explain how we can know even very basic mathematical truths, such as the proposition that two plus one is three. So Plato presents the challenge to explain human knowledge of numbers,[1] and Paul Benacerraf (1973) reformulates the long-standing challenge in contemporary terms. To meet the challenge, it is necessary to give a proper account of mathematical truths, including an account of the nature of numbers, and of what kinds of cognitive faculties enable us to know these truths. In this chapter, I aim to present an approach to explaining human knowledge of numbers that can meet the challenge. The approach rests on the view that natural numbers are properties. So I call it the *property approach*. This is essentially the approach Penelope Maddy (1980, 1990) and Jaegwon Kim (1981) present in response to Benacerraf. To articulate a viable version of the approach, however, I think it is necessary to make an important modification of the version they propose.

On the property approach, we humans have a limited yet fairly solid and reliable empirical foothold in the realm of numbers, and this provides a basis

---

1 In this chapter, I focus on knowledge of natural numbers and use 'number' and 'numeral' interchangeably with 'natural number' and 'numeral for a natural number', respectively. Although some systems include zero among natural numbers, I consider only positive natural numbers for convenience of exposition and use 'number' for (positive) natural numbers (see also note 5).

for gaining further knowledge about them by exercising cognitive faculties (e.g., reasoning) that can take us beyond the empirical foothold. When we see two pebbles, cows, or other sensible objects, we can count them to figure out that they are two, and sometimes simply *see* (i.e., *visually perceive*) them *as* being two or *as* being fewer than three.[2] The approach takes this to give us cognitive access to the realm of numbers because natural numbers (e.g., 2 or two) are *numerical properties* (e.g., being two), which counting and perception can relate to. So the approach rests on a view about the nature of numbers and a thesis about our cognitive access to them:

(A) The *property view of number*: Natural numbers (e.g., two) are numerical properties (e.g., being two).
(B) The *empirical access thesis*: Humans have empirical, and in some cases even perceptual, access to numerical attributes (e.g., being two, being fewer than).[3]

Kim and Maddy add to these a view about the nature of numerical properties:

The set property view of numerical property: Numerical properties (e.g., being two) are properties of *sets*, namely, their *cardinality properties* (e.g., being two-membered).

This commits them to the view that we can perceive some sets. And they argue that when we see some pebbles (e.g., Pebbie and Pebbo) as being two, for example, we see the *set* of the pebbles (e.g., the doubleton {Pebbie, Pebbo}). I think this is an implausible view. Moreover, I think the set property view is incoherent. The set {Pebbie, Pebbo}, for example, is *one* composite object with two members, but no one object (simple or composite) is two. So the numerical property of being two cannot be identified with the cardinality property of being two-membered (or having two members).[4]

We can avoid the difficulties of the Kim–Maddy approach by replacing the set property view with an adequate view of numerical property. To do so, it is necessary to consider what are *bearers* of numerical properties or

---

2 In this chapter, I use the construction "*see* the so-and-so's *as* being such-and-such" for veridical perception and interchangeably with "*visually perceive* the so-and-so's *as* being such-and-such" (not with "*regard* the so-and-so's *as* being such-and-such").

3 By 'attribute', I understand both properties and relations. Properties are one-place attributes, and relations are multi-place attributes.

4 In her earlier works—e.g., Maddy (1980, 1990)—Maddy holds realism about sets and pursues what I call the *Kim–Maddy approach*. In her later works—e.g., Maddy (1992, 1997, 2007, 2014, 2018)—she rejects realism about sets and departs considerably from the approach. In particular, she rejects the indispensability argument for the existence of sets and no longer holds the set property view of numerical property. So the Kim–Maddy approach can be properly attributed only to the early Maddy. I leave it for another occasion to discuss her later views about mathematical knowledge.

what the property of being two, for example, is *instantiated* by. In holding the set property view, Kim and Maddy assume a constraint ingrained in traditional views of attribute:

> The *singular instantiation thesis*: No property can be instantiated by many things *collectively* or *taken together* (in short, *as such*).

On this thesis, whatever can instantiate a property must be a single object or some one thing. So the numerical property of being two (if instantiated) would have to be instantiated by a single object, just as the property of being gray or that of being a pebble is instantiated by a single object (e.g., Pebbie). This directly yields an incoherent view about numerical property: some *one* thing (e.g., a set with two members) must also be *two* to instantiate the property of being two. So I think one must reject the singular instantiation thesis to have an adequate view of numerical property. And I hold that being two, for example, is a nonstandard property that violates the thesis. In this view, being two is a property instantiated by any *two* things, collectively or taken together. For example, Pebbie and Pebbo (which are two pebbles) instantiate this property, although neither of them alone does.

Unlike standard properties constrained by the singular instantiation thesis, the nonstandard properties that violate the thesis are properties that in a sense relate to *many* things taken together. I call the former properties *singular* properties and the latter *plural properties*. While Kim and Maddy hold that numerical properties are a kind of singular properties, I hold that they are plural properties.[5] I call this view the *plural property view* (*of numerical property*)[6] and the version of the property approach that results from adding the view to the two pillars of the approach (viz., (A) and (B)) the *plural property approach*. This is the approach I aim to motivate and articulate.

2.   The Plato–Benacerraf Challenge

We know that two is smaller than three, that two plus two is four, that three is a prime number, that there is no greatest prime number, and so on. But it seems hard to see how we can know such truths. How can we get to know, for example, that two plus two is four? This is one of the oldest problems

---

5 In this view, being one is also a plural property. Although it is not *instantiated* by many things as such, it in a sense *relates* to many things as such; its complement (*not being one*) is instantiated by many things as such (Pebbie and Pebbo, for example, are not one, but two). In this respect, it differs from *being a pebble*, a singular property whose complement (not being a pebble) also fails to be instantiated by any two or more things as such (see Section 4.4, especially note 45). (To add zero among natural numbers, one might take it to be an uninstantiated plural property. Its complement, not being zero, is instantiated by any one or more things.)

6 Combining this view with the property view of number yields the *plural property view of number*: natural numbers (e.g., two) are plural properties (e.g., being two).

in philosophy. Plato articulates the difficulties in solving the problem in a dialogue named after Parmenides that examines his Theory of Forms. In the dialogue, he has the eponymous character argue that "it does not even pertain to the forms to be known if they are such as we say they must be", and he takes the argument to pose "the greatest difficulty" to the theory (*Parm.*/1996, 133b).[7] While the difficulty concerns knowledge of forms, he would argue that giving an account of knowledge of numbers faces the same difficulty. Benacerraf (1973) essentially reiterates the difficulty in an influential article that formulates the contemporary challenge to explain mathematical knowledge. He argues: "If . . . numbers are the kind of entities they are normally taken to be. . . . It will be impossible to account for how *anyone* knows *any* properly number-theoretic propositions" (1973, 414; my italics).[8]

A key assumption underlying this impossibility thesis is what he calls "a platonist view of the nature of numbers" (1973, 406). In this view, as William D. Hart puts it, natural numbers are "abstract objects" (1977, 125 & 156). While we can see or touch this piece of paper[9] and the Eiffel Tower, we cannot see, touch, taste, or smell the number two or four. And numbers are "causally inert" (1977, 124). We cannot have any causal interactions with them, while we can have causal interactions with not only medium-sized material objects (e.g., cows) but also those too small for us to see or feel (e.g., atoms, photons, quarks). Moreover, they are "energetically inert" (1977, 125); they do not emit or transfer energy. Hart argues that this makes it hard to explain how our beliefs or knowledge (e.g., the belief that 2 + 2 = 4) can be *about* numbers (e.g., 2, 4):

> When you learn something about an object, there is change in you. Granted conservation of energy, such a change can be accounted for only by some sort of transmission of energy from, ultimately, your environment to, at least approximately, your brain. And I do not know how what you learned about that object can be *about* that object (rather than some other) unless at least part of the energy that changed your state came from that object. . . . [But] numbers . . . cannot emit energy.
>
> (1977, 125; original italics)

We can have no *de re* belief (or knowledge) about numbers on this argument because (a) our having *de re* beliefs (or knowledge) of an object requires transfer of energy from the object to us, but (b) numbers are energetically inert because they are abstract (1977, 125).[10]

---

7 See, e.g., Yi and Bae (1998) for an analysis of, and response to, the argument.
8 See also, e.g., Hart (1977), Field (1982, 1989), and Dehaene (2001, 30).
9 Imagine that I am pointing to the piece of paper you are holding while you are reading a hard copy of this chapter.
10 Benacerraf suggests a similar argument in accepting "a causal theory of *reference*" (1973, 412; original italics). See also Grandy (1973, 446).

3. The Property Approach

I agree that it would be hard, if not impossible, to give a plausible account of human knowledge of numbers while assuming platonism about numbers. But I do not think the difficulty arises from the thesis that numbers are *abstract*. In addition to this, contemporary platonism holds that numbers are *objects*, not features or attributes thereof, such as colors, shapes, or energy. While material objects are concrete entities, their attributes are abstract entities. For example, the color and shape of a basketball are abstract although the basketball itself is concrete. But I do not think it is plausible to hold that it is a *mystery*, something extremely hard to explain or even accommodate, that we humans have knowledge of colors and shapes. By comparing natural numbers to colors and shapes rather than basketballs, cows, and quarks, I think we can develop a suitable response to the Plato–Benacerraf challenge to explain human knowledge of numbers.

To see this, it would be useful to compare arithmetic with another ancient branch of mathematics: geometry. We know a great deal of geometrical truths, including many truths about shapes: 'Triangularity is a shape', 'At least three regular polygons have more sides than the regular triangle', etc. But some might argue that our knowledge of shapes is a mystery as follows:

> *The Shape Argument*: How can we know anything about shapes? Shapes (e.g., triangularity or the triangular shape) are *abstract* entities. So they must be *causally inert* (in particular, they emit no energy). If so, we cannot see, touch, or smell them, nor can we causally interact with them. (Surely, we can see and causally interact with material objects that have shapes, but this is beside the point.) So it seems impossible to explain how we can have any belief or knowledge about shapes themselves.

To assess this argument, it would help to compare it with the parallel argument about colors:

> *The Color Argument*: How can we know anything about colors? Colors (e.g., red or the red color) are *abstract* entities. So they must be *causally inert* (in particular, they emit no energy). If so, we cannot see, touch, or smell them, nor can we causally interact with them. (Surely, we can see and causally interact with material objects that have colors, but this is beside the point.) So it seems impossible to explain how we can have any belief or knowledge about colors themselves.

Are these compelling arguments that show that there is a fundamental mystery about our knowledge of colors or shapes? I think not.

Surely, our knowledge of colors has much to do with our color experience. When we see material objects, we often see their colors as well. That is, we can see that they are red, yellow, blue, etc. The visual experience we have of a red apple, for example, features not only the *object* with a color (the apple) but also its *color* (the red color). So our perception gives us epistemic access to colors as well as material objects, and this allows us to have reliable beliefs about colors. If so, there seems to be no fundamental difficulty in explaining our knowledge of colors. It is the same with shapes. For example, we can see a basketball as having a certain shape (e.g., as being spherical). The visual experience we have when we see this represents not only the basketball but also its shape. This enables us to have empirical access to the shape and to have some belief or knowledge about it.

We can now see the main problem of the shape and color arguments. There is no denying that shapes and colors are, as Hart puts it, "energetically inert" (1977, 125). Unlike material objects, they are not the kind of entities that can emit or reflect light or emit or absorb energy. But this does not mean that they are "causally inert" (1977, 124). Nor does this follow from the fact that shapes and colors, unlike material objects, are abstract. Figuring in causal relations are not only objects that emit or absorb energy but also their features, states, movements, actions, etc.[11] When the chair Bob was sitting on broke down because he was heavy (or he was sitting on the chair while he was too heavy for it), the cause of the collapse (or one of its causes) was his being heavy (or his having been sitting on the chair while he was too heavy for it). So it is one thing to say that the heaviness Bob has cannot emit energy, but quite another to say that the quality or property has no role in causality. It is the same with shapes and colors. They are as causally efficacious as weights. You might be cut by a piece of paper because it has a sharp edge; a wall with a bright yellow color might irritate your eyes because of its color. Moreover, shapes and colors can figure in causal processes that lead to our shape or color perception. We can see an apple as being red and a basketball as being spherical, as noted earlier. When we do so, that is in part because the apple is red or because the basketball is spherical.

The color and shape arguments break down, we have seen, because we can see some colors and shapes of some material objects. Although colors and shapes are abstract because they are attributes, this does not bar them from playing causal roles or being perceptually accessible. So, as Jaegwon Kim aptly puts in an article that inspires much of the discussion in this section (and beyond), "the inference we must resist is 'Abstract; therefore, causally inert; therefore, unperceivable and unknowable'" (1981, 347). Abstract as they are, colors and weights can figure in causality and perceptual experience. It is the same with shapes. If so, there is no profound mystery about our knowledge of geometry that separates it from our knowledge of colors or weights.

---

11 Incidentally, Hart would not hold that the *energy* that an atom has is causally inert because energy, unlike atoms, is not something that can absorb or emit energy.

The same, I think, holds for arithmetical knowledge, knowledge of natural numbers. Numbers are not like cows, pebbles, quarks, etc., but like colors, shapes, etc. That is, numbers are not objects but properties. They are *numerical properties*: being one, being two, being three, etc. For example, the number *two* is a property of *two* cows, pebbles, etc., and the number *three* is a property of *three* cows, pebbles, etc. And we have a reliable, if meager, epistemic foothold in the realm of numbers because we have empirical access to some of the numerical attributes figuring in situations involving objects we can perceive. When we see two horses and three cows, for example, we can see the former as *being two* and the latter as *being three*, and we can see the latter as *outnumbering* the former. This gives us an epistemic entry into the realm of numbers.

This is the point that Maddy (1980, 1990) and Kim (1981) make in response to the Plato–Benacerraf challenge. Kim sums it up as follows:

> As objects of perceptual discrimination and judgment, there is nothing unusual, uncommon or mysterious about numerical properties and relations. Seeing . . . that these are three green dots, that the dots over here are more numerous than those over there . . . are as common, and practically and psychologically unproblematic, as seeing that these dots are green, the dot on the left is larger . . . than the one on the right. . . . Numerical properties do not differ in respect of perceptual accessibility from sundry physical properties such as colors, shapes, odors, warmth and cold.
>
> (1981, 345)

This lays out the approach to meeting the challenge that I call the *property approach* because it rests on the *property view of number*—the view that natural numbers are properties of a special kind.

Aristotle holds this view when he says "number is one of the discrete quantities" (*Cat.*/1963, 4b32). And as Kim points out, modern empiricists regarded number as "a 'primary quality' of objects" (1981, 345). For example, John Locke lists numbers, together with shapes (or "Figure"), among "*original* or *primary* Qualities of Body*", which he thinks "we may observe to produce simple *Ideas* in us" (*Essay*/1975, Bk. II, Ch. VIII, §9; original italics).

Locke seems to suggest that we can have perceptual access to all natural numbers, but one cannot plausibly hold that we can perceptually recognize or discriminate the number of a large number of objects. When we see 1,000 cows, for example, we cannot see them as being 1,000 rather than 1,001.[12] But there is no denying that we have *empirical*, if not perceptual, access to

---

12 Hume denies that we have even an *idea* of 1,000. He says, "When we mention any great number, such as a thousand, the mind has generally no adequate idea of it, but only a power of producing such an idea by its adequate idea of the decimals, under which the number is comprehended" (1888/1978, Bk. I, Pt. 1, sc. 7).

the number of some objects: cows, pebbles, eggs, etc. We can *count* them, one by one, to find out how many there are. With much care and sustained attention, we can even count 1,000 pebbles (or cows) to conclude that they are 1,000, not 1,001. Moreover, psychological studies of numerical cognition suggest that humans, and some species of animals, have a more direct access to small numbers.[13] I think they provide a strong support for the view that we have perceptual access to numerical attributes.

In defending the view that we have perceptual access to small numbers, Kim (1981, 345) and Maddy (1990, 60) refer to Kaufman, Lord, Reese, and Volkmann (1949). In their study of discrimination of numbers of randomly arranged dots flashed on a screen for a short period (1/5 of a second), Kaufman et al. (1949) conclude that humans have a psychological mechanism for visually discriminating the number of dots that enables them to recognize the number without counting when it is small (up to five or six). They propose the term *subitize* (which derives from the Latin adjective *subitus*, which means *sudden*) for the recognition of number resulting from exercising the hypothetical mechanism (1949, 520).[14] Now, studies of perceptual discrimination of numbers of objects when the numbers are small date further back.[15] And Tobias Dantzig (1930/1954) proposes that humans have a mental faculty that enables them to attain perceptual recognition of numerical attributes of some objects when they are small in number:

> Man, even in the lower stages of development, possesses a faculty which, for want of a better name, I shall call *Number Sense*. This faculty permits him to recognize that something has changed in a small collection when, without his direct knowledge, an object has been removed or added to the collection.
>
> (1930/1954, 1)

He distinguishes the faculty he calls *number sense* from the ability to count, and adds that it is possessed by some animals (e.g., birds) as well (1930/1954, 1–5).[16] The number sense, in his view, has a biological basis and gives us the beginning of our knowledge of numbers.

---

13 See, e.g., Dehaene (1997) and Carey (2009a, 2009b).

14 They distinguish subitization from estimation as well as counting. They say, "Subitizing is, on the average, more accurate and more rapid than estimating, and it is done with more confidence" (1949, 520). They take estimation to be due to another mechanism—one that is exercised when an estimate is given of the number of dots when they are large in number (six or more). Incidentally, McCulloch seems to appeal to their studies when he says, "The numbers from 1 through 6 are perceptibles; others, only countables" (1960, §2).

15 For an overview of the studies, see Dehaene (1997, Chapter 3).

16 He notes the limitation of the number sense to small numbers (up to three or four) in both humans and animals. He says, "The direct *visual* number sense of the average civilized man rarely extends beyond four, and the *tactile* sense is still more limited in scope" (1930/1954, 4; original italics). See, e.g., Dehaene (1997, 17ff) for a survey of early studies of numerical abilities of animals.

This view receives strong support from recent psychological studies of numerical cognition. In a book that includes an overview of those studies, *The Number Sense* (1997), Stanislas Dehaene puts their import as follows:

> One of the brain's specialized mental organs is a primitive mental processor that prefigures . . . the arithmetic that is taught in our schools . . . numerous animal species . . . possess a mental module, traditionally called the "accumulator", that can hold a register of various quantities . . . [for example] rats exploit this mental accumulator to distinguish series of two, three, or four sounds or to compute approximate additions of two quantities. The accumulator mechanism opens up a new dimension of *sensory perception* through which the cardinal of a set of objects can be perceived just as easily as their *color, shape,* or *position.* This "number sense" provides animals and humans alike with a direct intuition of what numbers mean.
>
> (1997, 4f; my italics)[17]

Here he accepts and develops Dantzig's number-sense hypothesis. While calling the faculty (the number sense) a "mental module", "mental processor" or "one of the . . . mental organs", he mentions a recent model of the faculty he has helped to develop (the accumulator model).[18] And he takes the faculty to enable us to have perception of the number of some objects as we have perception of their colors and shapes.[19]

Now, some might object that this does not help at all to meet the Plato–Benacerraf challenge because seeing some pebbles as being two, for example, does not help to know any arithmetical truth or, as Benacerraf puts it, "any properly number-theoretic proposition" (1973, 414). I do not think this is correct. One can use cognition of numbers of pebbles to attain knowledge of some arithmetical truths. Suppose that one can see some pebbles (e.g., Pebbie and Pebbo) as *being two* but as *not being three.* The experience can lead to the belief that they are not three but two, and one can use this to conclude that *two is not three* (i.e., being two is not identical to being three); otherwise, any things that are two would also have to be three (and vice versa). This procedure for reaching the arithmetical truth ($2 \neq 3$) is parallel to (a) the procedure for concluding that Pebbie is not Pebbo from observing that Pebbie is white while Pebbo is not, and comparable to (b) that for concluding that white is not gray (or whiteness is not grayness) from observing

17 By "numbers" in the last sentence, he means *numerals* (or number words), not what numerals refer to or represent. His failure to clearly distinguish numbers from numerals, I think, mars some of the discussions in the admirable book.
18 See Dehaene (1997, 68ff) about the accumulator model.
19 For the present purpose, we can take Dehaene to use 'the cardinal of a set of objects' for the number of the objects in question. But see Section 5.1 for more about his talk of *set* or *cardinality.*

that Pebbie is not gray but white. In the arithmetical case, unlike in (a), one needs to shift the primary focus from the perceived objects (e.g., pebbles) to the attributes they possess or lack (e.g., two and three). But the same shift of focus is required in the color case, (b), as well.

So I think we have empirical foothold in the realm of numbers as in the realm of colors and shapes. Although the foothold falls far short of covering all arithmetical truths, the meager foothold can provide a reliable basis from which one can embark on gaining further knowledge of the realm. If so, the skeptical tone shrouding the Plato–Benacerraf challenge that suggests that our knowledge of numbers is a profound mystery stems from mistaken, if popular, views about the nature of numbers and the reach of our epistemic capacities. Needless to say, there remains much more to do to implement the approach sketched earlier, the *property approach*, to give a full account of our knowledge of numbers. But I do not think there is a *fundamental* difference between the difficulties in doing so and those one would meet in giving a full account of our knowledge of shapes or colors.

The approach, as noted earlier (Section 1), is essentially the one proposed and developed by Maddy (1980, 1990) and Kim (1981). But I think it is necessary to make a substantial modification of their approach by basing it on a viable version of the property view of number. In the next section, I discuss the main difficulty in their approach and formulate a suitable modification thereof.

Let me complete this section by noting that it is one thing to say that we can resort to experience to know some truths (e.g., $2 \neq 3$), but quite another to say that they are not a priori truths. Roughly, a truth is *non-empirical* (or *a priori*) if one *can* know it without relying on experience (or any particular experience), and *empirical* (or *a posteriori*) otherwise. This does not require all who know a priori truths to know them non-empirically, i.e., without relying on experience. So some might plausibly hold that Fermat's Theorem, for example, is an a priori truth although most people who know the truth know it only empirically (they have only second-hand knowledge that rests on hearsay). And most of us came to know that $7 \times 8 = 56$ by learning it from others in the process of memorizing the multiplication table, but this does not settle the epistemic status of the truth. In this connection, it is notable that Kim assumes that arithmetical truths are a priori truths (or truths of which we can have "a priori knowledge") in attempting to delineate "the epistemic role of perception" in our knowledge thereof (1981, 353).[20] I leave it for another occasion to discuss the epistemic status of arithmetical truths.

---

20 Note that the title of Kim's (1981) article is "The role of perception in a priori knowledge: Some remarks". He argues that the role of experience in our knowledge of, e.g., the truth that $3 + 2 = 5$ is not to provide instances of an inductive generalization but to give "a *cue* that prompts our apprehension of the truth" (1981, 343f & 350ff; original italics).

4.   Natural Numbers as Plural Properties

The property approach rests on the property view of number. In this view, natural numbers are properties that numerals or number words (e.g., 'two') refer to: the number two, for example, is the property of being two. This is a natural view that, as noted earlier, dates at least back to Aristotle. But there is a serious difficulty in articulating the view. The difficulty lies in stating what are bearers of numerical properties (e.g., being two). To do so, most proponents of the view identify numerical properties (or those except being one) with properties instantiated by *composite objects*, objects that in some sense comprehend other objects: aggregates, classes, sets, etc. Kim (1981) and Maddy (1980, 1990) identify them with properties of *sets*: their *cardinality* properties (e.g., being two-membered).[21] In their view, then, being two is a property instantiated by any set with two members (e.g., {Pebbie, Pebbo}). I think this view has a common problem with all the other versions of the usual property view that identify numbers with properties of a kind of composite objects (e.g., having two parts). A composite object cannot have the property of being two, that of being three, or the like, for it is *one* thing, not *two or more*, no matter how many parts, members, elements, etc., they might have.

The problem of the usual property view, I think, arises from assuming a constraint ingrained in the standard views of attribute:

> The *singular instantiation thesis*: No property can be instantiated by many things as such (or taken together); that is, no property can occur just once to be instantiated by many things.

To reach an adequate account of number and numerical property, we must depart from the standard views of attribute by rejecting this thesis. By doing so, we can revive the natural view of numerical property, the *plural property view*. In this view, numerical properties are properties relating to *many* things as such. Being two, for example, is a property instantiated by any two things (e.g., Pebbie and Pebbo) taken together. Let me elaborate on this view.

4.1.   Objects and Properties

The platonist view of number holds that numbers are abstract objects. In this view, numbers are unperceivable and causally inert. They cannot figure in perception or causality in the way that concrete objects (e.g., apples, electrons) can, nor can they do so in the way that attributes of concrete objects (e.g., colors, shapes) can. Frege holds this view in defending logicism about arithmetic, the view that all arithmetical truths are reducible to logical (or

---

21 Kim (1981) holds that numbers are properties of sets or classes. Unlike Maddy, however, he does not distinguish classes from sets and uses 'class' interchangeably with 'set'.

analytical) truths. To defend logicism, he argues that "Every individual number is a self-subsistent object" (1884/1980, 67), and identifies numbers as certain classes or "extensions of concepts" (1884/1980, 79f),[22] which he considers logical objects. His definition or identification of numbers as classes has a well-known problem (the so-called Russell's paradox). Nevertheless, platonism remains an influential, and arguably the standard, view in contemporary metaphysics and philosophy of arithmetic. So Benacerraf (1973) assumes the view in articulating the difficulty in explaining knowledge of numbers. And Hart tersely asserts that "platonism is the only adequate theory of mathematical truth" (1977, 125), and adds that "it is a metaphysical axiom that natural numbers are causally inert" (1977, 124).

Are they right to do so? Does platonism yield an adequate, and the only adequate, analysis of arithmetical truth? I think not. Much of the case for the view rests on the fact that numerals (e.g., '2' or 'two') figure as singular terms in arithmetical statements (e.g., '2 < 3' or 'Two is smaller than three').[23] But this does not mean that they must refer to objects, not attributes. I do not think we can take all singular terms of natural languages[24] to refer to objects.[25]

While distinguishing between singular and general terms, John Stuart Mill (1891/2002, Bk. I, Ch. II) distinguishes two kinds of singular terms:

(a) the concrete: 'John', 'London', etc.
(b) the abstract: 'whiteness', 'triangularity', etc.[26]

He says, "A concrete name is a name which stands for a thing; an abstract name is a name which stands for an attribute of a thing" (1891/2002, Bk. I, Ch. II, §4). For example, 'whiteness', in his view, is an abstract singular term—a term that refers to a property (viz., whiteness). The abstract term

---

22 He says, "The Number which belongs to the concept *F* is the extension . . . of the concept 'equal to the concept *F*'" (1884/1980, 79f). See also Russell (1919/1920), who says "*The number of a class is the class of all those classes that are similar to it*" (1919/1920, 18; original italics).

23 Benacerraf (1973, 405ff) argues that we must accept the platonist analysis because the mathematical statement 'There are at least three perfect numbers greater than 17' draws syntactic parallels with statements about ordinary objects, such as 'There are at least three large cities older than New York'. (This argument also rests on the fact that '17', like 'New York', figures as a singular term in the arithmetical sentence.) But consider 'There are at least three chromatic colors brighter than black'. This also draws syntactic parallels with the statement about New York, but this does not mean that 'black' must refer to an object or that black is not a property but an object.

24 Or all expressions that can figure as singular terms.

25 The reason I deny this is independent of the existence of vacuous singular terms, such as 'Pegasus'; we might consider this, so to speak, an 'object-referential' term, albeit not one that succeeds in referring to an object.

26 The two distinctions are independent; general terms are also divided into the concrete and the abstract.

'whiteness' is related to its adjectival cousin, the general term 'white', which he says is a "connotative" term—one that "implies, or . . . *connotes*, the attribute *whiteness*" (1891/2002, Bk. I, Ch. II, §5; original italics). In his view, then, an abstract singular term refers to (or, as he says, "denotes") the attribute that its general term cousin "connotes". While he uses different words (viz., 'denote' and 'connote') for the semantic relations that the two terms have to the same attribute, we might take the relations to be the same. Using 'refer' for the common relation, we can say that the abstract singular 'whiteness' refers to the same attribute that the general term 'white' does.[27] And the two terms can be considered different *forms* of the same underlying expression—a *predicable* expression that does not necessarily figure as a predicate. The underlying expression can figure as a predicate, as in 'This marble is white'; it can also figure as an argument of a higher order predicate,[28] as in 'Whit*eness* is a bright color', where it takes the singular noun form. And some predicable expressions might figure as abstract singular terms with no (apparent) morphological change. In '*White* is a bright color', for example, 'white' figures as a singular term referring to an attribute: whiteness (or the white color).

We can now see that abstract singular nouns for shapes (e.g., 'triangularity') refer to attributes, not objects, because shapes are attributes. Like abstract singular terms for colors, they inherit the reference of their adjectival cousins (e.g., 'triangular') or the underlying predicable expressions. It is the same with numerals figuring in arithmetical statements. In 'Two is smaller than three', the numerals 'two' and 'three' figure as singular terms. But they can also figure predicatively, as in 'These pebbles are two' and 'Those apples are three'. The two uses of numerals are as related as the two uses of the color word 'white' in '*White* is brighter than black' and 'This marble is *white*'. And we can see logical relations between the two uses. We can correctly argue:

> These are *two* pebbles but those are *three* apples. So the former are fewer (or less numerous) than the latter, for *two* is smaller than *three*.

So I think numerals are predicable expressions that refer to attributes. Like 'white', they can figure as arguments of higher order predicates without morphological change.

On this analysis, the number two, for example, is the property that the numeral 'two' refers to whether it figures predicatively or as a singular term.

---

27 Mill also assigns denotations to general terms: the general term 'white', for example, denotes any white object, i.e., anything that has the attribute it connotes. I do not think the relation a general term has to its denotations is the same relation as the reference relation.

28 Higher order predicates refer to attributes of attributes, not those of objects.

To highlight its predicable nature, we might call it *twoness* or, as I prefer, *being two*. It is the property referred to by the predicate 'be two' that figures in, e.g., 'The pebbles over there *are two*'. This yields the alternative to (contemporary) platonism that I propose: the property view that, as we have seen, helps to resolve the apparent difficulties in explaining our knowledge of numbers. Like platonism, the property view yields a *realist* analysis of arithmetic by taking numerals to refer to real entities (viz., numbers) that exist independently of human epistemic access. And the view agrees with platonism in holding that numbers are not concrete but abstract entities. But it diverges from platonism in holding that numbers are not objects but properties.

4.2. The Problem of the One and the Many

In the property view, natural numbers are properties: being one, being two, being three, and so on. This should strike one as a compelling view, I think, when one compares talk of numbers with talk of colors or shapes. But there is an important difference between numbers, on the one hand, and colors and shapes, on the other. This gives rise to a serious difficulty in formulating a viable version of the property view, which might help to explain the prevalence of platonism. The difficulty concerns *bearers* of numerical properties. What would instantiate, for example, the property of being two? The usual answers proponents of the property view give to this question have a common problem because they all rest on the singular instantiation thesis.

Mill holds that a number is a property of an "aggregate", a whole composed of parts, such as an aggregate of pebbles, apples, etc.; he says that a number is "some property belonging to the agglomeration of things" or "the manner in which single objects of the given kind must be put together, in order to produce that particular aggregate" (1891/2002, Bk. III, Ch. XXIV, §5). While motivating the view of numbers as certain classes of classes,[29] Bertrand Russell suggests the view that numbers are properties of *classes* (or *collections*): "The number 3 is something which all trios have in common, and which distinguishes them from other collections. A number is something that characterizes certain collections" (1919/1920, 12).[30] As noted earlier, Maddy (1980, 1990) and Kim (1981) hold a variant of this view: numbers are properties of *sets*.

In all these views, whatever can instantiate the number two (or being two), for example, must be some one thing, a single composite object comprehending two objects: an aggregate with two pebbles as parts, a class with

---

29 In this view, numbers are objects of a special kind. See note 22. See Yi (2013) for a discussion of the development of Russell's views of number and class.

30 And he continues, "Instead of speaking of a 'collection', we shall as a rule speak of a 'class', or sometimes a 'set'" (1919/1920, 12).

two pebbles as elements, a set with two pebbles as members, etc. No matter how many parts, members, elements, etc., such an object has, however, the object itself is only one thing just as a water molecule is only one object, albeit one composed of three atoms. Thus the composite objects that the views identify as bearers of the number two fail to instantiate the property of being two. The aggregate of two pebbles has the property of being one, not that of being two, for it is only one object. Its having two parts does not make the aggregate itself two anymore than my having two parents makes me two parents, for it is one thing to say that *the parts* of the aggregate are two, but quite another to say that *the aggregate* itself is two. It is the same with the set, the class, and the like.[31]

In *Principia Mathematica*, Whitehead and Russell call this "the ancient problem of the One and the Many" (1962, 72). They formulate the problem as a problem of the theory of classes:

> If there is such an object as a class, it must be in some sense *one* object. Yet it is only of classes that *many* can be predicated. Hence if we admit classes as objects, we must suppose that the same object can be both one and many, which seems impossible.
>
> (1962, 72; original italics)

Although this puts it as a problem of the view of classes as objects, it is a common problem of views that take numerical properties to be instantiated by composite objects.[32] A composite object of any kind whatsoever is one object, but no one object (composite or not) can be many (i.e., two or more). So no composite object can instantiate the property of being two.

And we can see that the usual versions of the property view violate the logic of numerical statements. Consider, e.g., the set property view: numerical properties (e.g., being two) are cardinality properties of sets (e.g., being two-membered). This view violates logical relations among (1a)–(1d):

(1)   a. Pebbie is not identical with Pebbo.

     b. Pebbie and Pebbo are two.

---

31 Sorin Bangu suggests that any material object (e.g., Socrates) is two because it has two parts (e.g., his left and right halves). But it is one thing to say that the two halves of Socrates, for example, are two, but quite another to say that Socrates himself is two. To draw this conclusion, it is necessary to assume what I call the *many-one identity thesis*: one composite object (e.g., a set or aggregate) is identical with its many components (e.g., its members or parts). I think this thesis is incoherent. See Yi (1999a, 2014), where I argue against the thesis and its close cousin, the thesis that composition is a kind of identity.

32 For more about their objection to classes, which leads to the no-class theory, see my (2013, §5).

c. The set of which Pebbie and Pebbo are members (i.e., {Pebbie, Pebbo}) has two members.

d. There is something of which both Pebbie and Pebbo are members.

In the view, (1c) gives an analysis of (1b). If so, (1b) must imply (1d) because (1c) does so. But (1b) does not imply (1d), for (1a) implies (1b) but not (1d). We can give the same argument, *mutatis mutandis*, against views that identify numerical properties as properties of other kinds of composite objects (e.g., aggregates, classes).[33]

The usual versions of the property view of number, we have seen, are incoherent. Some might invoke this to return to the platonist view. I do not think this is a correct response, for essentially the same problems arise for the platonist view.[34] The problems of the usual versions of the property view noted earlier arise from ignoring the natural answer to the question about bearers of numeral properties. By holding onto the answer, we can give an adequate account of numerical properties. To do so, however, it is necessary to depart radically from traditional views of attribute.

4.3. *The Singular Instantiation Thesis and Plural Properties*

What can instantiate numerical properties, such as being two? Whatever can instantiate this property must be *two* things. And any two things instantiate it. For example, Pebbie and Pebbo, the two pebbles, instantiate it. This, to be sure, does not mean that both of them do so. *Neither* pebble instantiates the property; Pebbie is one, not two, and the same holds for Pebbo. Thus being two is instantiated by Pebbie and Pebbo (taken together), although it is not instantiated by either of them. It is the same with being three, being four, etc. Being three is instantiated by any three things (albeit not by any one of them), being four by any four things (as such), and so on. Being one, to be sure, is instantiated by any one thing.

---

33 This argument helps to clarify the view that a numerical property is in a sense *intrinsic* to any objects that instantiate the property. In this view, the property of being two is intrinsic and internal to any two things (taken together). For example, Pebbie and Pebbo are two as long as they are different objects. No other condition is relevant to their being two; as long as they differ from each other, it does not matter whether they are nearby, how similar they are, what kind of objects they are, etc. In particular, it is irrelevant whether they combine, so to speak, to compose a whole or aggregate, a set or class, etc. See Yi (1999b, 188–190).

34 The view must help to yield a suitable analysis of the predicative use of numerals as well, and the usual versions of the view carry implicit analyses of the predicative use. For example, those who take numbers to be classes of classes must take the predicate 'be two' to refer to a property of certain classes (viz., the complex property of *being an element of the class of classes* identified with the number two or that of *being equinumerous with* {∅, {∅}}). As we have seen, however, such a property cannot be what the predicate refers to, because it is a property of classes.

This, I think, is the natural and correct answer to the question about bearers of numerical properties. But most discussions of the question ignore it altogether. The reason is that it conflicts with a central, if implicit, component of standard views of attribute. To see this, it would be useful to consider Maddy's argument for the set property view.

In response to the Plato–Benacerraf challenge, as noted earlier, Maddy (1980, 1990) and Kim (1981) hold the set property view. While Kim (1981) merely assumes the view, Maddy presents an argument for it. She holds that when we humans (e.g., Steve) see two eggs (e.g., Eggie and Eggo) in a carton, for example, we can see that they are two and gain the matching perceptual belief.[35] And she argues that this numerical belief is a belief *about a set*, namely, the doubleton {Eggie, Eggo}.[36] This argument depends on the view that "the bearers of number properties are sets" (1990, 61), and she defends this view by ruling out alternative answers to the question "What is the subject of a number property?" (1990, 60). To do so, she begins by listing several possible answers to the question (1990, 60f).[37] But her list does not include the natural answer: it is *the eggs in the carton* (i.e., Eggie and Eggo) that have the property of being two. In presenting the candidates in the list, she does mention this answer but only to replace it right away with other answers clearly different from it. She says, "The easiest answer would be that Steve's belief is about *the eggs*, the physical stuff there"; she also says, "Some would say, 'the eggs', meaning by this the physical stuff as divided up by the property of being an egg, what I call an 'aggregate'" (1990, 60; my italics). But *the eggs* (in the carton), which are two objects, are not the same things as *the physical stuff* in it. Although the stuff constitutes the two eggs by providing their matter, this does not mean that the stuff itself *is* two eggs. And Eggie is one of the two eggs but not one of the stuff. Similarly, one cannot identify the two eggs (taken together) with the aggregate composed of them. Eggie is one of the eggs, but not one of the aggregate.

Why does Maddy then dismiss or distort the natural and "easiest" answer (1990, 60)? She assumes that "what we perceive [when we see two eggs] is simply *something* with a number property" (1990, 61; my italics). And she explains, "When I say he gains a perceptual belief about a set, I mean he gains a perceptual belief about *a something* with a number property, which we theorists know to be a set" (1990, 63, no. 70; my italics).

---

35 She discusses a case of seeing three eggs, but I change it to a case of seeing two eggs for convenience of exposition.

36 Moreover, she argues that those who see the eggs as being two see the doubleton (the set perceivability thesis). See Section 5.1 and Appendix.

37 According to them, the bearers of numerical properties are: (i) physical stuffs, e.g., "the physical stuff there in the carton"; (ii) what Maddy calls *aggregates*, physical stuffs as divided by properties, e.g., "the physical stuff as divided up by the property of being an egg"; (iii) Fregean concepts, e.g., "the concept of 'egg in the carton'"; (iv) classes; and (v) sets (1990, 60f).

By "something" or "a something", she means some *one* thing or a single object. The numerical belief (if true) must be a belief about some one thing, she thinks, because the numerical property the belief attributes *cannot* be instantiated by *many* things (viz., two or more objects) taken together. This is so, she assumes, because *no* property *whatsoever* can be instantiated by many things (as such)—a property can be instantiated only by one thing, so to speak, at a time. This, to be sure, is not to deny that a property (e.g., being white) can in a sense be instantiated by both Eggie and Eggo. It is to say that for the property to be instantiated by the two in this sense is just for it to be instantiated, so to speak, twice (once by Eggie and once by Eggo), that is, to be instantiated by *each one of them*. If so, Maddy rules out the natural answer because she implicitly assumes the singular instantiation thesis: no property can be instantiated by many things as such (i.e., no property can be instantiated by many things while occurring just once). This thesis bans the view that the property of being two is instantiated by the two eggs (as such).

Is there a good reason to accept the thesis to rule out this view? I think not. Although the thesis is a central, if implicit, component of standard views of attribute, I think it is a dogma that encapsulates a *bias against the many* in favor of the one. The thesis directly yields the incoherent view that bearers of most numerical properties must be both one and many: what instantiates the property of being two, for example, must be *one* (for they instantiate a property) as well as *many* (for they must be two). Accordingly, views of numerical properties constrained by the thesis, as we have seen, violate the logic of numerical statements. For example, 'Pebbie and Pebbo are two' does not imply the existence of anything except Pebbie and Pebbo, but the set property view analyzes the statement as one that implies the existence of a set (viz., {Pebbie, Pebbo}), which is an object that differs from both Pebbie and Pebbo. So I think the singular instantiation thesis is the main stumbling block to reaching an adequate view of the nature of number.[38]

Removing the blindfolds imposed by the long-standing dogma, we can revive the natural view of number and of the many. It is two things, not one thing, that are two; it is many things, after all, that are many. In this view, there are properties of a special kind that violate the singular instantiation thesis. Such properties are instantiated by many things as such and thus relate in a sense to many things as such. In addition to numerical properties I identify as natural numbers, they include being two logicians, being many, and collaborating. For example, Russell and Whitehead instantiate all these properties (as well as being two) while neither of them does so.

---

38 See Yi (1999b, 167–169) for discussion of the singular instantiation thesis and its general-ization, the *principle of singularity*; see also Yi (1998, 105).

We need to distinguish such nonstandard properties from the usual, standard properties that conform to the singular instantiation thesis: being white, being spherical, being a human, etc. I call the latter *singular properties*, and the former *plural properties*.[39] And I call the liberal conception of attribute that accepts the existence of plural properties the *plural conception*. By accepting this conception, we can reaffirm the natural view of numerical property: being two, for example, is a plural property instantiated by any two things (as such). I call this view the *plural property view of numerical property*, and the view of number that results from combining this view with the property view of number the *plural property view of number*.[40]

4.4.   Plural Attributes and Plural Languages

The plural conception of attribute provides the metaphysical framework for giving a natural account of the semantics of *plural constructions* of natural languages. A sketch of the guiding idea of the account would help to clarify some features of the plural property view of number.

Consider the following sentences:

(2)   a. Pebbie and Pebbo *are two*. (=(1a))
      b. Russell and Whitehead *are two logicians*.
      c. Russell and Whitehead *collaborate*.

(3)   a. Venus and Serena *won* a Wimbledon doubles title.
      b. Russell and Whitehead *wrote* "Principia Mathematica".
      c. The boys *lifted* a piano (by working together).
      d. The girls *are fewer than* the boys.
      e. The boys *outnumber* the girls.
      f. Russell *is one of* the authors of "Principia Mathematica".

---

39 Not all plural properties, which in a sense relate to many things as such, violate the singular instantiation thesis (they all violate the principle of singularity mentioned in note 38). Being one is a plural property that does not violate the thesis although its complement (not being one) does (see note 5). For an elaboration of the notion of plural property, see Section 4.4.

40 For more about the plural property view, see Yi (1998, 1999b, 2002, Chapter 5). In proposing the notion of *class as many*, Russell (1903/1937) comes close to rejecting the singular instantiation thesis. But he ends up retaining the thesis by holding that even a class as many must in some sense be one. He holds that the class as many is "a complex" (1903/1937, 76) that is one in an attenuated sense: "In a class as many, the component terms . . . have *some kind of unity*", albeit one that is "less than is required for a whole" (1903/1937, 69; my italics). So his view of class as many turns out to violate his guiding idea that "the many are only many, and not also one" (1903/1937, 76). If so, he falls short of accepting the plural conception. See Yi (2013, 110).

The underlined phrases in these sentences are *plural terms,* and the italicized expressions are predicates that can take plural terms (in short, *plural predicates*). In (2a)–(2c), the plural terms combine with one-place predicates to yield the sentences. In (3a)–(3f), which have two-place predicates, the underlined plural terms combine with the predicates by filling one or the other of their two argument places. In (3d), for example, 'the girls' and 'the boys' fill the first and second argument places of the predicate 'be fewer than' (or its plural form), respectively.

We may contrast plural terms with singular terms (e.g., 'Pebbie', 'the boy'), and sentences involving plural constructions with sentences involving no plural constructions (e.g., 'Venus is a tennis player'). As the earlier examples suggest, plural constructions (in short, *plurals*) are as prevalent in natural languages as singular constructions (in short, *singulars*). But the usual account of the logic and semantics of natural languages show a strong *prejudice against plurals* in favor of their singular cousins. They regard plurals as redundant devices used merely to abbreviate their singular cousins.[41] This prejudice against *plurals* is the linguistic cousin of the bias against the *many* that is encapsulated in the singular instantiation thesis. Those who assume this bias cannot take (2a) to involve the plural predicate 'be two' (which can combine with 'Pebbie and Pebbo'), for that would require a property, being two, that can be instantiated by two things (e.g., Pebbie and Pebbo) as such. But the usual accounts of plurals contravene logical relations among them and this calls for shedding the prejudice against plurals, as we have seen.[42]

The alternative approach I take to explain the logic and meaning of plurals is implicit in the rejection of the singular instantiation thesis. The approach takes plurals not as mere abbreviation devices but as devices serving a distinct semantic function. In this view, plurals are by and large devices for talking about many things as such while singulars are mostly devices for talking about one thing, so to speak, at a time. While the singular term 'Pebbie' refers to one thing (viz., one pebble), the plural term 'Pebbie and Pebbo' refers to two things (viz., two pebbles) taken together. So a typical plural term is a *plural referential* term, a term that refers to many things as such.[43] How about predicates?

Predicates that can combine with only singular terms (in short, *singular predicates*) (e.g., 'is white', 'is a human', 'is identical with') refer to the usual, singular properties: being white, being a human, being identical

---

41 For example, some accounts of plurals regard plural terms (e.g., 'Pebbie and Pebbo') as devices for abbreviating singular terms for sets (e.g., 'the set {Pebbie, Pebbo}') and consider (2a) an abbreviation of "The set {Pebbie, Pebbo} is two-membered."

42 See the argument about logical relations given in Section 4.2. For related arguments, see, e.g., Yi (1999b, 2005, 2006).

43 Some plural terms (e.g., 'Cicero and Tully', 'the Romans identical with Cicero') are not plural referential terms.

with, etc. By contrast, plural predicates (e.g., 'be two', 'outnumber') refer to attributes with a special kind of argument places that I call *plural attributes*. They are attributes with at least one argument place that can admit many things taken together. For example, the one-place plural predicate 'be two' refers to a one-place plural attribute (i.e., a plural property): being two. This property can be instantiated by many things as such, for the one argument place it has can admit many things (e.g., Pebbie and Pebbo) taken together. Call such an argument place a *plural argument place*. Similarly, the predicates in (2b)–(2c) refer to plural properties (viz., being two logicians and collaborating) whose argument places can admit Russell and Whitehead (taken together). And the two-place predicate 'wrote' (which admit the plural term 'Russell and Whitehead' in (3b)) refers to the two-place plural relation *writing*, whose first argument place is plural; the predicate of (3e), 'outnumber', refers to a two-place relation both of whose argument places are plural. Similarly, the other predicates in (3a)–(3f) refer to two-place plural relations.

Typical plural properties (e.g., being two) are *plurally instantiated*, that is, instantiated by many things (e.g., Pebbie and Pebbo) as such. But some plural properties are not plurally instantiated. Although 'collaborate' refers to a plural property (viz., collaborating), whether the property is plurally instantiated or not depends on whether or not there are any humans or agents who collaborate (e.g., whether or not (2c) is *true*). And the predicate 'be one' (which figures in 'Pebbie and Pebbie *are* not *one*'[44]) refers to a plural property, being one, but this cannot at all be instantiated by any two or more things as such.[45] This means that all the numerical properties identified as natural numbers (being one, being two, etc.) are plural properties. Among them, being one is not plurally instantiated but all the others are (if there are infinitely many things).

Using this analysis of the semantics of plural terms and plural predicates, we can characterize the truth and falsity of *plural predications*, i.e., sentences that result from combining plural predicates with a suitable number of singular or plural terms:

> For example, (2a) is true if and only if the things that the term 'Pebbie and Pebbo' refers to (taken together) instantiate the property that the predicate 'be two' refers to.

---

44 Or 'It is not the case that Pebbie and Pebbo *are one*'.

45 Although being one is not instantiated by any two or more things (taken together), its argument place must be plural because its complement, *not being one*, is instantiated by any two or more things, such as Pebbie and Pebbo (taken together)—'Pebbie and Pebbo *are not one*' is true. (Assuming that Eleatic monism is false, we can see that the complement of a plural property that is not plurally instantiated must be plurally instantiated.) For more about plural attributes, see Yi (1999b, §2 & §4).

We can use this characterization to see that (2a) is true: the plural term refers to Pebbie and Pebbo (as such), and these (taken together) instantiate being two.[46]

5. Numerical Cognition

We can now turn to the role of numerical cognition in mathematical knowledge. The property approach to explaining mathematical knowledge, as we have seen, rests on two pillars:

(A) The *property view of number*: Natural numbers are numerical properties.
(B) The *empirical access thesis*: Humans have empirical, and in some cases even perceptual, access to numerical attributes.

As noted earlier (Section 3), I think psychological studies of numerical cognition provide a strong support for the empirical access thesis. In stating the results and imports of the studies, however, most psychologists seem to assume traditional views of number that presuppose the singular instantiation thesis, such as the set property view. This gives rise to substantial problems for their statements of the results and epistemic significance of the studies. The plural property view, I think, provides a suitable framework for formulating their results and stating their significance.

5.1. The Number Sense and the Set Perceivability Thesis

Recent studies of numerical cognition, as noted earlier (Section 3), support the *number-sense hypothesis*: humans (including infants) and some species of animals (e.g., rats, pigeons, chimpanzees) have a mental faculty for an approximate representation of the numbers of certain objects or individuals (e.g., physical objects, sounds, jumps, syllables in a word). Stanislas Dehaene (1997) calls the faculty the *number sense*, borrowing the term from Dantzig (1930/1954), and Susan Carey calls it "a system of analog magnitude representations of number" (2009a, 118). The faculty or cognitive system yields pre-linguistic representations that differ from the *discrete* representations of numbers that use numerals or their mental counterparts. It represents numbers of individuals as *analog* magnitudes or quantities "roughly proportional" to the numbers (2009a, 118). The mark or "signature" of an analog representational system is that the discrimination of quantities the system represents is governed by Weber's law: "The discriminability of any two magnitudes is a function of their ratio" (2009a, 118). Thus it is easier to use the number sense to distinguish 2 dots from 3 dots than to distinguish 3 dots from 4 dots or even 9 dots from 12 dots (for two-thirds is smaller

---

46 See, e.g., Yi (1999b, 2002, 2005, 2006) for more about the logic and semantics of plurals. See also, e.g., Linnebo (2003), McKay (2006), Oliver and Smiley (2001), and Rayo (2002).

than three-fourths or nine-twelfths). And one can use the faculty, without counting, to have a fairly accurate discrimination of a small number of individuals (e.g., 1, 2, or 3 dots) and tell, e.g., that 30 dots outnumber 20 dots.[47]

So Dehaene compares recognition and comparison of numbers due to the number sense to perception of colors or shapes. He says, as noted earlier (Section 3), that the number sense "opens up a new dimension of sensory perception through which the *cardinal of a set* of objects can be perceived just as easily as their color, shape, or position" (1997, 4f; my italics). I agree with the gist of this statement: using the number sense, we can have perceptual access to the *number* of some objects. This elaborates on the empirical access thesis. But he states a far less plausible thesis by assuming the identification of the *number* of objects with the *cardinality* of the *set* thereof. This thesis is only a distant cousin of the empirical accessibility thesis for numbers. And I think it has a serious problem. It implies that when we see two pebbles (e.g., Pebbie and Pebbo) as being two, for example, we see one composite object different from both pebbles, i.e., the *set* of the pebbles (e.g., the doubleton {Pebbie, Pebbo}).

Although some might use 'cardinal' for certain sets (e.g., Zermelo or von Neumann ordinals), I think Dehaene uses it for some *properties* of sets: their cardinality properties (e.g., being two-membered or having two members), which he seems to identify with numbers.[48] So he assumes the set property view: numbers or numerical properties are cardinality properties of sets. This view, like other views that rest on the singular instantiation thesis, has serious problems (the many-one problem, violation of the logic of numerical statements), as we have seen (Section 4.2). Moreover, the view gives rise to an additional problem when it is combined with the empirical access thesis. The view and the thesis (taken together) imply the *set perceivability thesis*:

> The *set perceivability thesis*: Some *sets* (e.g., a set of two pebbles, a set of three sounds), as well as their members (e.g., the pebbles, the sounds), are perceivable.

When we see two pebbles, on this thesis, we can *see* the *set* comprehending the pebbles as well as the pebbles, and when we hear three sounds, we can *hear* the *set* comprising the sounds as well as the sounds. But I do not think one who sees some pebbles as being two sees the set thereof.

As his formulation of the empirical accessibility thesis suggests, Dehaene seems to accept or presuppose the set perceivability thesis. He flatly says, "In the early 1990s. . . . Some [laboratories] focused on how infants perceived

---

47 For more about the number sense, see Dehaene (1997, Chapter 3) and Carey (2009a, 118f).
48 He concludes a discussion of a deficiency in children's understanding of number by saying, "Up to a relatively advanced age, children . . . [have] the maxim 'Number is a property of sets of discrete physical objects' . . . deeply embedded in their brains" (1997, 61). In his view, I think, the correct thesis is that a number is a property of sets of some entities (discrete or not).

sets of objects" (2011, ix).[49] In his view, then, human infants and some animals alike can see (and hear) some sets as well as their members, for they (like human adults) have the number sense. Some psychologists who study numerical cognition take note of a problem with this view. Although Bloom and Wynn (1997) assume the set property view,[50] they say:

> Sets are notoriously abstract entities. One can see and hear cats, but nobody has ever been wakened in the middle of the night by the yowling of a set.
>
> (1997, 512)

By this, I think, they deny that humans have ever perceived a set (e.g., a set of cats). I agree. Although I have occasionally seen *two cats* and seen them as being two, I have never seen a *set* of two cats—one composite object comprising two cats. I think other humans and animals have the same limitations in perception.[51]

Some philosophers (e.g., defenders of Benacerraf) might go a step further in rejecting the set perceivability thesis. They might argue that it is *in principle* impossible to perceive sets because sets are abstract objects and thus are causally inert. I think defenders of the set perceivability thesis might object to this argument by rejecting the usual view that sets are abstract. But this does not mean that they have a good reason to hold the thesis.

Kim and Maddy, who accept the set property view, hold the set perceivability thesis. Kim holds, "We have perceptual access to some sets" (1981, 349); and Maddy holds, "We can and do perceive sets" (1990, 58). And they respond to the objection that it is in principle impossible to perceive sets by denying that they are abstract in the relevant sense. That is, they deny that sets are not "spatio-temporal objects", which would make them "logically incapable of entering into causal relations" (Kim, 1981, 348). Kim holds, "The metaphysical assumption that sets . . . are abstract in the present sense is by no means obvious or compelling" (1981, 348f). And he suggests that "sets are concrete if their members are concrete" so that "the class [or set] of these dots, as well as the dots themselves, is right here on this piece of paper . . .

---

49 And he says, "As humans, we are born with multiple intuitions concerning numbers, *sets*, continuous quantities", etc. (2001, 31; my italics).

50 They say, "*Two* is a predicate that applies to the *set* of cats. More generally, as Frege . . . has argued, numbers are predicates of sets of individuals" (1997, 512; original italics).

51 I do not have a compelling *a priori* argument against the *possibility* that some of them have a superior sensory or mental faculty that enables them to see sets of cats as well as cats, but I have no good reason to think they actually do. (I have heard some philosophers flatly claim that they could see sets, but I think they were misled by their philosophy in construing the contents of their perception.) In any case, the number sense I share with them does not enable me to perceive sets.

moves when this piece of paper is moved" and "goes out of existence when the paper is burned" (1981, 349). Similarly, Maddy holds:

> There is no real obstacle to the position that the set of eggs [for example] comes and goes out of existence when they [i.e., the eggs] do . . . and . . . spatially as well as temporarily, it is located exactly where they are.
>
> (1990, 59)

And she proposes to "adopt this view" (1990, 59).

I agree that this is a coherent view that helps to meet the objection they address. But the view does not help to meet the initial objection that there is no good reason to accept the set perceivability thesis (the Bloom–Wynn objection). The thesis holds not just that it is in principle possible for us to see sets but also that we humans do have the capacity for perceiving some sets and actually see them in suitable conditions. That is, it holds that we see the doubleton {Pebbie, Pebbo}, for example, whenever we get to see the pebbles as being two. Although I have sometimes seen some pebbles as being two, however, I have never seen a doubleton comprising pebbles, as noted earlier. One cannot reply to this objection by simply holding that sets of pebbles are located where their members are, for we might still not be able to see them because we have no faculty that enables us to see those sets.

Proponents of the set perceivability thesis might hold that I, like others, have actually seen a set of pebbles *without realizing it*. Maddy says, as we have seen, "When I say he gains a perceptual belief about *a set*, I mean he gains a perceptual belief about *a something* with a number property, which we *theorists* know to be a set" (1990, 63, no. 70; my italics). Similarly, she would hold that whenever I see Pebbie and Pebbo, I see "a something", a single composite object (with a number property) although I cannot identify it as a set or distinguish it from a class, an aggregate, etc. (we need a mathematical theory to do so, for our vision does not have enough power to uncover some latent or theoretical features of sets). This assumes that when I see the two pebbles, I see something else: one composite object that is different from both. This is what I deny. Suppose the pebbles are in two adjacent locations in front of us. We can then see them both without shifting attention, and see them as being two. But this does not mean that we also see something else, a third object that is in both locations by having different components in them.

Kim and Maddy hold the set perceivability thesis because they think it follows from the empirical accessibility thesis. But this thesis, which concerns numbers or numerical properties, is different from its cousin about cardinality properties of sets formulated by Dehaene. The former, unlike the latter, does not imply the set perceivability thesis. To draw this thesis from the empirical accessibility thesis (about numerical properties), Kim (1981, 348f) simply assumes the set property view. But this view is incoherent, as we have seen (§4.2). In arguing for the set perceivability thesis,

Maddy (1990, 58ff) does not simply assume but gives a defense of the view. She argues that it is superior to other views about bearers of numerical properties. In doing so, however, she overlooks the *plural property view*, and compares her view only with other views that are equally incoherent because they also rest on the singular instantiation thesis, as we have seen (Section 4.3).[52]

By accepting the plural property view, we can accept empirical accessibility of numbers while rejecting the set perceivability thesis. On the resulting account, which underwrites the plural property approach, we have perceptual access to some numerical attributes of concrete objects. For example, we can see some pebbles (e.g., Pebbie and Pebbo) as *being two* and see some cows (e.g., seven cows) as *being more than two* or as *outnumbering* those pebbles. When we do so, what we see as being two are the pebbles themselves, not a set thereof, and what we see as outnumbering them or as being more than two are the cows themselves, not a set. And the perceptions are veridical, for the pebbles have the *plural property* of being two, the cows (which are, e.g., seven) have the plural property of being more than two, and the cows (taken together) have the *plural relation* of outnumbering with regard to the pebbles (taken together). On this account, neither sets of pebbles (or cows) nor their cardinality attributes (e.g., being two-membered, having a larger cardinality) figure in the perceptions, and we have no mental faculty that enables us to have perceptual access to those sets or attributes.

5.2. Numerical Cognition and Cognition of the Many

Susan Carey (2009a, 2009b, 2011) holds that the *number sense*, which she calls "a system of analog magnitude representations of number" as noted earlier (2009a, 118),[53] is one of "several systems of core cognition", which include the "core object cognition" system that yields "representations of objects individuated from the background" (2009a, 67ff).[54] And she argues, "The analog magnitude number representations . . . require representations of *sets*" (2009a, 136; my italics), which implies that human infants and some animals (as well as human adults) have mental

---

52 See Appendix for more about Kim's and Maddy's arguments for perceivability of sets.
53 She maintains that humans have another core system with numerical content, the "parallel individuation" system (2009a, 137ff). See also Dehaene (2011, 256ff).
54 She holds that "core cognition resembles perception" but differs from perception in that the "representations in core cognition" have "conceptual content" (2011, 113). It is not clear whether this is meant to deny that we can literally see (i.e., visually perceive) a red apple or see it as being red or that we can literally see some apples as being two. In response to Tyler Burge's objection that "there are object perceptions" (2011, 125), she agrees that her cases against perception of physical objects "do not rule out that object representations in infancy are perceptual" (2011, 155).

representation of sets because they (like human adults) have the number-sense system. I agree that the system's representation of a number (e.g., two) must usually be linked to a representation of *the things* that activate it to give rise to the number representation (e.g., Pebbie and Pebbo). But I do not think it requires a representation of the *set* of those things (e.g., {Pebbie, Pebbo}).

Suppose that when someone, Ariel, looks at two pebbles (viz., Pebbie and Pebbo), the pebbles activate her number sense, and she gets to see them as being two and gains a numerical belief that she expresses by asserting (4) while using 'they' as a demonstrative for the pebbles:

(4) *They* are two.

In (4), the predicate 'be two' (or the general term 'two') *is predicated of* the demonstrative symbol 'they'. And the assertion is true because the pebbles (which she uses the symbol to refer to) have the property the predicate refers to (i.e., being two). Similarly, in holding the belief she expresses by asserting (4), Ariel predicates the mental cousin of 'be two' (or 'two') of the mental cousin of 'they' that represents or refers to the pebbles. And the belief is true because the pebbles (taken together) have the property of being two (which the mental cousin of 'be two' represents or refers to). I think her perception of the pebbles as being two that involves a number-sense representation has essentially the same structure. When the two pebbles activate her number-sense system to give rise to a representation of two, the representation does not arise in her mind as a free-floating device with no anchorage. It must be anchored to, or predicated of, a perceptual representation of the two pebbles, a perceptual cousin of the linguistic symbol 'they' that refers to the pebbles. Otherwise her perception cannot help her to answer the question whether *they* (i.e., Pebbie and Pebbo) are two, and there would be no reason to take her (putative) perception involving the number representation to be a correct one about the two pebbles rather than an incorrect one about some others (e.g., Pebbie alone). So her seeing the pebbles as being two must involve *predicating* a number representation *of* a representation of the pebbles, and the perception is correct because the pebbles have the property, being two, that the number representation refers to.

So I think both humans and some animals (viz., those with the number sense) have mental cousins of *plural referential* terms (e.g., 'they', 'these', 'those', 'this and that', 'this, this, and this', 'Pebbie and Pebbo', 'the pebbles over there'), as well as mental cousins of *singular referential* terms (e.g., 'it', 'this', 'Pebbie'). While one can use the singular 'it', for example, to refer to only *one* thing (e.g., Pebbie) at a time, one can use the plural 'they' to refer to *many* things (e.g., Pebbie and Pebbo), collectively or taken together. And one needs to use a plural referential term (e.g., 'they', 'Pebbie and Pebbo') to say of some pebbles that *they* are two. Similarly, one needs to have a *plural referential representation* (e.g., the mental cousin of 'they') to see some pebbles as being two, namely, to see that *they* are two. To do so, one needs to predicate a mental representation of the number two of that of the pebbles. This means that the number-sense system requires the capacity for plural referential representation.

Does this mean that the number sense requires the capacity for representing sets? Must one who activates the system to see some pebbles as being two have a representation of the *set* of those pebbles? I think not. The plural term 'Pebbie and Pebbo' that refers to Pebbie and Pebbo (taken together) is not a term referring to the set comprising them (i.e., {Pebbie, Pebbo}), for the two pebbles (taken together) are not identical with the set. Similarly, a mental representation of the pebbles is not a representation of the set. But what one needs in order to see the pebbles as being two is a representation of *the pebbles*, not a representation of the set.

As noted earlier, however, Carey argues that number-sense representations require representations of sets (call this the *set representation thesis*). To do so, she assumes a thesis about representation that rests on the set property view:

> *The cardinality property representation thesis*: The number-sense representations of numbers (e.g., two or being two) are representations of cardinality properties of sets (e.g., being two-membered or having two members).

She says that representations of the number sense, which represents the *number* of some individuals,[55] are "representations of *cardinal values of sets*of individuals" (2009a, 136; my italics).

Some might use this thesis to argue for the set representation thesis as follows:

> When Ariel sees some pebbles (viz., Pebbie and Pebbo) as *being two* using the number sense representation of two (or being two), she must link the number sense representation of the cardinality property of *being two-membered* to a representation of something with this property (for the perception is veridical). And this representation must be a representation of a set, i.e., {Pebbie and Pebbo} (for only sets have the cardinality property).

The problem with this argument lies in invoking the cardinality representation thesis to identify the number-sense representation of being two with a representation of being two-membered. One cannot identify the representation of being *two* as a representation of being *two-membered* without assuming the set property view: numbers or numerical properties are cardinality properties of sets. This is a widely held view, and I think Carey assumes it to hold the thesis. But the view is incoherent, as we have seen. The property of being two cannot be identified with that of being two-membered; the set

---

55 She says that it represents a number "by a physical magnitude that is roughly proportional to *the number of individuals* in the set being enumerated" (2009a, 118; my italics).

{Pebbie, Pebbo}, for example, has the latter but not the former. This means that representations of the number property cannot be considered representations of the cardinality property.

Now, Carey gives a more sophisticated argument for the set representation thesis. But her argument equally rests on the cardinality representation thesis or the set property view.

She argues for the set representation thesis by highlighting the *attention* one needs to direct to some *particular* things (rather than others) to have a number-sense representation:

> At any given moment, an indefinite number of possible sets to enumerate are in a visual field; attentional mechanism must pick out a particular set [among the many possible sets] to enumerate. [So] Analog magnitude representations [of numbers] are predicated of particular sets of individuals, and thus demand that the animal or infant [who exercises the number sense] represent which set is being assigned which approximate cardinal value. Thus, in addition to the analog magnitude symbols themselves . . . the baby must at least implicitly represent {box} or {red object} or {object on the table}, where { } designates a set and the symbol therein represents the kind of entity contained in the set.
>
> (2009a, 136)

I would agree that when one gets a number-sense representation by looking at some things, the mental process that yields the representation involves directing attention to *some* particular *things* (rather than others), and that the attention involves or leads to a representation of *the things* attended to (i.e., a mental counterpart of 'they', 'the pebbles', 'the boxes', etc.). By assuming the set representation thesis, however, Carey goes a step further. She thinks that the selective attention mechanism involves selecting one particular *set* among many, which implies that many *sets* to select among are located *in* the visual field, and that we (including human infants and some animals) have a kind of pre-attentive and pre-perceptual cognitive access to them. I disagree. I do not think it is necessary to take the attention mechanism to involve cognitive access to sets.

Consider, again, the case of Ariel, who sees Pebbie and Pebbo as being two. In this case, those pebbles activate her number sense to give rise to its representation of two. If so, must she also get a representation of *one*? When she looks at Pebbie and Pebbo (or in their direction), her visual field includes Pebbie, and this is *one*. (The same holds for Pebbo.) If so, would she not get a representation of one as well? We might consider two possible proposals about this issue:

(a) One's number sense can represent only the number of *all* of the things in one's visual field.

(b) One's number sense must represent the number of any of the things in one's visual field.

Suppose for the sake of simplicity that Ariel's visual field has nothing other than the two pebbles. Then she will have only one number-sense representation (viz., that for the two pebbles) on proposal (a); on proposal (b), by contrast, she will have three separate representations (one for Pebbie and Pebbo, one for Pebbie alone, and one for Pebbo alone). Carey thinks that neither proposal is correct. Both assume that what and how many number-sense representations one gets is determined by the things in one's visual field. But she holds that whether Ariel, for example, has a number-sense representation of the number of some things in her visual field depends on whether she *selects* them for her *attention* among all the things in the field, and that she might direct attention to some of those (e.g., all the pebbles) but not to others (e.g., Pebbie alone).[56] If so, directing attention to some things (e.g., all the pebbles) might be necessary to have a mental representation of them (taken together), and this might in turn be necessary to have a representation of their number. Then those who have a number-sense representation, it seems, would get the representation as predicated of or at least linked to the representation of the things they have attended to in order to eventually attain the number-sense representation.

I think this captures the gist of Carey's argument for the set representation thesis. If so, we can formulate its gist without assuming the cardinality property representation thesis or the set property view, as we have seen. That is, there is no need to assume either of these implausible doctrines to state the role of selective attention in the number-sense representation. Moreover, we can do so without implying that human infants and some animals alike have mental representations of sets or that sets of pebbles (as well as pebbles themselves) exist in the visual field that includes pebbles. To see some pebbles as being two, one needs only to have the pebbles in the visual field and possess the capacity for attending to them (as taken together) to build a representation thereof to feed to the number sense. Thus I do not think representations of sets figure in the minds of animals or human infants or adults when they exercise the number sense to get representations of the number of some pebbles. Sets have representations only in the minds of those who give theories of their minds that assume standard yet incoherent views of numbers and of their representations.

6. Concluding Remarks

Warren S. McCulloch (1960) raises two issues one must address to attempt to understand or explain how we humans can know something about numbers:

(a) What is a number?
(b) What kinds of epistemic faculties do humans have?

---

56 She (2009a, 136f) refers to Halberda, Sires, and Fergenson (2006) to support this view.

The plural property approach is based on a view of number, the *plural property view*, that gives an answer to the first question. In this view, the number two, for example, is the property of being two, and this is a property instantiated by any two things (as such): two pebbles, two cows, a pebble and a cow, and so on. If so, we gain cognitive access to the realm of numbers when we get to know, for example, that Pebbie and Pebbo are two, just as we gain cognitive access to colors or shapes when we get to know that Pebbie is gray or that it is oval. And as we can empirically get to know that Pebbie is gray (or oval), we can empirically get to know that Pebbie and Pebbo are two. We can count the pebbles using numerals to find out that they are two. Moreover, we can see them as being two as we can see them both as being gray or oval, as Kim and Maddy hold. Recent psychological studies of numerical cognition provide a strong support for this view. They suggest, as noted earlier (§5), that humans and some animals alike have a mental faculty, the *number sense*, that enables them to make, without counting, reliable (if not error-free) numerical discriminations between two and three pebbles and between 20 and 30 cows. This justifies the partial answer to the second question, the *empirical access thesis*, that underwrites the property approach to explaining human knowledge of numbers: humans have faculties akin to perception they share with some animals that provide them with a solid and reliable, if modest, epistemic foothold in the mathematical realm.

Surely, humans have other cognitive faculties, including probably some unique to them, that enable them to reach far beyond the modest foothold. It is necessary to consider those faculties as well to give a full account of human knowledge of arithmetic, let alone various other branches of mathematics. And the property approach calls for examination of such faculties as well. But the aim of this chapter is not to give a full account of mathematical knowledge by implementing the approach. Its aim is to formulate the approach in order to dispel the long-standing and influential view that we cannot even embark on the project because we have no clue about how to begin. I leave it for other occasions to discuss cognitive faculties that complement the empirical faculties that secure us a modest foothold in the mathematical realm.

Acknowledgments

The work for this chapter was partially supported by SSHRC research grants (410–2011–1971 and 435–2014–0592), which is hereby gratefully acknowledged. I presented its ancestors in the First Veritas Philosophy Conference and the Philosophy of Mathematics, Experimental Approaches to Philosophy, and Color Perception conference, and in meetings of Canadian Philosophical Association, American Philosophical Association, and Society for Exact Philosophy. I also presented them at Keio University, Kyoto University, Oxford University, Seoul National University, SUNY at Buffalo, Kyungpook National University, University of Oslo, Central European University, Korea University, and Postech. I wish to thank the audiences for useful comments and discussions. I am especially grateful to

T. Williamson, C. Shields, T. Iida, Y. Deguchi, T. Kouri, P. Hovda, J. Clarke-Doane, F. Terseman, N. Pedersen, F. Tremblay, Ø. Linnebo, H. Ben-Yami, I. Chung, J. Ha, and C. Lee for useful discussions, and M. Durand and S. Bangu for helpful comments on the penultimate version of this chapter. I am especially indebted to J. Kim, P. Maddy, and T. Martin for their help and encouragement. Needless to say, none of the aforementioned is responsible for any infelicity of this chapter.

References

Aristotle. (*Cat.*/1963). *Aristotle's Categories and de interpretatione* (J. L. Acrill, Trans. with notes and glossary). Oxford: Oxford University Press.

Benacerraf, P. (1973). Mathematical truth. *Journal of Philosophy, 70*, 661–679. Reprinted in Benacerraf & Putnam (1983), 403–420. [Page numbers are to the reprint].

Benacerraf, P., & Putnam, H. (Eds.). (1983). *Philosophy of mathematics* (2nd ed.). Cambridge, MA: Cambridge University Press.

Bloom, P., & Wynn, K. (1997). Linguistic cues in the acquisition of number words. *Journal of Child Language, 24*, 511–533.

Burge, T. (2011). Border crossings: Perceptual and post-perceptual object representation. *Behavioral & Brain Sciences, 34*, 125–126.

Carey, S. (2009a). *The origin of concepts*. Oxford: Oxford University Press.

Carey, S. (2009b). Where our number concepts come from. *Journal of Philosophy, 106*, 220–254.

Carey, S. (2011). Précis of *The origin of concepts*. *Behavioral & Brain Sciences, 34*, 113–167.

Chihara, C. (1982). A Gödelian thesis regarding mathematical objects: Do they exist? And can we perceive them? *Philosophical Review, 91*, 211–227.

Dantzig, T. (1930/2007). *Number: The language of science*. New York, NY: Macmillan, 1930. Masterpiece Science ed. (J. Mazur ed., B. Mazur Forward). New York, NY: Plume, 2007.

Dehaene, S. (1997). *The number sense: How the mind creates mathematics*. Oxford: Oxford University Press.

Dehaene, S. (2001). Précis of *The number sense*. *Mind & Language, 16*, 16–36.

Dehaene, S. (2011). *The number sense: How the mind creates mathematics* (revised and updated ed.). Oxford: Oxford University Press.

Field, H. (1982). Realism and anti-realism about mathematics. *Philosophical Topics, 13*, 45–69. Reprinted in Field (1989), 53–79.

Field, H. (1989). *Realism, mathematics & modality*. Oxford: Basil Blackwell.

Frege, G. (1884/1980). *Die Grundlagen der Arithmetik*. Breslau: Köbner, 1884. Translated by J. L. Austin as *The foundations of arithmetic* (2nd rev. ed.). Oxford: Basil Blackwell, 1980.

Grandy, R. (1973). Reference, meaning, and belief. *Journal of Philosophy, 70*, 439–452.

Grice, H. P. (1961). The causal theory of perception. *Proceedings of the Aristotelian Society: Supplementary Volumes, 35*, 121–152.

Halberda, J., Sires, S. F., & Fergenson, L. (2006). Multiple spatially overlapping sets can be enumerated in parallel. *Psychological Science, 17*, 572–576.

Hart, W. H. (1977). Review of Steiner, *Mathematical knowledge*. *Journal of Philosophy, 74*, 118–129.

Hume, D. (1888/1978). *A treatise of human nature* (L. A. Selby-Bigge, Ed. with an analytic index, 2nd ed. with text revised & variant readings by P. H. Nidditch). Oxford: Oxford University Press, 1978.

Kaufman, E. L., Lord, M. W., Reese, T. W., & Volkmann, J. (1949). The discrimination of visual number. *American Journal of Psychology, 62,* 498–525.

Kim, J. (1981). The role of perception in a priori knowledge: Some remarks. *Philosophical Studies, 40,* 339–354.

Lavine, S. (1992). Review of Maddy, *Realism in mathematics. Journal of Philosophy, 89,* 321–326.

Linnebo, Ø. (2003). Plural quantification exposed. *Noûs, 37,* 71–92.

Locke, J. *(Essay/1975). An essay concerning human understanding.* Oxford: Oxford University Press.

Maddy, P. (1980). Perception and mathematical intuition. *Philosophical Review, 89,* 163–196.

Maddy, P. (1990). *Realism in mathematics.* Oxford: Oxford University Press.

Maddy, P. (1992). Indispensability and practice. *Journal of Philosophy, 89,* 275–289.

Maddy, P. (1997). *Naturalism in mathematics.* Oxford: Oxford University Press.

Maddy, P. (2007). *Second philosophy.* Oxford: Oxford University Press.

Maddy, P. (2014). *The logical must: Wittgenstein on logic.* Oxford: Oxford University Press.

Maddy, P. (2018). Psychology and the a priori sciences. In S. Bangu (Ed.), *Naturalizing logico-mathematical knowledge: Approaches from philosophy, psychology and cognitive science,* 15–29 New York, NY and London, UK: Routledge.

McCulloch, W. S. (1960). What is a number, that a man may know it, and a man, that he may know a number? *General Semantics Bulletin, 26/27,* 7–18. In J. Paul (Ed.), *www.vordenker.de* (Winter ed. 2008/09). Retrieved from www.vordenker. de/ggphilosophy/mcculloch_what-is-a-number.pdf.

McKay, T. J. (2006). *Plural predication.* Oxford: Oxford University Press.

Mill, J. S. (1891/2002). *A system of logic: Ratiocinative and inductive* (reprinted from the 1891 ed.). Honolulu, HI: University Press of the Pacific, 2002.

Oliver, A., & Smiley, T. (2001). Strategies for a logic of plurals. *Philosophical Quarterly, 51,* 289–306.

Pitcher, G. (1971). *A theory of perception.* Princeton, NJ: Princeton University Press.

Plato. *(Parm./1996). Parmenides* (M. L. Gill & P. Ryan, Trans.). Indianapolis, IN: Hackett.

Putnam, H. (1971). *Philosophy of logic.* New York, NY: Harper.

Quine, W. V. (1948). On what there is. Reprinted in Quine (1980), pp. 1–19.

Quine, W. V. (1951). Two dogmas of empiricism. Reprinted in Quine (1980), pp. 20–46.

Quine, W. V. (1980). *From a logical point of view* (2nd ed. rev.). Cambridge, MA: Harvard University Press.

Rayo, A. (2002). Word and objects. *Noûs, 36,* 436–464.

Russell, B. (1903/1937). *Principles of mathematics.* Cambridge: Cambridge University Press. 2nd ed. London, UK & New York, NY: Norton, 1937.

Russell, B. (1919/1920). *Introduction to mathematical philosophy.* London: George Allen & Unwin. 2nd ed. London: George Allen & Unwin, 1920. (Dover ed. New York, NY: Dover, 1993).

Whitehead, A. N., & Russell, B. (1962). *Principia mathematica to *56.* Cambridge: Cambridge University Press.

Yi, B.-U. (1998). Numbers and relations. *Erkenntnis, 49,* 93–113.

Yi, B.-U. (1999a). Is mereology ontologically innocent? *Philosophical Studies, 93,* 141–160.

Yi, B.-U. (1999b). Is two a property? *Journal of Philosophy, 96,* 163–190.

Yi, B.-U. (2002). *Understanding the many.* New York, NY: Routledge.

Yi, B.-U. (2005). The logic and meaning of plurals: Part I. *Journal of Philosophical Logic, 34,* 459–506.

Yi, B.-U. (2006). The logic and meaning of plurals: Part II. *Journal of Philosophical Logic, 35,* 239–288.

Yi, B.-U. (2013). The logic of classes of the no-class theory. In N. Griffin & B. Linsky (Eds.), *The Palgrave centenary companion to Principia mathematica* (pp. 96–129). New York, NY and Basingstoke, UK: Palgrave Macmillan.

Yi, B.-U. (2014). Is there a plural object? In A. J. Cotnoir & D. Baxter (Eds.), *Composition as identity* (pp. 169–191). Oxford: Oxford University Press.

Yi, B.-U., & Bae, E. (1998). The problem of knowing the forms in Plato's *Parmenides. History of Philosophy Quarterly, 15*, 271–283.

Appendix
Can We See Sets?

Kim argues, "We have perceptual access to some sets and their properties" (1981, 349). To do so, he assumes the set property view and argues that we can see, for example, the set of some dots because we can see some dots as being five. The argument is implicit in his discussion of the objection that his views imply that we can see something unperceivable:

> If ascribing a numerical property to these dots on the paper, that they are five in number, is to have the consequence that we are attributing a property to some suprasensible object . . . inaccessible to human perception, then something is wrong with the descriptive apparatus being used, and it is this apparatus . . . that need to be changed, not our belief that the number five is visibly instantiated by these dots right in front of us.
>
> (1981, 348)

Here he assumes that when we see some dots as being five, we perceptually attribute the numerical property of being five to *those dots*, and takes this to mean that we perceptually attribute the property to *the set* of the dots (and thus see the set).[57] But it is one thing to say that we attribute the property to the five dots and yet another to say that we attribute it to the set. One cannot identify the five dots with the set thereof (the dots are five but the set is not; it is one thing comprising five dots). This also means that the set property view is incoherent, as discussed earlier (Section 4.2).

While Kim assumes the set property view, Maddy (1980, 1990) defends the view in her argument for the set perceivability thesis. But her argument has essentially the same problem as Kim's. Her defense of the set property view implicitly assumes the singular instantiation thesis. But this is an incoherent thesis that implies that some one thing (e.g., a doubleton) is also many (e.g., two). It would be useful to see the role of the thesis in her argument.

---

57 But he argues that this does not mean that we see something unperceivable by rejecting the assumption that sets are abstract and have no spatiotemporal locations.

She argues that those who see two or three eggs, for example, can see the *set* of those eggs. Suppose that Steve opens a carton with two eggs (viz., Eggie and Eggo), sees the eggs in the carton, and gains a numerical belief that he might express by asserting '*They* are two' while using 'they' as a demonstrative referring to the eggs. She argues that in that case, Steve has seen or visually perceived a set of eggs: the doubleton {Eggie, Eggo}.[58] She does so by applying Pitcher (1971)'s analysis of perception. On this analysis, as she puts it,

> . . . for Steve to perceive a tree before him is for there to be a tree before him, for him to gain perceptual beliefs, in particular that there is a tree before him, and for the tree before him to play an appropriate causal role in the generation of these perceptual beliefs.
>
> (1990, 51)[59]

So she argues that Steve's case satisfies three conditions (1990, 58):

(a) The set {Eggie, Eggo} exists in the carton.
(b) Steve acquires perceptual beliefs about the set.
(c) The set plays an appropriate causal role in the generation of those beliefs.

In defense of (a), Maddy presents her version of the indispensability argument for the existence of mathematical entities (including sets)[60] and she holds that the set {Eggie, Eggo} "comes into and goes out of existence when they [i.e., the eggs] do . . . and . . . spatially as well as temporarily, it is located exactly where they are" (1990, 59). In defense of (b), she argues that the numerical belief Steve gains when he sees the eggs (i.e., the belief that *they* are two) is a perceptual belief about the set. In defense of (c), she holds that the set plays the same sort of causal role in Steve's gaining the belief as the role that my hand plays in my gaining the perceptual belief that there is a hand before me.

---

58 See Maddy (1980, 178; 1990, 58) for her presentations of Steve's case. Her case involves three eggs, but I change it to a case of seeing two eggs for convenience of exposition. In her (1980), she argues that the existential "belief . . . that there are three eggs" (1980, 58) is a belief about a set by holding that numerical beliefs are "beliefs that something or other has a number property" (1980, 179). In her (1990), she does not make it clear what kind of numerical beliefs are at issue, but she argues that Steve "gains perceptual beliefs about the set of eggs, in particular, that this set is three-membered" (1990, 59), which suggests that her focus is on beliefs whose contents can be expressed by '*They* are two' and its ilk.

59 The causal role played by the things perceived is said to be *appropriate*, if they play "the same sort of role in the causation of the perceiver's perceptual state as my hand plays in the generation of my belief that there is a hand before me when I look at it in good light" (Maddy 1990, 51). This way of characterizing the appropriateness in question is due to Grice (1961).

60 See Quine (1948, 1951), Putnam (1971), and Maddy (1990, Chapter 1) for the indispensability argument.

I think the main problem of her argument lies in her defense of (b).[61] To defend this thesis, she argues that true numerical beliefs must be beliefs about sets because numerical properties are cardinality properties of sets. But this view assumes the set property view, which is incoherent as we have seen (Section 4.2). She defends the view by arguing that it is superior to other views about bearers of numerical properties. But this argument ignores the natural view, the plural property view, on which it is the eggs (i.e., Pebbie and Pebbo) that Steve sees as being two and that instantiate the numerical property of being two. She implicitly rejects this view by assuming the singular instantiation thesis and compares the set property view only with views that are equally incoherent because they conform to the thesis. As we have seen (Sections 4.3–4.4), however, this thesis is incoherent and we must reject it to reach an adequate view of number and numerical property.

---

61 I do not accept (a), which is presupposed in Maddy's argument for (b). But my objection to this argument is independent of whether or not (a) is true. In her later works (1992, 1997), Maddy rejects the indispensability argument, which leads her to reject (a) and undermines her argument for (b). Lavine (1992, 323) objects to (c). See also Chihara (1982).

5 Intuitions, Naturalism, and Benacerraf's Problem

Mark Fedyk

1. Introduction

Mathematical knowledge does not arise in the mind spontaneously. Both rationalists and naturalists in the philosophy of mathematics need a plausible theory that explains either how accurate mathematical beliefs are formed and justified or, if not that, then why skepticism about mathematical knowledge is warranted. And it is most usually held by rationalists that mathematical knowledge enters the mind by way of the operation of intuition; consequently, intuitions play a central role in rationalist mathematical epistemologies. But what can a *naturalist* say about the connection between intuition and mathematical knowledge? Indeed, given that naturalists frequently are skeptical about a priori knowledge and the evidential role that intuitions play in establishing such knowledge, *should* a naturalist try to say anything substantive about intuition and mathematical knowledge?

There are, actually, at least three arguments that ought to lead a naturalist to answer 'yes' to the later question. The first argument is itself an implication of the view that inquiry can only produce truths that are, at most, approximate (Boyd, 1990; Wimsatt, 1972; Teller, 2014)—the implication, specifically, is that it is a sign of progress when researchers working within a single disciplinary matrix construct multiple definitions of roughly equivalent concepts and where, importantly, each of these definitions can help in some way to satisfy some of the matrix's epistemic ends. This pertains to the role that intuitions play in the epistemology of mathematics because, as Philip Kitcher wryly notes, "'Intuition' is one of the most overworked terms in the philosophy of mathematics" (Kitcher, 1984, 49). To a naturalist, this is an auspicious sign. The emergence of a plurality of partially overlapping technical concepts means, generally speaking, that none of them should be interpreted as *the* exact, final, absolutely true definition, and that conclusion holds despite the fact that their respective individual ability to bear some inductive or explanatory weight is evidence that many of the concepts are, each in their own way, representing something of interest with a non-trivial degree of accuracy. Thus, a profusion of concepts with varying but overlapping definitions within a discipline can be seen as the emergence of a novel conceptual framework, the nodes of which are different approximations that, when taken together, convey more information about the referent of the concepts at each node than any single concept can when it is taken in isolation. This means that the proliferation of rationalist conceptions of intuition in the philosophy of math should be interpreted by naturalists— perhaps with a bit of irony—as evidence that a theory of intuition is integral to the epistemology of mathematics. It may consequently be productive for naturalists to explore definitions of 'intuition' that are both compatible with a generally naturalistic mathematical philosophy and which capture at least some of the insights motivating rationalist treatments of intuition.

The second argument begins from the observation that nothing could be more counterintuitive than the claim that intuitions do not exist. We all have them, and they play an important role in most aspects of our cognitive lives. Young children must rely almost entirely on certain very basic intuitions to make sense of their physical and social worlds (Wellman, Kushnir, Xu, & Brink, 2016), as well as elementary mathematical phenomena (Xu & Carey, 1996). Not unrelatedly, philosophers have made the practice of collating and analyzing intuitions into an area of professional competency (Deutsch, 2010, 2015; De Cruz, 2015; Cappelen, 2012). There simply is overwhelming evidence that intuitions are both a real and an important part of the human cognitive system: it is much more likely than not that intuitions support rather than hinder the formation of knowledge. So, the second argument supporting a naturalistic investigation of the relationship between intuition and mathematical knowledge is that this project is a component of the larger epistemological undertaking that consists of seeking an understanding of the role that intuition plays in supporting reliable cognition in all its forms. The ubiquity of intuition forces us to take seriously questions about what kinds of beliefs intuition can justify, including whether intuition can be a source of justification for mathematical knowledge.

Finally, there is substantial ex ante plausibility that there is some kind of an important relationship between intuition and mathematical belief; it is not only rationalists who must believe that intuition and mathematical belief have something to do with one another. Accordingly, the third argument is that an epistemology of mathematics that can make sense of the connection between intuition and mathematical belief would be better, most other things being equal, than one that cannot. Yet it has proven difficult to bring this connection into sharper relief. Richard Tieszen is referring to the history of failed attempts to make this connection explicit when he writes,

> "Intuition" has perhaps been the least understood and the most abused term in philosophy. It is often the term used when one has no plausible explanation for the source of a given belief or opinion.

7. Intuitions have a (perhaps distinctive) phenomenological character: their mode of presentation in consciousness has a distinctive 'seeming-ness' or 'apparently-ness'.
8. Intuitions are, often, not factive. An intuition that p (much of the time) does not mean that p is the case. In this respect, intuitions are typically unlike everyday perceptions.

So far, so good, but we still need to ask at this point, what makes this account of intuitions a naturalistically plausible account? The short answer is that this account does not entail that intuitions are a source of a priori knowledge.

But there is a deeper reason why the ideas listed earlier form a naturalistically friendly theory of intuition: the ideas are each projectable on the basis of contemporary cognitive science. Here it is important to avoid the common mistake of assuming that consistency with cognitive science is sufficient to establish the naturalistic bona fides of an epistemological theory. The reason this is a mistake is that even the very best theories in cognitive science are not themselves mutually consistent with one another, and so it makes no sense at all to ask whether some epistemological theory is consistent with cognitive science in toto. A more reasonable question to ask, instead, is whether the main concepts used in the explanation of, in this case, the nature of an intuition are themselves projectable—according to the sense given to that term by Nelson Goodman (Goodman, 1955)—within some projects in cognitive science. So, for illustration, while there are scientifically plausible arguments for skepticism about mental representations (Smolensky, 1988; Bechtel & Abrahamsen, 1991), concepts (Machery, 2009), and even intentionality or *aboutness* itself (Chomsky, 2000; Mcgilvray, 1998), both of the other conceptual ingredients in my account of intuitions are highly projectable in modern cognitive science. For example, Susan Carey's theory of cognitive development (Carey, 2011) is an instance of a highly projectable psychological theory that uses all of the elements of our definition; Joseph Perner's earlier work in the same area is another example (Perner, 1991).

3. Meaning-Directed and World-Directed Uses of Intuitions

An intuition is, by itself, a single mental occurrence. However, intuitions are frequently used as evidence by philosophers and nonphilosophers alike. This happens when the propositional content of an intuition is taken to increase the warrant or credence of a belief; frequently, though not necessarily, the relevant belief has the same propositional content of the intuition. But can we be more specific about what types of learning are possible when intuitions are used as evidence?

To answer this question, we should distinguish between *world-directed* and *meaning-directed* uses of intuitions as evidence. This distinction is grounded in the observation that the reliability of an intuition is partly a matter of the purpose to which its use is being put and thus partly a matter of the epistemic context in which the intuition occurs. Accordingly, the distinction here is between the contexts in which we can use intuitions to learn about the representational content of the concept implicated in a person's intuition (*meaning-directed* uses) and contexts in which we can learn about whether or not situations in the world contain properties (objects, events, kinds, etc.) that are instances of the properties defined by a concept implicated in an intuition (*world-directed* uses).

Examples are the best way to make the distinction between meaning-directed and world-directed uses of intuitions clear. Suppose I want to use one of my intuitions to find out something about the world—specifically, whether some of the things that I can recall seeing or imagine existing fall into a specific category. When appealing to my intuitions, I first bring into my imagination a representation of a certain situation, and I then apply the concept I intend to 'implicate' to some aspect or feature of the situation that I have before my mind. Doing this produces the intuition itself, the propositional content of which is a by-product of whether what I have brought to mind satisfies, or does not satisfy, the implicated concept. To make this more concrete, suppose that I have an intuition that implicates my concept of a spark plug, and suppose also that I know a lot about internal combustion engines. In this case, it is easy enough for me to imagine a new combustion process based on a hypothetical build of a turbulent jet ignition system (Mauß, Keller, & Peters, 1991), and I may then appeal to my intuitions to determine whether the jet ignition system itself is a spark plug. Since I know a lot about engines, my concept of a spark plug can be expected to reliably track spark plugs, and so my intuition about whether the turbulent jet ignition system I have imagined counts as a spark plug will be accurate most of the time. In this case, it is appropriate to treat my intuition as a world-directed intuition: it can be used, defeasibly, as evidence about what things are and are not spark plugs, so long as these things are close approximations of what I can imagine.

Suppose we change the example. Rather than engines and their components, let us imagine now that we are interested in cosmological questions related to gravitational waves. Since I know almost nothing about cosmology, if I were asked to remember the waveform data generated by the Laser Interferometer Gravitational-Wave Observatory experiments and then asked for an intuition about whether the data indicates the existence of gravitational waves, whatever intuitions come to mind about the relevant data should not be interpreted at world-directed intuitions and thereby as defeasible evidence of the existence of such waves. Why? My concept of gravitational waves is too inaccurate for it to be a reliable guide about whether anything that my mind finds itself able to fit under the concept is, in fact, a gravitational wave. My intuition in this case should not be used as a world-directed intuition. Nevertheless, the intuition itself is still evidence of something: it is evidence about the representational (or intensional) content of my concept of gravitational waves, such as it is. We might not be able to find out anything about whether this or that data would be indicative of gravitational waves by appealing to my intuitions; still, we can find out something about the content of any number of my concepts and doing so is to make use of a meaning-directed intuitions.

So, the distinction we have is this: in normal circumstances, all intuitions are reliable evidence about something: all intuitions provide accurate information about the content of the concept implicated in an intuitor's intuitions. To the extent that we want to learn about only the meaning of a person's concept, we can rely upon meaning-directed uses of intuitions. But if we want to instead use intuitions to learn something about the world, then we must have independent evidence that the implicated concept is accurate enough for the task. When we are in possession of this additional evidence, an intuition initially used as meaning-directed evidence can also then be used as world-directed evidence.

4. A Different Take on Benacerraf's Epistemological Problem

For contemporary naturalists, there is the possibility of a novel response to Benacerraf's epistemological problem as it was expressed in (Benacerraf, 1973). There, Benacerraf argued,

> Accounts of truth that treat mathematical and nonmathematical discourse in relevantly similar ways do so at the cost of leaving it unintelligible how we can have any mathematical knowledge whatsoever; whereas those which attribute to mathematical propositions the kinds of truth conditions we can clearly know to obtain, do so at the expense of failing to connect these conditions with any analysis of the sentences which shows how the assigned conditions are conditions of their truth.
> (Benacerraf, 1973, 662)

Benacerraf operates with the desiderata that we should aim for a single comprehensive theory of truth that applies to all cases of mathematical and nonmathematical belief alike. Importantly, he also presupposes, like almost all subsequent commentators in philosophy, that the appropriate concept of truth for such a theory is *exact* truth.

But since the publications of Benacerraf's seminal article, work done by philosophers of science on the problem of how the representational by-products of scientific inquiry—that is, theories stated in both natural and mathematical language, but also models, graphs, data tables, and so on—represent various bits and pieces of nature has opened a new avenue for meeting Benacerraf's challenge. According to this alternative, we should reject ideas such as the view that the natural world consists of facts that can be identified with, or placed into some form of homomorphism with, sets of propositions. Consequently, the notion of an *exactly* true representation—as, for instance, one frequently finds in model theory for logic—does not tell us how true statements correspond to the world. At best, the theory of truth familiar to most philosophers is an idealized and greatly simplified (that is, inexact) picture of how language achieves correspondence (cf. (Boyd, 2016; Teller, 2012)). The implication is that facts are not the sort of thing that can be described with perfect fidelity, whether or not a scientist is using a mathematical vocabulary (cf. (Levins, 1966)). But the idea that truth should be cashed out as exact truth is nevertheless implicit in Benacerraf's challenge: he requests a treatment of mathematical and nonmathematical discourse in which the exact truth of the relevant statements is adequately explained. At least for scientific discourse, however, that is no longer a reasonable demand (Boyd, 2001; Giere, 2006; Cartwright, 2017, 1999). So, a naturalist inspired by this work in philosophy of science can look for a theory of how mathematical statements can be inexactly or approximately true. In the next section, I will argue that the distinction between meaning-directed uses and world-directed uses of intuitions allows for exactly that.

First, though, one more clarifying comment about Benacerraf's problem. In suggesting a shift away from asking how both mathematical and nonmathematical statements (or beliefs) can be exactly true, I am not also suggesting that we abandon the other component of the challenge—namely, the request that a causal relation exists between the relevant discourse and its object. About this, Benacerraf writes, "For X to know that S is true requires some causal relation to obtain between X and the referents of the names, predicates, and quantifiers of S" (Benacerraf, 1973, 671). Thus, for the purposes of this chapter, I will understand Benacerraf's problem to be the problem of explaining how (a) mathematical statements can be approximately true in (b) the same way that nonmathematical statements can be approximately true, while also requiring that (c) in order for X to know that S is approximately true there must be some causal relation between X and the referents of the names, predicates, and quantifiers of S.

5.   Meaning-Directed Intuitions and Mathematical Concepts

The first part of a solution to this formulation of Benacerraf's problem is to interpret intuitions, the propositional content of which are mathematical statements, as instances of *meaning-directed* uses of intuitions that implicate specific mathematical concepts. On such an interpretation, these intuitions are a kind evidence, but only about something very specific and limited, as they do nothing more than to make explicit the content of whichever mathematical concepts are implicated in the relevant intuitions. Consequently, it would be a mistake to ask, of the propositional content of intuition itself, whether it is true or false.

An example of how this proposal works can make it clearer. Consider the concept DENSE, as expressed in the following theorem:

(t) The rational numbers are dense in the reals: if a and b are real numbers such that a is less than b, then there is a rational number p/q such that a is less than p/q, which is less than b.

This theorem gives a definition of DENSE, and understanding (t) seems to be sufficient to grasp (but, importantly, not justify) the concept. However, the proof of (t) helps establish a firmer understanding of the concept DENSE. It is also short enough to hold in one's mind 'all at once'. To wit, the Archimedean property tells us that if p and e are positive, then ne is greater than p for some integer n. So, setting $p = 1$ and $e = b - a$, there is a positive integer q for which $q(b - a)$ is greater than 1. The Archimedean property also tells us that, setting $e = 1$ and $p = qa$, where a is greater than 0, there is also an integer j such that j is greater than qa. From these two steps, we can reason as follows:

Let p be the smallest integer such that $p > qa$.
It follows that $p - 1 \leq qa$.
Thus $qa < p \leq qa + 1$.
But since $1 < q(b - a)$, it also follows that $qa < p < qa + q(b - a) = qb$.
So $qa < p < qb$.
Thus $a < p/q < $ b.

The proof, partly because of its brevity, can operate as an explication of the intensional content of the concept DENSE.

Of course, it is customary to interpret proofs as not only ways of developing or refining your grasp of a mathematical concept but also as justification for, in this case, (t). The point here, though, is that, without denying that proofs confer justification, proofs—or, rather, conscious apprehensions of each of all of the steps of a proof—can act as the representations that the implicated concept is applied in order to generate meaning-directed intuitions. Indeed, that is the utility of having the concept of a meaning-directed intuition in our mathematical epistemology: it makes room for distinguishing between using proofs as a vehicle for acquiring and refining mathematical concepts, on the one hand, and proofs as establishing certain mathematical truths, on the other. We can interpret intuitions implicating DENSE as evidence *only* about the intensional content of the concept DENSE—on this view, running through the proof in one's imagination is a way of eliciting the intuition that the rational numbers are *dense* in the reals without necessarily taking on an ontological commitment to the real numbers (cf. (Lakatos, 1976)).

However, for longer proofs, it might not be possible to hold all the steps of the proofs in one's imagination and thereby generate the appropriate meaning-directed intuitions so immediately. Nevertheless, it is still possible to interpret these proofs as eliciting meaning-directed intuitions that can, in turn, be used to develop or refine a particular mathematical concept. The suggestion here is to see the concept implicated by steps 1 through 10 in proof Z as the concept $STEPS_{1-10 \ in \ z}$, the concept implicated by steps 10 through 12 as $STEPS_{1-12 \ in \ z}$, and so on. These concepts may never acquire a public name, but complexes of them would form the content of imaginative experience that the relevant implicated mathematical concept is meant to be applied to. Of course, a certain amount of training is required to learn how to 'chunk' together all of the steps, or even the smaller chunks, of a complicated proof; nevertheless, this example illustrates how chains of meaning-directed intuitions can facilitate the acquisition of mathematical concepts much more complicated than DENSE.

So, according to the position that I am advancing, meaning-directed intuitions that implicate mathematical concepts are *procedurally instrumental* for the production of mathematical knowledge. This does not mean that the proofs involved in eliciting the relevant intuitions must also be viewed as entailing the accuracy of either the mathematical concepts implicated by them or statements formulated using some of the same concepts. This distinction holds because, as a more general truth, the fact that some act or occurrence might be procedurally instrumental in the formation of certain types of belief does not mean that whichever fact this fact happens to be must also play a central role in justifying the relevant beliefs. For example, it is procedurally instrumental that light refract off the surface of the table at which I am sitting in order for me to form the belief *the table is brown*, but this fact does not justify my belief all on its own. Likewise, working through a mathematical proof, and thereby using it to elicit meaning-directed intuitions, is a process that may be procedurally instrumental for acquiring certain mathematical concepts—still, it does not follow that the mathematical concepts so learned are thereby either true or false.

Charles Parsons draws a distinction between intuitions-*that* and intuition-*of* (Parsons, 1979). Intuitions-*that* are intuitions that, because of the putative obviousness of their propositional content, are readily believed, while intuitions-*of* are more like perceptual experiences that do not immediately generate beliefs. Parson's distinction is similar to, but not the same as, the two uses of intuitions that we have been examining. Meaning-directed intuitions are most like intuitions-*of*, because it is a mistake to subject their propositional content to factive interpretation. However, if there is independent evidence that a concept implicated in an intuition is accurate, then we can shift our interpretation of an intuition as an intuition-*of* to something like seeing it as an intuition-*that*. However, in the example used previously, the intuition <if a and b are real numbers such that a is less than b, then there is a rational number p/q such that a is less than p/q which itself is less than b> is an intuition-*of* a mathematical proposition: it would be a mistake to interpret this intuition as evidence of anything other than the meaning of a mathematical concept.

6. Realism, Complexity, and Mathematical Concepts in Science

When can mathematical intuitions be interpreted as world-directed intuitions? To answer this question in a manner compatible with the response to Benacerraf's problem that we are developing, we must use the same criteria that would justify treating a nonmathematical proposition as approximately or inexactly true. Doing this will give us the second and final part of our solution to Benacerraf's problem.

But before proceeding any further, a clarification is necessary. Indispensability arguments in mathematics often have as their target substantive ontological conclusions (Maddy, 1992; Resnik, 1995; Baker, 2001; Colyvan, 2001; Bangu, 2012)—for instance, that certain mathematical objects such as the real numbers exist. I mention this only to make it clear that the argument in this section does not offer an indispensability argument, or at least not an argument in the traditional form of such arguments, because the only ontological commitment that falls out of the argument to follow is that the natural world contains something *like* mathematical objects. The reason for this is that standard arguments for scientific realism, when applied to the mathematical and nonmathematical elements of scientific vocabularies alike, yield the conclusion that we should see all elements of the respective vocabularies as, at most, approximate truths. And yet that is exactly what is called for by way of this chapter's solution to Benacerraf's problem. In fact, we can move our focus even further toward standard issues in the philosophy of science by proceeding from a plausible answer to Philip Kitcher's question of *why* mathematical concepts are necessary components of successful science (Kitcher, 1984). The answer I propose here is the same as for the parallel question of why technical concepts are necessary components of successful science: in both cases, we construct proprietary concepts, models, formulas, and theories as a way of dealing with the world's inherent complexity—a complexity that renders the natural world largely illegible to both common sense and those who would attempt to see it from the perspective only of mathematical platonism.

Indeed, the natural world—which includes of course the social worlds of humans and corresponding mental worlds—is *exceedingly* complex. It really is something of a miracle that, for example, we are able to build transistors with gate widths of approximately one nanometer. But the word 'approximately' in that last sentence is important. We have not, and we cannot ever, build transistors that are all and only *exactly* a single nanometer in width, and this is not just because of the relevance of Heisenberg's uncertainty principle to examples at this scale. Paul Teller offers us a distinction between accuracy, precision, and exactness (Teller, 2012, 2014). A statement is *accurate* to the extent that it is true, *precise* to the extent that it is free of vagueness, *exact* to the extent that it is simultaneously both highly accurate and highly precise. The chemical models and equations that tell us that, in this case, carbon nanotubes and molybdenum disulfide can be put together to form extremely small transistors are accurate, but not precise. Using recently developed techniques, it is possible to build hundreds of these extremely small transistors, but at that scale, even small shifts in local gravitational fields will alter the width of the gates in the transistors: their width, importantly, is never *precisely* a single nanometer. Still, that does not matter; it is accurate enough to say that we can build transistors with nanometer-wide gates.

Considerations such as this example illustrate the important lesson that rarely, if ever, are scientific statements *exactly* true: the complexity of the world means that we constantly face a trade-off between precision and accuracy. That is partly why precision is an important criterion to care about, because it is a way of handling the world's complexity in an epistemically constructive fashion. The precision we create by introducing both technical jargon and mathematical concepts into scientific vocabularies allows us to idealize the world and talk about its processes and systems *as if* they were simple enough for us to reason about directly. To illustrate, no one would take, for instance, a linear model of the relationship between immigration and a nation's gross domestic product (GDP) growth as the exact truth; everyone knows that the relationship between these two phenomena is much, much more complicated than that. Nevertheless, the actual relationship between immigration and GDP is *too* complex for anyone to model and thus to represent exactly. So, we must model it *as if* it is a linear relationship and thereby recognize that the mathematical precision of our model is true only of an idealized and fictional world.

Nancy Cartwright has this to say about one component of the epistemic processes that I am describing here:

> A Ballung concept is a concept with rough, shifting, porous boundaries . . . Many of our ordinary concepts of everyday life are just like this. Ballung concepts also can, and often do, play a central role in science and especially in social science. But they cannot do so in their original form. To function properly in a scientific context they need to be made more precise. . . . I sometimes use the ugly word "precisification" to describe the process by which a Ballung concept is transformed into one fit for science.
>
> (Cartwright, 2017, 136)

Something like the opposite of precisification occurs when concepts from pure mathematics are turned into, for instance, the components of scientific models about complicated causal processes, interactions, forces, and so on. Instead of taking Ballung concepts and regimenting their meaning in service of some inductive or explanatory end, scientists often also take mathematical concepts and use them in a (compared to their use in pure math) more rough, shifting, and porous manner. To see this, think about what the numerals in an equation expressing a linear relationship between immigration and GDP growth are representing in the messy, complicated real world.

The point of using mathematical concepts and formulas in scientific practice, then, is not to describe how the world is *exactly*. We introduce the mathematical vocabulary we need in order to establish terminological frameworks that are precise enough to allow us to, for instance, specify the contrast classes that drive some forms of scientific explanation (Salmon, 1984), or talk about causal interactions as if they were simple Boolean relationships, or even frame research questions with just enough specificity to allow such questions to be tested by empirically generated data sets. The mathematical statements that get used for these purposes are *inaccurate*—though not false. This is in fact crucial to their inductive and explanatory utility. If they were perfectly exact—that is, if they were both maximally accurate and maximally precise—the formulas we would try to write down would be too complicated for anyone to reason with. Indeed, there is nothing but the dictates of epistemic prudence and philosophical humility standing in the way of trying to write down a formula with thousands and thousands of variables that attempts to describe *exactly* all the actual interactions through which immigration and GDP influence one another. But as Dani Rodrik says of the use of models in scientific inquiry,

> What makes a model useful is that it captures an aspect of reality. What makes it indispensable, when used well, is that it captures the most relevant aspect of reality in a given context. Different contexts [. . .] require different models.
>
> (Rodrik, 2015)

And so, what justifies using a mathematical concept, whether in a scientific theory or a model or any other of the many different kinds of representational tools used in service of scientific inquiry, is its ability to support induction or explanation. But it is a mistake to think that, for the concept in question to do this, it must be used exactly.

These considerations mirror the parallel argument for the conclusion that uses of jargon within science should likewise be treated as inexact, and therefore as approximate, truths. As I noted earlier, to a naturalist, a proliferation of jargon can be a sign of progress, to the extent that the new technical terms support reasoning about different aspects of some complex natural process or system. Rather than worry about which is the one true definition of, say, 'mass', scientists will introduce as many different technical definitions as is necessary to understand something like molecular bonding. Different contexts require different jargon.

This passage from Thomas Kuhn combines our observations about mathematical concepts and jargon into a description of how these observations play out in scientific practice:

> Consider . . . the quite large and diverse community constituted by all physical scientists. Each member of that group today is taught the laws of, say, quantum mechanics, and most of them employ these laws at some point in their research or teaching. But they do not all learn the same applications of these laws, and they are not therefore all affected in the same ways by changes in quantum mechanical practice. On the road to professional specialization, a few physical scientists encounter only the basic principles of quantum mechanics. Others study in detail the paradigm applications of these principles to chemistry, still others to the physics of the solid state, and so on. What quantum mechanics means to each of them depends on what courses he has had, what texts he has read, and which journals he studies.
>
> (Kuhn, 1970, 49)

What explains the respective indispensability of mathematical and technical concepts to scientific practice is the need to idealize to handle the world's complexity, and this involves either regimenting or distorting the concepts in order that they are able to serve very specific inferential aims.

Ultimately, the outcome of these processes of conceptual repurposing is a menagerie of lexical and mathematical concepts that are put to any number of different referential uses in service of satisfying a scientific discipline's distinctive epistemic ends. And it needs to be observed explicitly here that the only philosophically plausible explanation of how it is that these menageries of inexact concepts can do this epistemic work is that users of the relevant terms have established sustained causal feedback loops running between uses of the concepts and certain of the world's properties, processes, and structures. Indeed, we should extend Kuhn's list of what can affect the meaning of the concepts of quantum mechanics to include the experimental phenomena with which a physicist spends most of her or his time interacting. It is only because of sustained causal interactions between users of the relevant concepts and the referents of the concepts themselves that the concepts can be used to construct assertions that are approximately true.

Thus, an intuition, the propositional content of which reflects the application of a mathematical concept to an imaginary representation of a causal process that the intuitor knows quite a bit about, can be interpreted as a world-directed intuition. The reason for this is that the implicated concept has been shaped by whatever processes lead it to its inferential home in the messy world of scientific reasoning. Consequently, it will almost always be a mistake to treat the propositional content of such an intuition as an exact truth. But at the same time, the reason why such an intuition can be treated as world-directed evidence is that there is a causal connection between uses of the concept implicated in the intuition and the referent of the same concept.

7. Conclusion

Let us take stock, then. We have a naturalistically friendly theory of intuitions that allows us to explain the connection between intuitions and mathematical beliefs. According to this theory, intuitions are procedurally instrumental for the formulation of mathematical beliefs, as it is through uses of meaning-directed intuitions that a person can acquire the mathematical concepts out of which such beliefs are constructed. But it is not using these meaning-directed intuitions to grasp and refine any number of mathematical concepts that subsequently justifies beliefs constructed out of these concepts. What's more, because this theory does not require that we ask whether the propositional contents of the relevant mathematical intuitions are true or false, we avoid the difficult question of finding in the natural world the objects to satisfy an exact, literal interpretation of statements of pure mathematics.

Nevertheless, this conclusion does not imply that we should strive to reformulate scientific sentences in order to remove their mathematical content. Rather, by interpreting mathematical intuitions as meaning-directed intuitions we open space to treat the use of mathematical concepts imported into scientific reasoning using the same theory of truth and reference we use for the rest of the representational output of scientific practice. Both the mathematical and nonmathematical content of science can be interpreted as, at most, a set of approximate truths. To that end, I have argued that there is an important parallel between the use of jargon and the use of mathematical concepts in scientific reasoning: in both cases, scientifically useful concepts are frequently borrowed from contexts and vocabularies exogenous to those of scientific practice, after which the concepts are put to a variety of uses that depend onthe concepts not being used to formulate exact truths. But at the same time, it is the causal feedback loops running between uses of the relevant concepts and the world's own causal processes that sustains the epistemic utility of the relevant terms.

When these ideas are put together, then, they form, at least in outline, a solution to Benacerraf's epistemological problem.

References

Baker, A. (2001). Mathematics, indispensability and scientific progress. *Erkenntnis*, 55(1), 85–116.

Bangu, S. (2012). *The applicability of mathematics in science: Indispensability and ontology*. London: Palgrave Macmillan.

Bealer, G. (1998). Intuition and the autonomy of philosophy. In M. DePaul & W. Ramsey (Eds.), *Rethinking intuition: The psychology of intuition and its role in philosophical inquiry* (pp. 201–240). London: Rowman & Littlefield.

Bechtel, W., & Abrahamsen, A. (1991). *Connectionism and the mind*. Oxford: Blackwell.

Benacerraf, P. (1973). Mathematical truth. *The Journal of Philosophy*, 70(19). *Journal of Philosophy, Inc.*, 661–679.

Boyd, R. (1990). Realism, approximate truth, and philosophical method. In W. Savage (Ed.), *Scientific theories* (Vol. 14). Minnesota Studies in the Philosophy of Science. Minneapolis: University of Minnesota Press.

Boyd, R. (2001). Truth through thick and thin. In R. Schantz (Ed.), *What is truth?* Berlin, NY: De Gruyter.

Boyd, R. (2016). *How not to be afraid of correspondence truth*. Unpublished Manuscript, Sage School of Philosophy, Cornell University, Ithaca, NY.

Cappelen, H. (2012). *Philosophy without intuitions*. Oxford: Oxford University Press.

Carey, S. (2011). *The origin of concepts* (Reprint ed.). Oxford: Oxford University Press.

Cartwright, N. (1999). *The dappled world: A study of the boundaries of science*. Cambridge: Cambridge University Press.

Cartwright, N. (2017). How mechanisms explain? In *Making a difference: Essays on the philosophy of causation*. Oxford: Oxford University Press.

Chomsky, N. (2000). *New horizons in the study of language and mind*. Cambridge: Cambridge University Press.

Clarke-Doane, J. (2016). What is the benacerraf problem? In F. Pataut (Ed.), *Truth, objects, infinity* (pp. 17–43). Logic, Epistemology, and the Unity of Science 28. Dordrect: Springer International Publishing.

Colyvan, M. (2001). *The indispensability of mathematics*. Oxford: Oxford University Press.

De Cruz, H. (2015). Where philosophical intuitions come from. *Australasian Journal of Philosophy*, 93(2). Routledge, 233–249.

Deutsch, M. (2010). Intuitions, counter-examples, and experimental philosophy. *Review of Philosophy and Psychology*, 1(3). Springer Netherlands, 447–460.

Deutsch, M. (2015). *The myth of the intuitive: Experimental philosophy and philosophical method*. Cambridge, MA: MIT Press.

Devitt, M. (2013). Linguistic intuitions are not 'the voice of competence'." In M. C. Haug (Ed.), *Philosophical methodology: The armchair or the laboratory?* (p. 268). Abingdon, UK: Routledge.

Fedyk, M. (2009). Philosophical intuitions. *Studia Philosophica Estonica*, 2(2), 54–80.

Giere, R. N. (2006). Perspectival pluralism. In *Scientific Pluralism* (pp. 26–41). Minneapolis: University of Minnesota Press.

Goldman, A. I. (2013). Philosophical naturalism and intuitional methodology. In *The a priori in philosophy*. Oxford: Oxford University Press.

Goodman, N. (1955). *Fact, fiction, and forecast*. Cambridge, MA: Harvard University Press.

Huemer, M. (2005). *Ethical intuitionism* (Vol. 78). London: Palgrave Macmillan.

Kant, I. (1996). *Critique of pure reason*. Indianapolis: Hackett Publishing.

Kitcher, P. (1984). *The nature of mathematical knowledge*. Cambridge: Cambridge University Press.

Kornblith, H. (2014). Naturalism and intuitions. In *A naturalistic epistemology*. Oxford: Oxford University Press.

Kuhn, T. (1970). *The structure of scientific revolutions* (2nd ed.). Chicago: University of Chicago Press.

Lakatos, I. (1976). *Proofs and refutations: The logic of mathematical discovery*. Cambridge: Cambridge University Press.

Levins, R. (1966). The strategy of model building in population biology. *American Scientist, 54*(4). Sigma Xi, The Scientific Research Society, 421–431.

Machery, E. (2009). *Doing without concepts*. Oxford: Oxford University Press.

Maddy, P. (1980). Perception and mathematical intuition. *The Philosophical Review, 89*(2), 163–196.

Maddy, P. (1992). Indispensability and practice. *Journal of Philosophy, 89*(6), 275–289.

Mauß, F., Keller, D., & Peters, N. (1991). A lagrangian simulation of flamelet extinction and re-ignition in turbulent jet diffusion flames. *Symposium on Combustion, 23*(1). Elsevier, 693–698.

Mcgilvray, J. (1998). Meanings are syntactically individuated and found in the head. *Mind & Language, 13*(2). Blackwell Publishers Ltd, 225–280.

Molyneux, B. (2014). New arguments that philosophers don't treat intuitions as evidence. *Metaphilosophy, 45*(3), 441–461.

Parsons, C. (1979). Mathematical intuition. In *Proceedings of the Aristotelian society* (pp. 145–168). New Series, 80. London: The Aristotelian Society.

Perner, J. (1991). *Understanding the representational mind*. Cambridge, MA: The MIT Press.

Resnik, M. (1995). Scientific vs. mathematical realism: The indispensability argument. *Philosophia Mathematica. Series III, 3*(2), 166–174.

Rodrik, D. (2015). *Economics rules: The rights and wrongs of the dismal science*. New York: W. W. Norton and Co.

Ross, W. D. (2000). *Foundations of ethics*. Oxford: Oxford University Press.

Salmon, W. C. (1984). Scientific explanation: Three basic conceptions. *PSA: Proceedings of the Biennial Meeting of the Philosophy of Science Association, 1984*(2), 293–305.

Shieber, J. (2012). A partial defense of intuition on naturalist grounds. *Synthese, 187*(2), 321–341.

Smolensky, P. (1988). On the proper treatment of connectionism. In *The behavioral and brain sciences*. Cambridge: Cambridge University Press.

Teller, P. (2012). Modeling, truth, and philosophy. *Metaphilosophy, 43*(3), 257–274.

Teller, P. (2014). *Language and the complexity of the world*. Manuscript.

Tieszen, R. L. (1989). *Mathematical intuition*. New York: Springer.

Wellman, H. M., Kushnir, T., Xu, F., & Brink, K. A. (2016). Infants use statistical sampling to understand the psychological world. *Infancy: The Official Journal of the International Society on Infant Studies, 21*(5), 668–676.

Wimsatt, W. (1972). *Complexity and organization. PSA: Proceedings of the Biennial Meeting of the Philosophy of Science Association, 1972*, 67–86.

Xu, F., & Carey, S. (1996). Infants' metaphysics: The case of numerical identity. *Cognitive Psychology, 30*(2), 111–153.

6   Origins of Numerical Knowledge

Karen Wynn

Introduction

In this chapter, I will make three central arguments. First, I will argue that human infants possess extensive numerical competence. Empirical findings show that young infants are able to represent and reason about numbers of things. Infants' ability to determine number is not based on perceptual properties of displays of different numbers of items, nor is it restricted to specific kinds of entities such as physical objects. Rather, it spans a range of ontologically different kinds of entities. Infants, therefore, have a means of determining and representing number per se. Furthermore, infants are able to reason numerically about these representations; they possess procedures for operating over these representations in numerically meaningful ways, and so they can appreciate the numerical relationships that hold between different numerical quantities. They can thus be said to possess a genuine system of numerical knowledge. These early capacities suggest the existence of an unlearned basic core of numerical competence.

Second, I will propose a mechanism that underlies this early numerical competence, the accumulator mechanism, and discuss the structure of the numerical representations that it produces. In particular, I will argue that our initial representations of number have a magnitudinal structure in which the numerical relationships that hold between the numbers are inherently specified.

Finally, I will address the issue of how the numerical competence arising from the accumulator mechanism relates to more advanced numerical and mathematical knowledge. There are limits to the mathematical knowledge the proposed mechanism could give rise to, and evidence, both developmental and historical, of humans' acquisition of aspects of mathematical knowledge beyond these limits lends further support for this mechanism. Furthermore, there are significant differences in the representations of number provided by the accumulator mechanism, and those instantiated in linguistic counting. Thus, despite possession of an early appreciation of numbers, learning the system of the counting routine and learning the meanings of the number words are difficult and complex achievements for children.

1. Numerical Capacities of Infants

1.1 Infants' Sensitivity to Numerosity

'Subitization' is the term for the ability to 'recognize' small numbers of items automatically without having to engage in conscious counting. In typical experiments, adult subjects are presented briefly with displays containing items such as black dots and asked to name as quickly as possible the number of items displayed. Subitization is characterized by a very shallow reaction time slope and by near-perfect performance in the range of one to three or four items. Beyond this, the reaction time (RT) slope typically increases steeply, and errors become more frequent, indicating a much slower, nonautomatic mental process applies to enumerating items in displays outside of this range (e.g., Chi & Klahr, 1975; Kaufman, Lord, Reese, & Volkmann, 1949; Mandler & Shebo, 1982).

Findings suggest that the ability to subitize is inborn. It has been reliably demonstrated that human infants can distinguish between different numbers of items. When infants are repeatedly presented with displays containing a certain number of items until their looking time at the displays drops off (indicating a decrease in interest), they will then look longer at new displays (evidencing renewed interest) containing a new number of items (e.g., Antell & Keating, 1983; Starkey & Cooper, 1980; Strauss & Curtis, 1981; van Loosbroek & Smitsman, 1990). This kind of habituation paradigm has found that infants can distinguish two from three and, in certain conditions, three from four (e.g., Strauss & Curtis, 1981, van Loosbroek & Smitsman, 1990) and even four from five (van Loosbroek & Smitsman, 1990).

Given that the upper limits of numerical discrimination in infants correspond to the limits of the subitizable number range in adults, it seems likely that the same quantification process underlies both capacities. This suggests that a better understanding of the subitization process can inform our understanding of the nature of infants' early numerical concepts and, conversely, that improved knowledge of infants' numerical discrimination abilities may help shed light on the nature of subitization.

Different proposals have been made as to the nature of subitization. One view is that at root, it is not a process that actually determines numerosity; rather, it is a pattern-recognition process that holistically identifies perceptual patterns that correlate with arrays containing a certain number of items (e.g., Mandler & Shebo, 1982). On this view, children must somehow 'bootstrap' themselves up into an ability to represent numbers of things from an initial ability to recognize and represent the perceptual patterns that piggyback along with small-sized collections.

However, there is now ample evidence that subitization is not a pattern-recognition process. For one thing, habituation studies have varied the perceptual attributes of the items in the displays presented to infants. In one experiment, infants were presented with a slide picture of either two or three household objects arranged in random locations (Starkey, Spelke, & Gelman, 1990). The objects within and between trials were all different—no object was presented more than once. Objects were of different sizes, colors, and textures, and differed in complexity of contour (e.g., an orange, a glove, a keychain, a sponge, sunglasses). Following habituation, infants were alternately presented with new pictures of two and of three such objects. Despite the lack of perceptual similarity, infants successfully discriminated pictures on the basis of their number, looking longer at pictures showing the novel number of objects.

Furthermore, infants' numerical discrimination is not restricted to visual arrays. Infants can also enumerate sounds and can recognize the numerical equivalence between arrays of objects and sounds of the same number (Starkey et al., 1990). In one experiment, infants were habituated to slide pictures of either two or three objects like those described earlier. Following habituation, infants were presented with a black disk that emitted alternately two drumbeats and three drumbeats. Infants looked longer at the disk when it emitted the same number of drumbeats as the number of items they had been habituated to visually, indicating that (a) infants were enumerating the drumbeats, and (b) the numerical representations resulting from their enumeration of sounds and of objects is a common one in that it can be matched across these two ontologically distinct kinds of entities.

Experiments recently conducted in my infant cognition laboratory indicate that infants' enumeration capacities apply even more broadly. We have been investigating 6-month-olds' ability to enumerate another ontologically different kind of entity: physical actions (Wynn, 1996). A sequence of actions such as those used in these experiments—the repeated jumps of a single puppet—differs from displays of objects typically presented to infants in several respects. Objects in an array have an enduring existence, exist together in time, and occupy distinct portions of two-dimensional (in the case of photographs or pictures) or three-dimensional space. Actions in a sequence, on the other hand, have a momentary existence, occupy distinct portions of time, and may or may not occur at precisely the same location in space. Perhaps most importantly, perceptual access is available to the entire collection of objects at once, but to just one element at a time of a collection of sequential actions so that one cannot anticipate the final element. Actions also differ from sounds in several ways. There are specific material objects or 'agents' associated with actions, whereas sounds are disembodied entities—though a sound may emanate from a specific physical object, it is perceived independently of the object in a way that an action cannot be. Because of this, sounds are (psychologically, if not physically) inherently temporal entities, whereas actions are inherently spatiotemporal; both spatial and temporal information together specify an action, making the identification of an action a complex task requiring the integration of spatial information over time.

Furthermore, actions are interesting to study because Frege's (1884/1960) point that number is not an inherent property of portions of the world applies every bit as much to actions as it does to physical matter. Just as a given physical aggregate may accurately be described as *52 cards*, *1 deck*, or *10^{28} molecules*, if I walk across a room it counts as 1 "crossing-the-room" action, but may count as 5 "taking-a-step" actions. Number of actions is only determinate relative to a sortal term that identifies a specific kind of action. There is no inherent, objective fact of the matter as to where, in the continuously evolving scene, one 'action' ends and the next 'action' begins. The individuation of discrete actions from this continuous scene is a cognitive imposition.

Six-month-old infants were presented with a display stage containing a puppet. Half the infants were habituated to a puppet jumping two times, the other half to a puppet jumping three times. During the habituation phase of the first experiment, on each trial, the puppet jumped the required number of times, pausing briefly between the jumps. Upon completion of the jump sequence, the infant's looking time at the now-stationary puppet was recorded. In the test phase of the experiment, infants were presented with trials in which the puppet alternately jumped two times and three times. The structure of the jump sequences ruled out the possibility that infants could be responding on the basis of tempo of jumps or overall duration of jump sequences. The test sequences containing the old number of jumps always had both a different tempo and a different overall duration from the habituated sequence. The test sequences containing the novel number of jumps differed from the habituated sequence on only one of these dimensions (for half the infant's tempo, for half duration), having the same value on this dimension as the test sequence with the old number of jumps. Therefore, with respect to duration and tempo, the test sequence with the old number of jumps was the most different from the habituated sequence.

Infants looked significantly longer at the puppet after it jumped the new number of times than after it jumped the habituated number of times (see Figure 6.1). To exhibit this pattern of preferences, infants must have been sensitive to the number of jumps in each sequence.

In order to begin investigating what cues infants use to individuate the events—how infants decide where one action ends and the next begins— a second experiment was conducted, exploring the possibility that infants were segmenting the sequence on the basis of the presence and absence of motion. In the first experiment, infants may have been detecting and enumerating those periods of time in which the puppet was in motion, which were bounded in time by periods in which the puppet was motionless. In the next experiment, therefore, we presented infants with sequences of jumps

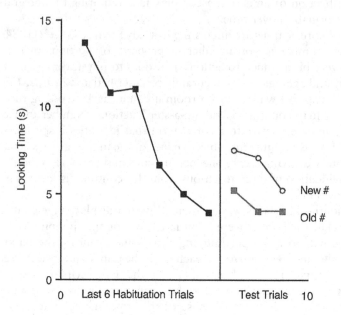

Figure 6.1 Six-month-olds' looking to puppet following novel versus habituated number of jumps

Experiment 1 of Wynn, 1996: motionless pause between jumps

in which the puppet was in constant motion—between jumps, the puppet wagged from side to side. As in the first experiment, one group of 6-month-olds was habituated to sequences of two jumps and another to sequences of three jumps. Following habituation, infants were presented with trials in which the puppet alternately jumped two and three times (Figure 6.2).

Again, infants were sensitive to the number of jumps, looking significantly longer at the puppet following the new number of jumps (see Figure 6.2). Thus infants are able to parse a sequence of continuous motion into discrete segments and are able to enumerate the individuals so arrived at.

These findings have two implications. First, they show that infants have some concept of 'individual' that can apply to physical action; they possess a cognitive process for imposing some as-yet-undetermined criteria for individual-hood over continuous spatial and motion input. Whether this notion of 'individual', and the accompanying principles of individuation that specify such individuals, are the same as those that specify individual physical objects, individual sounds, and possibly other kinds of individuals as well, remains to be determined (see Bloom, 1990, 1994b for evidence that a single notion underlies the conception of 'individual' across different domains).

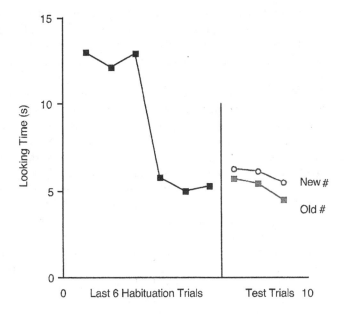

Figure 6.2 Six-month-olds' looking to puppet following novel versus habituated number of jumps

Experiment 2 of Wynn, 1996: puppet in constant motion between jumps

Second, the fact that infants can enumerate different ontological kinds of entities ranging from objects, to sounds, to physical actions indicates that the mechanism for determining number is quite general in the kinds of entities it will take as input to be counted. It may even take in *anything* the cognitive/conceptual system specifies as an individual. This generality, in turn, indicates that the enumeration mechanism cannot be operating over physical or perceptual properties of the entities, recognizing certain perceptual patterns and the like; it must be determining number of entities per se.

1.2 Do Infants Possess a System of Numerical Knowledge?

The studies reviewed earlier show that human infants are sensitive to number itself; they have representations of number that apply over a wide range of perceptually different situations and different kinds of entities.

But possessing genuine numerical knowledge entails more than simply the ability to represent different numbers. A numerical system is composed not only of numbers but also of procedures for the manipulation of these numbers to yield numerical information—for example, information on how two numerical values relate to each other. Infants may be able to determine numbers of things without being able to reason about these numbers or to make numerical kinds of inferences on their basis. If so, we would not want to say that infants' representations of these small numbers play numerically meaningful roles in the infant's cognitive system, nor to credit infants with an actual numerical understanding with a system of numerical knowledge.

Cooper (1984), for example, has proposed that human infants' initial ability to discriminate small numbers of items does not include any ability to relate these numbers to each other; infants' concepts of, say, 'oneness', 'twoness', and 'threeness' are initially as unrelated to each other as are our adult concepts of, say, 'clocks' and 'chickens'. That is, they have no superordinate concept of *number* (or of *numbers of things*) to unite their concepts of individual numbers. On this view, infants gradually come to learn that the numbers belong to a single category by repeatedly observing situations where one or more objects are added to or removed from small collections of objects. From such observations, infants see that the actions of addition and removal reliably result in a change from one number to another, and they learn the orderings of the numbers and the relationships that hold between them. Thus they build up a superordinate concept, *number*. Quite similar views have also been expressed in the philosophy of mathematical epistemology; Kitcher (1984), for example, has proposed that "Mathematical knowledge arises from rudimentary knowledge acquired by perception" and that "children come to . . . accept basic truths of arithmetic by engaging in activities of collecting and segregating".

Studies in my laboratory show that, in contrast to these kinds of learning accounts, 5-month-old infants are able to engage in numerical reasoning; they possess an appreciation of the numerical relationships that hold between different small numbers. Thus they have procedures for manipulating these numerical concepts to attain numerical information.

In these experiments, infants are shown a small collection of objects, which then has an object added to or removed from it. The resulting number of objects shown to infants is either numerically consistent with the events or inconsistent. Since infants look longer at outcomes that violate their expectations, if they are anticipating the number of objects that should result, they will look longer at the inconsistent outcomes than the consistent ones.

In the first experiments (Experiments 1 and 2 of Wynn, 1992a), 5-month-old infants were divided into two groups. One group was shown an addition situation in which one object was added to another identical object, while the other group was shown a subtraction situation in which one object was removed from a collection of two objects (see Figure 6.3). Infants in the '1 + 1' group saw one item placed into a display case. A screen then rotated up to occlude the item, and the experimenter brought a second item into the display and placed it out of sight behind the screen. The '2−1' group saw two items placed into the display; the screen rotated up to hide the two items; then the experimenter's hand reentered the display, went behind the

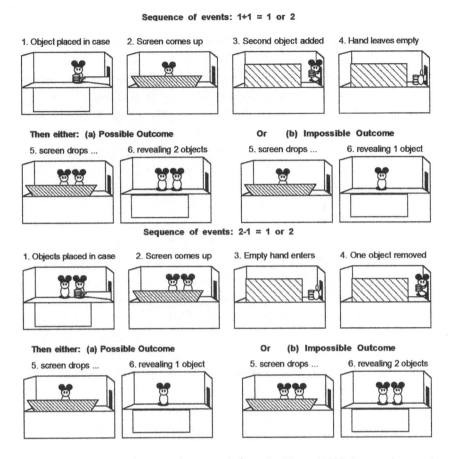

Figure 6.3 Sequence of events shown to infants in Wynn (1992a), experiments 1 and 2

courtesy of *Nature*

screen, and removed one item from the display. For both groups, the screen then dropped to reveal either one or two items. Infants' looking time at the display was then recorded. Infants received six such trials in which the consistent and inconsistent outcomes were alternately revealed.

Pretest trials, in which infants were simply presented with displays of one and two items to look at, revealed no significant preference for one number over the other and no significant difference in preference between the two groups. But there was a significant difference in the looking patterns of the two groups on the test trials; infants in the 1 + 1 group looked longer at the one-item result than at the two-item result, while infants in the 2–1 situation looked longer at the result of two items than the result of one item (see Figure 6.4).

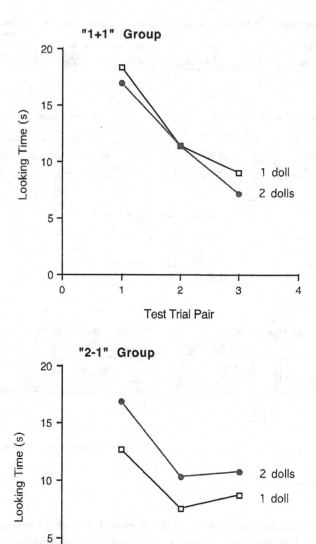

Figure 6.4 Five-month-olds looking to one versus two objects, following either '1 + 1' or '2–1' sequences of events

These results show that 5-month-olds know there should be a change in the display as a result of the operation. In both situations, infants looked longer when the screen dropped to reveal the same number of objects as there were before the addition or subtraction. But these results do not show that they are anticipating the precise nature of the change. They may simply be expecting the display to be changed in some way from its initial appearance, or expecting there to be a different number of objects without expectations of a *specific* number. To address this issue, another experiment was conducted (Experiment 3 of Wynn, 1992a). Five-month-olds were shown an addition of one item to another in which the final number of objects revealed was either two or three. Since both outcomes are different from the initial display of one object, then if infants do not have precise expectations of how the display should be changed, they should be equally unsurprised by both outcomes and so look equally. However, if infants are computing the exact result of the addition, they will expect two items to result and so should look longer at the result of three items than of two items.

This was the pattern observed: infants looked significantly longer at the inconsistent outcome of three objects than at the consistent outcome of two objects (see Figure 6.5). (Pretest trials revealed no baseline preference to look at three items over two items.) These results show that infants are computing the precise result of the operation; they know not only that the display of objects should change but also exactly what the final outcome should be.

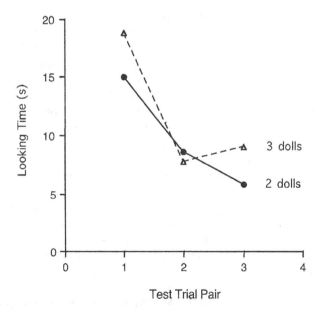

Figure 6.5 Five-month-olds' looking to two versus three objects, following a '1 + 1' sequence of events

The findings obtained from these studies are quite robust; they have been obtained in a number of other laboratories, using different stimuli and with variations in the procedure. Baillargeon, Miller, and Constantino (1994; see Baillargeon, 1994) showed 10-month-olds an addition situation in which a hand deposited first one, and then another, item out of sight behind a screen. The screen was then lowered, revealing either two or three items. The infants looked longer when shown three items behind the screen than when shown two, showing that they had been expecting only two. (They had previously demonstrated equal preference to look at two versus three items.) Similarly, Uller, Carey, Huntley-Fenner, and Klatt (1994) obtained longer looking at the numerically incorrect result when they presented 8-month-olds with '1 + 1' situations that resulted in either one or two items.

In another study, Moore (1995) presented one group of 5-month-olds with '1 + 1' situations and another group with '2–1' situations; for both groups, the outcome alternated between one and two items. All sequences of events were presented on a computer display. The 'objects' were computer-generated rectangular random checkerboards, and the 'screen' was a red square, which descended from the top of the computer display to obscure the checkerboard(s), and ascended to reveal the 'outcomes'. Despite the shift from actual three-dimensional objects to two-dimensional, nonrepresentational, computer-generated patterns, the effect was still obtained: each group of infants looked longer at the impossible outcome.

In another variation of this paradigm, Simon, Hespos, and Rochat (in press) obtained results that rule out the possibility that infants might be computing over visual aspects of the display, expecting an outcome scenario containing particular visual elements rather than expecting a specific number of objects. They presented one group of 5-month-olds with '1 + 1' situations and another group with '2–1' situations; for both groups, the outcome alternated between one object and two objects. But in addition to manipulating numerical possibility, they manipulated 'identity possibility'—on some trials, the items in the outcome display were the kind of object that should be expected, while in other trials, one of the items in the outcome display had an unexpected identity. For example, in a situation in which one Ernie doll was added to another Ernie doll, the outcome display might be two Ernie dolls (numerically and identity correct), one Ernie doll and one Elmo doll (numerically correct but identity incorrect), one Ernie doll (numerically incorrect but identity correct), or one Elmo doll (numerically and identity incorrect). Infants looked longer at the outcomes that were numerically incorrect, regardless of the identity of the dolls (despite a control condition showing that infants could perceptually distinguish Elmo from Ernie). Therefore, they could not have been responding on the basis of expectations about the precise visual aspects of the display.

In a particularly elegant study, Koechlin (1994) tested the possibility that infants were anticipating certain spatial locations to be filled and others empty rather than anticipating number. In the original Wynn (1992a) studies, infants may have been able to determine the approximate location behind the screen where the second object was placed (in the '1 + 1' situations) or removed from (in the '2–1' situations). In the impossible outcomes, therefore, they might have been surprised not because there was an incorrect number of objects, but because a location was empty when it should be filled ('1 + 1' impossible outcome), or filled when it should be empty ('2–1' impossible outcome). In this experiment, 5-month-olds were presented with '1 + 1' and '2–1' situations as in Wynn (1992a). However, all objects were placed on a large revolving plate that was located behind the screen, so no object retained a distinct spatial location throughout the experimental operation. Nonetheless, infants looked reliably longer at the numerically incorrect outcomes, showing that they were computing over number of objects, not over the filled/empty status of different spatial locations.

Finally, an intriguing extension of this paradigm has obtained similar results in nonhuman primates (Hauser, MacNeilage, & Ware, 1996). In these studies, rhesus monkeys were presented with the following situation: one eggplant was placed in an open box where the monkey could see it; a partition was then placed in front of the eggplant obscuring it from view; the monkey then saw a second eggplant placed in the box; the partition was then removed to reveal either one or two eggplants in the box. Although the monkeys showed no preference in control situations to look at one eggplant over two, in this situation, they looked significantly longer at the one-eggplant outcome, indicating that they had been expecting two eggplants to be in the box.

We have recently extended these results to further numerical situations, exploring whether infants have expectations about the results of situations in which one item is added to two items or one is taken away from three. In this experiment, we modified the methodology somewhat. Previously, we had shown infants the same sequence of events on all trials, alternately resulting in two different outcomes. Here we presented infants with exactly the same outcome on all trials, but with two different sequences of events that led up to the outcome—on half of the trials, the events were consistent with the outcome and on the other half, they were inconsistent.[1] The motivation for this modification was to remove any influence on test-looking times of possible subtle preferences for looking at one number of objects more than another.

Infants were alternately presented with '2 + 1' and '3–1' trials. In the '2 + 1' trials, infants saw two items initially in the display, the screen rotated up to hide them, and then a hand entered the display with a third item, placed it behind the screen along with the other two, and left the display empty. In the '3–1' trials, the display initially contained three items, the

1 I thank Renee Baillargeon for suggesting this methodological modification to me.

screen rotated up to hide them, and then an empty hand entered the display and removed one of the objects. For half of the infants, the result of these two sequences was always two items—the correct outcome for the 3–1 sequence, but incorrect for the 2 + 1 sequence. For the other infants, the result was always three items—the correct result for the 2 + 1 sequence but not the 3–1 sequence.

Infants in the two groups should give different patterns of looking if they are anticipating the outcomes of the events. Infants for whom the presented outcome is always two objects should look longer on the '2 + 1' than on the '3–1' trials if they have expectations about what the result of one object added to two should be. However, this result would not suggest that they appreciate that one object removed from three results in two objects, since they might be looking for a lesser length of time during 3–1 trials *either* because they appreciate that the result is numerically correct, *or* because they have no clear expectations about the result in this case. Infants for whom the outcome is always three objects, on the other hand, should look longer during the 3–1' than on the '2 + 1' trials, if they have expectations about what the outcome of one object removed from three should be. Again, however, this result would tell us only about their expectations in the '3–1' situation; a lack of surprise in the '2 + 1' trials could be either because the outcome of three matches their expectation, or because they have no clear expectations about the result in this case.

Infants in the two groups differed significantly in their looking patterns in the predicted direction. Infants for whom the outcome was two looked longer following the '2 + 1' than the '3–1' situations, while those for whom the outcome was three showed the reverse pattern (see Figure 6.6). However, separate analyses of each group's looking times revealed that only the three-outcome infants significantly distinguished between the consistent and inconsistent events; though the two-item infants looked longer following the sequence of events inconsistent with the outcome, this difference was not significant. In this experiment, therefore, infants showed an appreciation of the outcome of three objects with one removed, but not of the outcome of two objects with one added. Inferring the results of the addition in these experiments appears to be more difficult for some reason than inferring the results of the subtraction.

The many experiments reviewed earlier show that 5-month-olds appear to be sensitive to the numerical relationships between small numbers of objects. This implies that (a) infants' representations for these numbers possess a structure that embodies this kind of numerical relational information, and (b) infants possess a means of operating over their representations of these numbers to extract this information. I will now present a theory of an innate mechanism for determining number that entails such a structure to the numerical representations it yields and is amenable to straightforward processes by which information of the numerical relationships between these representations could be obtained.

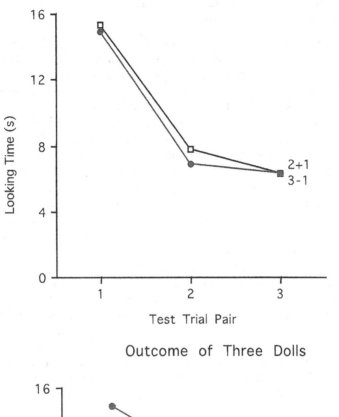

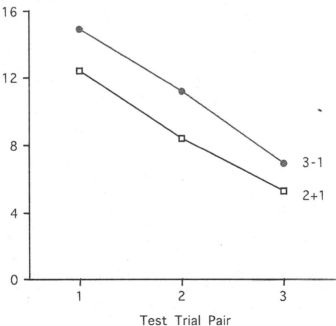

Figure 6.6 Five-month-olds looking to either two objects or three objects, following '2 + 1' versus '3 – 1' sequences of events

2.   A Mechanism for Determining and
Reasoning About Number

This model, the "accumulator mechanism", was originally proposed to account for rats' ability to discriminate number (Meck & Church, 1983). It has recently been extended to account for human infants' preverbal numerical competence and for nonverbal numerical processes such as subitization (Gallistel & Gelman, 1991, 1992; Wynn, 1990a, 1992c).

Meck and Church (1983) suggest that a single mechanism underlies both animals' ability to determine number and their ability to measure duration. The model operates as follows: a pacemaker puts out pulses at a constant rate, which can be passed into an accumulator by the closing of a switch. When this mechanism is in its counting mode, every time an entity is experienced that is to be counted, the switch closes for a fixed brief interval, passing the pulses into the accumulator during that interval. Thus the accumulator fills up in equal increments, one for each entity counted. In its timing mode, the switch closes at the beginning of the event to be timed and remains closed for the duration of the interval, passing energy into the accumulator continuously. The final fullness level of the accumulator represents the number of items counted or how much time has gone by. The mechanism contains several accumulators and switches so that the animal can count different sets of entities and measure several durations simultaneously. The fullness value in the accumulator can be passed into working memory and compared with previously stored accumulator values to allow the animal to evaluate whether a number of items or events (or a duration) is greater, smaller, or the same as a previously stored number (or duration).

Note that there will be differences in the exact fullness of the accumulator on different counts of the same number of items, resulting from inherent variability in the rate at which the pacemaker is generating pulses and in the amount of time the switch closes for each increment. This variability is normally distributed around a mean fullness value for each number and obeys Weber's law; the variance is greater for larger numbers than it is for smaller ones and increases in proportion to the numerosity represented. Thus as number increases, the discriminability of adjacent numbers decreases (see Gallistel & Gelman, 1991 for discussion). This accords with data showing decreasing discriminability with increasing numerosity in animals (e.g., Mechner & Guevrekian, 1962; Platt & Johnson, 1971), in human adults (e.g., Chi & Klahr, 1975; Mandler & Shebo, 1982), and in human infants (e.g., Starkey & Cooper, 1980; von Loosbroek & Smitsman, 1990; Wynn, 1995).

Meck and Church (1983) obtained experimental support for the claim that the same mechanism underlies animals' counting processes and their timing processes. First, the administration of methamphetamine increases rats' measure of duration and of number by precisely the same factor, suggesting that it is a single mechanism being affected. This effect can be explained in the model by positing that the drug causes an increase in the rate of pulse generation by the pacemaker, leading to a proportionate increase in the final value of the accumulator regardless of its mode of operation. Second, they tested the following prediction: if a rat's decision (e.g., when or whether to press a lever for food reward based on presented stimuli) is based on comparing the final value of the accumulator with a previously stored value, then it might not matter whether the previously stored value is the result of a timing process or a counting one. For example, a count that results in the same final fullness value in the accumulator as a previously trained duration might be responded to as if it were that duration. This prediction was confirmed: when rats were trained to respond to a specific *duration* of continuous sound, they immediately generalized their response when presented with a certain *number* of one-second sound segments that had been calculated by the experimenters to fill up the accumulator to the same level as that for the duration the rats had been initially trained on. Similar transfer results were obtained when rats were initially trained to respond to a specific number of events and then tested on a duration of continuous sound. Meck and Church concluded that it was indeed the same mechanism underlying both counting and timing processes in rats.

Because it is the *entire fullness of the accumulator* that represents the number of the items counted, the magnitudinal relationships between the numbers are inherently specified in the accumulator's representations for them. For example, the number six is three more than the number three, or twice as large, and the accumulator's representation for six has three more increments than the representation for three, and so the accumulator is twice as full. Provided the infant or animal has certain procedures for operating over these representations, it will be able to extract some of these numerical relationships. Addition, for example, could be achieved by combining or concatenating two numerical representations, 'pouring' the contents from an accumulator representing one value into an accumulator representing another value (or, to avoid the loss of the original separate values, creating two accumulators matching in value to the two to be added together and 'pouring' the contents of these two into a third empty accumulator). Subtraction could be similarly achieved: suppose accumulator A represents the initial value and accumulator B the value to be subtracted from it. Successively removing increments of the contents of A—or, to avoid loss of information of the initial value of A, the contents of A', a 'copy' of A—and 'pouring' them into an initially empty accumulator C until C reaches the same fullness level as B (indicating the right amount has been removed) would result in A' representing the difference between the values represented by A and B.

3. Relationship of the Accumulator Mechanism to Later Numerical Knowledge

By virtue of yielding numerical representations in which magnitudinal information is inherently embodied, the accumulator mechanism, in conjunction with mental processes for operating over these representations, allows for the extraction of a rich body of numerical information. However, there are two sorts of limitations to the knowledge that could be obtained from the outputs of the accumulator model.

First, there will be practical limits on how the outputs of the mechanism can be manipulated. While an extensive body of numerical information is in principle accessible by virtue of the magnitudinal structure of the representations—for example, multiplicative relationships, division (with remainder), whether a given value is even or odd, or prime—access to these facts requires that the cognitive system has procedures for manipulating the representations in appropriate ways to obtain it. The kind of procedure required for determining the product of two values, for example, will be much more involved and complex than that for determining the sum of two values, requiring an additional accumulator to count the number of successive additions. A procedure for determining how many times one number goes into another will be more complex yet; one for computing an exponential value will be even more complex, and so on. What kinds of processes comprise this system of numerical competence is an empirical question; the accumulator theory does not predict that the animal or infant must be able to appreciate *all* these kinds of numerical relationships.

Second, there are limits to what the accumulator is *in principle* capable of representing. One limitation is that it can only represent numbers *of specific individuals* (be they objects, sounds, events, or whatever). The body of mathematical knowledge that has developed through history and been transmitted culturally, however, operates as an abstract, logical system independent of the physical world. Somehow, the mind must make the abstraction, from conceiving of numerical properties of collections of entities, to conceiving of abstract numbers that are entities in themselves.

Another principled limitation to the accumulator is the kinds of mathematical entities and concepts it can give rise to. It is unlikely that the notion of infinity could result from this mechanism, as all physical processes are limited, and there is presumably some point beyond which the accumulator cannot measure. Numbers other than the positive integers cannot be represented—by its nature, the accumulator mechanism does not measure fractional values, negative values, imaginary values, and so on. Interestingly, representation of these kinds of numerical entities emerges very gradually and with much effort, both ontogenetically and culturally. Children's acquisition of fractions as numerical entities is extremely difficult, requiring an expansion beyond their initial conceptualization of numbers as "things that represent numbers of discrete entities" (Gelman, Cohen, & Hartnett,

1989). More generally, this conceptual expansion can be seen in the historical development of mathematics in numerous domains. The number 'zero', for example, was initially not part of the number system. It was introduced as a 'placeholder' symbol representing an absence of values in a given position in place-value numeral notation (so as to reliably distinguish, say, 307 from 37) and eventually attained status as a number in its own right by virtue of becoming embedded in the system, with rules for its numerical manipulation (Kline, 1972). The emergence of negative numbers, irrational numbers, complex numbers, and so on follow similarly complex and gradual progressions.

The earliest numerical thought involved the positive integers (Ifrah, 1985)—the very values produced by the accumulator mechanism. Moreover, the earliest physical representations (and also, later on, the earliest written notations) for the numbers had a magnitudinal structure in which a given number (of sheep, for example) was represented by a collection of that same number of pebbles or other 'counters' (Ifrah, 1985; Schmandt-Besserat, 1986, 1987). All these facts suggest that the positive integers are psychologically privileged numerical entities, and these are precisely the numerical entities that the accumulator model is capable of producing and representing.

The accumulator mechanism provides mental representations of numbers and allows for manipulation of these representations in numerically meaningful ways. In this respect, the structure of these psychological foundations of numerical knowledge is similar to that of formal systems of mathematics. Such systems can be characterized by a body of mathematical entities (often defined by a set of axioms—e.g., the Peano axioms of arithmetic or Euclid's axioms of geometry) along with a set of operations performable upon these entities. To ultimately understand the role the initial numerical foundations supplied by the accumulator mechanism play in the development of further numerical knowledge, it will be critical to determine the numerical operations the accumulator mechanism can perform, as well as the 'axioms' that psychologically 'define' its outputs (see Wynn & Bloom, 1992), and to determine the ways in which these psychological foundations map onto the abstract system of arithmetic and the ways in which they differ.

3.1 The Accumulator Model and Linguistic Counting

Even within the domain of the positive integers, extensive conceptual development occurs. There are significant differences in the way the accumulator mechanism represents number and the way in which number is represented in culturally transmitted, linguistic counting systems. In linguistic counting, an ordered list of number words are applied to the items to be counted in a one-to-one correspondence, and the final word used in the count represents the total number of items counted (Gelman & Gallistel, 1978). Number is not inherently represented in the symbols for the numbers (the number words); these obtain their numerical meaning by virtue of their positional relationships with each other—the fifth word in the series represents the number *five*, and so on. The number words represent cardinalities through a system that is isomorphic to, but distinct from, the system of cardinals and their internal relations. For example, the number *six* is three units larger than, or twice as large as, the number *three*, while the linguistic symbol for *six* occurs three positions later in the number word list than the linguistic symbol for *three*, or twice as far along.

In order to master the linguistic counting system, then, children must learn the mapping between their own magnitudinal representations of number and the ordinal representations inherent in the linguistic counting system—surely not a trivial process. A series of studies examining children's developing understanding of the linguistic counting system suggests that a mastery of linguistic counting is in fact a complex achievement with an extensive period of learning.

In one experiment (Wynn, 1990b), 2-1/2- to 4-year-olds were asked to give a puppet from one to six toy animals from a pile (the 'give-a-number' task). If children understand how linguistic counting determines numerosity, they should be able to use counting to give the correct number. Children were also given a counting task in which they were asked to count sets of items ranging in number from two to six and were asked, after counting, 'how many' items there were.

In the give-a-number task, there was a bimodal distribution in response strategy. Some children (the 'Counters', who on average were about 3-1/2 years old) did use counting to solve the task, much as would an adult. They typically counted the items from the pile, stopping at the number word asked for, thus succeeding on the task in giving the right number. The other children (the 'Grabbers', who were typically under 3-1/2 years of age) did not use counting to solve the task. They never counted items from the pile to give the correct number, even though they were quite good at performing the counting routine (for example, children who could count six items perfectly well were unable to give, say, four items from the pile). When asked for larger numbers, these children just grabbed and gave a random number of items. When asked for smaller numbers (three and fewer), some children had apparently directly mapped some of these number words onto their correct numerosities (presumably as a result of an ability to subitize) and so could give the correct number just by looking, but not by using linguistic counting. A response made by a typical Grabber follows:

Experimenter: "Can you give Big Bird five animals?" (*The child grabs a handful that contains three animals and puts them in front of the puppet.*)

Experimenter: "Can you count and make sure there's five?"

Child:	(*Counting the three items perfectly*) "One, two, three— that's five!"
Experimenter:	"No, I really don't think that's five. How can we fix it so there's five?"
Child:	(*Pauses and then very carefully switches the positions of the three items*) "There. Now there's five".

Thus the Grabbers appear not to appreciate just how linguistic counting determines the number of a collection of items.

A second analysis examined the relationship between children's performance in the give-a-number task and their responses to the 'how many' questions following counting in the counting task. The 'how-many' task has often been considered an indicator of whether a child understands that the last number word used in a count represents the number of items counted—if a child understands this, it is sometimes argued, then when asked 'how many' immediately after counting a collection of items, he or she should give the last number word used in the count as the response to the question. But there are reasons to doubt the validity of this task: children might correctly answer a 'how-many' question without having the assumed knowledge, simply as a result of having observed adults modeling counting for them. Conversely, children might give a response other than the last number word used in the count, even if they do understand the significance of the last word. Asking about the number of items immediately after the child has just counted them (and hence provided information as to their number) might cause the child to interpret the request as an indication that the count was incorrect and to recount the items rather than state the last word. For these reasons, whether or not a child gives the last word used in the count when asked 'how many' may be not be a good indication of her grasp of the significance of the last number word in the count.

However, looking at *when* children give the last number word might be more valid. The last word used in a count is the correct response to 'how many' only when the count has been executed correctly—in an incorrect count in which the one-to-one correspondence of words to items has been violated, or in which the number words have been applied in an incorrect ordering, the last number word will not indicate how many items there are. If children understand how the counting system determines number, they might therefore be expected to give last-number-word responses less often after incorrect counts than after correct ones. If, however, they do not possess such an understanding, they should not differentiate correct from incorrect counts, but give last-number-word responses equally often after both kinds of counts. Accordingly, Grabbers' and Counters' responses to the 'how-many' questions were divided into those following correct counts by the child and those following incorrect counts. Results showed that the Counters, but not the Grabbers, gave significantly more last-number-word responses after correct counts than they did after incorrect counts; Grabbers

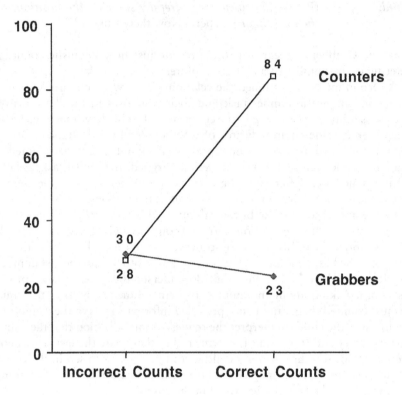

**Mean % Cardinality Responses Following
Correct versus Incorrect Counts**

Figure 6.7 Percentage of grabbers' versus counters' responses to 'how-many' question following correct and incorrect counts in which they gave the last number word used in the count

made no distinction between the two kinds of counts (see Figure 6.7), further strengthening the conclusion that they did not understand how the counting system represents number, despite their competence at performing the counting routine.[2]

2 Gelman (1994) presents evidence suggesting that children's early understanding of linguistic counting may be significantly greater than these studies indicate. The experimenter first showed children a card with one item on it (e.g., a cat), asking children, "What's on this card?" When children invariably responded "a cat", the experimenter would respond, "That's right; *one* cat". Children were then shown a card with two cats and were again asked, "What's on this card?", and, following their response, they were asked "Can you show me?" The same questions were then asked using cards with larger numbers. Using this procedure, Gelman obtained a high percentage of young children correctly stating the

3.2 Intermediate Stages in the Developing Understanding of Number Words

The earlier conclusion leads to the question of the precise nature of children's understanding of both the purpose of the counting routine and the meanings of the number words used in counting. In a longitudinal study (Wynn, 1992b), children's developing understanding of the meanings of number words was looked at in detail. Two aspects of knowledge of number words were examined: (1) whether children knew the semantic category that these words pick out—that is, that number words refer to numerosities—and (2) whether they knew exactly *which* numerosity each word picks out.

To address whether children knew that number words picked out numerosities, pairs of pictures were shown to children who already knew the meaning of the word 'one'. These children were asked to point out the picture depicting a certain number of objects. For example, children were shown one picture depicting a single blue fish and another depicting four yellow fish, and asked, "Can you show me the four fish?" If they know that 'four' refers to a specific number, they should (in accordance with the principle of contrast; see, e.g., Clark, 1987; see also Markman, 1989) infer that it does not refer to the number *one* since they already have a word for the numerosity *one*. They should thus choose the correct picture by a process of elimination. However, if they do not know that 'four' is a number word, they will not know to contrast it with 'one' and should respond as they do when asked a nonsense question, such as, "Can you show me the *blicket* fish?" (When children were asked this, they were equally likely to point to either picture.)

To address whether children knew the precise number a number word picks out, they were shown pairs of pictures, one of which contained a given number of items and the other of which contained that number plus one. For example, to test whether children knew the precise meaning of the word 'four', they were shown a picture of four green dogs and one of five red dogs, and asked, "Can you show me the four dogs?" on some trials and, "Can you show me the five dogs?" on others. (Again, they were asked nonsense questions as well, such as, "Can you show me the *blicket* dogs?", and chose each picture equally often on these questions).

Even the youngest children (2-1/2-year-olds) succeeded when one of the pictures in the pair contained a single item. They correctly pointed out the number asked for 94% of the time, showing a clear understanding that the number words pick out numerosities. However, despite this early knowledge, it took children nearly a full additional year to learn the general

---

number of items on the cards (at least for the numbers two and three) and using counting in response to the experimenter's "can you show me?" questions. Further research is needed to clarify children's understanding of the task and to explore the extent to which this finding affects the conclusions of the earlier discussion.

pattern for which words refer to which numerosities—i.e., that successive words in the number word list refer to successively higher numbers. The same pattern of results was found as with the give-a-number task earlier: when asked to distinguish small numbers (two from three and, sometimes, three from four), some children had directly associated that number with its correct number word. They thus succeeded at these smaller numbers, but never by using counting to determine the answer. When presented with larger numbers (e.g., pictures of four and five dogs), the only children who succeeded were children who counted the items in the pictures (these children tended to be 3-1/2 years of age or older). These children succeeded on all the numbers they were asked to identify and tended to use counting even when asked for the smaller numbers, showing that once they understand that linguistic counting is a method for determining number, they can then extend their use of this strategy even to situations for which they already have another strategy available.

These studies show that there appears to be a lengthy period of time during which young children know how to apply the counting routine and know that these words each refer to a specific numerosity, but do not know the system behind which numerosity each word picks out—the ordinal system by which the list of counting words represents numerosity. This protracted period of learning is consistent with the accumulator model, which posits initial numerical representations very different in form from those employed by the linguistic counting system. On the accumulator model, what children must do in order to learn the linguistic counting system is to map its ordinal representations of number onto children's own initial, magnitudinal representations of number by making a kind of analogical mapping: *larger* numbers are represented by symbols occurring *later* in the number-word list and *smaller* numbers by symbols occurring *earlier* in the list.

This raises a number of intriguing questions. If the linguistic counting system represents number in a very different way from the initial accumulator representations, then its acquisition should have a major influence on numerical thought. To acquire a new representational system, one that exploits different properties of the entities to be represented, is to acquire a new tool with which to think and reason about the entities under consideration. The counting system embodies a wealth of information about number not available from the accumulator mechanism—for example, information about the iterative structure of the number naming system, which may in turn provide information about numerical infinity (Bloom, 1994a). An understanding arrived at in acquiring the linguistic counting system that may be particularly important is that numbers can be represented by arbitrary symbols whose structures are independent of the values they represent. This allows the possibility of representing variables, which opens the door to an entirely new realm of mathematical concepts, representations, and entities, and is surely a major achievement in the development of mathematical reasoning, both in the individual and in the history of mathematics. This achievement would be impossible within any system in which the structures of the symbols representing numbers are inextricably dependent on the values of those numbers, such as the accumulator mechanism.

Acknowledgments

I would like to thank Renee Baillargeon, Tom Bever, Paul Bloom, Randy Gallistel, Rochel Gelman, Marcus Giaquinto, and Elizabeth Spelke for helpful discussion, and Marcus Giaquinto and especially Paul Bloom for their painstaking comments on preliminary versions of this chapter. Some of the work presented here was supported by an NICHD FIRST Award to the author.

References

Antell, S., & Keating, D. P. (1983). Perception of numerical invariance in neonates. *Child Development, 54*, 695–701.

Baillargeon, R. (1994). Physical reasoning in young infants: Seeking explanations for impossible events. *British Journal of Developmental Psychology, 12*, 9–33.

Baillargeon, R., Miller, K., & Constantino, J. (1994). *Ten-month-old infants' intuitions about addition*. Unpublished manuscript.

Bloom, P. (1990). *Semantic structure and language development*. Unpublished doctoral dissertation, MIT: Cambridge, MA.

Bloom, P. (1994a). Generativity within language and other cognitive domains. *Cognition, 51*, 177–189.

Bloom, P. (1994b). Possible names: The role of syntax-semantics mappings in the acquisition of nominals. *Lingua* 92(1–4), 297–329.

Chi, M. T., & Klahr, D. (1975). Span and rate of apprehension in children and adults. *Journal of Experimental Child Psychology, 19*, 434–439.

Clark, E. V. (1987). The principle of contrast: A constraint on language acquisition. In B. MacWhinney (Ed.), *Mechanisms of language acquisition*. Hillsdale, NJ: Erlbaum.

Cooper, R. G. (1984). Early number development: Discovering number space with addition and subtraction. In C. Sophian (Ed.), *Origins of cognitive skills*. Hillsdale, NJ: Erlbaum.

Frege, G. (1884). *Die Grundlagen der Arithmetik*. Translated as *The Foundations of Arithmetic*, by J. L Austin. 2nd revised edn. 1960, New York: Harper & Brothers.

Gallistel, C. R., & Gelman, R. (1991). The preverbal counting process. In W. E. Kesson, A. Ortony, & F. I. M. Craik (Eds.), *Thoughts, memories, and emotions: Essays in honor of George Mandler*. Hillsdale, NJ: Erlbaum.

Gallistel, C. R., & Gelman, R. (1992). Preverbal and verbal counting and computation. *Cognition, 44*, 43–74.

Gelman, R. (1994). Paper presented at the International Interdisciplinary Workshop on Mathematical Cognition, 'Concepts of Number and Simple Arithmetic', at the Scuola Internazionale Superiore di Studi Avanzati, Trieste, December 10–14.

Gelman, R., Cohen, M., & Hartnett, P. (1989). *To know mathematics is to go beyond the belief that 'Fractions are not numbers'*. Proceedings of Psychology of Mathematics Education, Vol. 11 of the North American Chapter of the International Group of Psychology. New Brunswick, New Jersey.

Gelman, R., & Gallistel, C. R. (1978). *The child's understanding of number*. Cambridge, MA: Harvard University Press.

Hauser, M. D., MacNeilage, P., & Ware, M. (1996). *Numerical representations in primates*. Proc. Natl. Acad. Sci. 93, 1514–1517.

Ifrah, G. (1985). *From one to zero: A universal history of numbers.* New York: Viking Penguin.

Kaufman, E. L., Lord, M. W., Reese, T. W., & Volkmann, J. (1949). The discrimination of visual number. *American Journal of Psychology, 62,* 498–525.

Kitcher, P. (1984). *The nature of mathematical knowledge.* Oxford: Oxford University Press.

Kline, M. (1972). *Mathematical thought from ancient to modern times* (Vol. 1). Oxford: Oxford University Press.

Koechlin, E. (1994). Paper presented at the International Interdisciplinary Workshop on Mathematical Cognition, 'Concepts of Number and Simple Arithmetic', at the Scuola Internazionale Superiore di Studi Avanzati, Trieste, December 10–14.

Mandler, G., & Shebo, B. J. (1982). Subitizing: An analysis of its component processes. *Journal of Experimental Psychology: General, 11,* 1–22.

Markman, E. M. (1989). *Categorization and naming in children.* Cambridge, MA: MIT Press.

Mechner, F. M., & Guevrekian, L. (1962). Effects of deprivation upon counting and timing in rats. *Journal of the Experimental Analysis of Behavior, 5,* 463–466.

Meck, W. H., & Church, R. M. (1983). A mode control model of counting and timing processes. *Journal of Experimental Psychology: Animal Behavior Processes, 9,* 320–334.

Moore, D. S. (1995). *Infant mathematical skills? A conceptual replication and consideration of interpretation.* Manuscript under review.

Platt, J. R., & Johnson, D. M. (1971). Localization of position within a homogeneous behavior chain: Effects of error contingencies. *Learning and Motivation, 2,* 386–414.

Schmandt-Besserat, D. (1986). Tokens: Fact and interpretation. *Visible Language, 20*(3), 250–273.

Schmandt-Besserat, D. (1987). Oneness, twoness, threeness: How ancient accountants invented numbers. *The Sciences, 27,* 44–48.

Simon, T. J., Hespos, S., & Rochat, P. (in press). Do infants understand simple arithmetic? A replication of Wynn (1992). *Cognitive Development.*

Starkey, P., & Cooper, R. G., Jr. (1980). Perception of numbers by human infants. *Science, 210,* 1033–1035.

Starkey, P., Spelke, E. S., & Gelman, R. (1990). Numerical abstraction by human infants. *Cognition, 36,* 97–128.

Strauss, M. S., & Curtis, L. E. (1981). Infant perception of numerosity. *Child Development, 52,* 1146–1152.

Uller, M. C., Carey, S., Huntley-Fenner, G. N., & Klatt, L. (1994). *The representations underlying infant addition.* Poster presented at the Biennial Meeting of the International Conference on Infant Studies, Paris, June 5–8.

van Loosbroek, E., & Smitsman, A. W. (1990). Visual perception of numerosity in infancy. *Developmental Psychology, 26,* 916–922.

Wynn, K. (1990a). *The development of counting and the concept of number.* Unpublished doctoral dissertation, MIT: Cambridge, MA.

Wynn, K. (1990b). Children's understanding of counting. *Cognition, 36,* 155–193.

Wynn, K. (1992a). Addition and subtraction by human infants. *Nature, 358,* 749–750.

Wynn, K. (1992b). Children's acquisition of the number words and the counting system. *Cognitive Psychology, 24,* 220–251.

Wynn, K. (1992c). Evidence against empiricist accounts of the origins of numerical knowledge. *Mind & Language, 7,* 315–332.

Wynn, K. (1996). *Infants' individuation and enumeration of sequential actions.* Psychological Science, 7, 3: 164–169.

Wynn, K., & Bloom, P. (1992). The origins of psychological axioms of arithmetic and geometry. *Mind & Language, 7,* 409–416.

7 What Happens When a Child Learns to Count?

The Development of the Number Concept

Kristy vanMarle

Developing an understanding of number is one of the most important achievements made during childhood. It is the foundation for the uniquely human ability to acquire formal, symbolic mathematical knowledge and skills (Gelman & Gallistel, 1978), which in turn is an important determiner of successful functioning in modern society. Children's mathematical skills at school entry predict their mathematics achievement in adolescence (Duncan et al., 2007; Geary, Hoard, Nugent, & Bailey, 2013), which impacts their employability and wage-earning potential (Bynner, 1997; Parsons & Bynner, 1997; Ritchie & Bates, 2013; Rivera-Batiz, 1992). But where do number concepts come from, and how do they develop?

As described by Frege (Frege, 1884/1980), formally, numbers are a special kind of sets. Because any set of individuals can be counted (e.g., objects, sounds, actions, ideas, events), children's experience of 'twoness' on one occasion (e.g., two apples) may share almost no phenomenal similarity with their experience of 'twoness' on another occasion (e.g., two elephants, or two bites of cereal, or two birthday parties, and so on). From a practical standpoint, this featural variability across exemplars poses a learnability problem (Fodor, 1983; Wynn, 1995; cf. Mix & Sandhofer, 2007). If a child is trying to learn what 'apple' refers to, the thing to do is to focus on the perceptual features of various exemplars of the category. Apples are small (relatively), red (or green, or sometimes yellow), round, smooth, sweet, and so on. In other words, one can immediately experience what it means to be an 'apple'. In contrast, in order to learn what 'two' means, the child must abstract away from the immediate perceptual features of the items in a set (color, size, shape, kind, etc.) in order to discover that 'two' refers to *any* set made up of exactly two individuals.

Because learning to count is the first step children take toward a symbolic understanding of number, researchers often focus on this milestone as a way to shed light on the process(es) underlying the development of the number concept. Nonetheless, after decades of research, there is still vigorous debate about exactly how the process unfolds. One aspect, however, is clear: it takes a remarkably long time for a child to learn how to count. And not only is the development protracted, it also appears to be stage-like with the initial stages taking several months each, followed by a rapid shift where the child seems to grasp the logic of counting suddenly, which may be a behavioral indicator of an underlying conceptual change.

Empirical evidence of this developmental progression was originally demonstrated by Karen Wynn in her seminal work (Wynn, 1990, 1992a) where she showed both cross-sectionally and longitudinally that although children can correctly recite the count list (one, two, three, . . . , ten) as early as 2 years of age, it can take up to one and a half to two additional years before they understand the meanings of the number words as well as the counting routine and how to apply it. Children's performance on the GiveN task, which measures children's cardinal knowledge, as well as their knowledge of the counting routine, was especially telling. The general pattern that emerged is robust and has been replicated many times in English-speaking children (Condry & Spelke, 2008; Le Corre & Carey, 2007; Le Corre, Van de Walle, Brannon, & Carey, 2006; Lee & Sarnecka, 2010, 2011; Slusser & Sarnecka, 2011) as well as Japanese-, Mandarin Chinese–, and Russian-speaking children (Li, Le Corre, Shui, Jia, & Carey, 2003; Sarnecka, Kamenskaya, Yamana, Ogura, & Yudovina, 2007), suggesting the developmental process is not language-specific.

At the earliest stage, despite knowing the number words and being able to recite them in order, children do not seem to understand their cardinal meanings (e.g., that 'three' refers to sets with exactly three individuals). If asked to give a puppet a particular number of toys, the child will be able to accurately give 'one' toy, but will give a random number for any number greater than 'one'. Such children are labeled 'one-knowers', following Le Corre and Carey (2007). After about six months, the child can accurately give 'one' and 'two' items (a 'two-knower'), but still gives a random handful for requests greater than 'two'. After about six more months, the child can give 'one', 'two', and 'three' accurately, but not more. Then, finally, after several more months, children seem to undergo a shift: once able to reliably give 'four' items, they seem to be able to give any number accurately up to the limit of their count list. Thus over the course of about two years, children learn the cardinal meaning of 'one', followed by the meaning of 'two', followed by the meaning of 'three', and then, suddenly, they seem to understand counting. Once children make this transition, they are deemed 'CP-knowers' because they can properly apply the counting principles (described next). With this knowledge, children presumably understand (either implicitly or explicitly) how to implement the counting routine, and in doing so, they can determine the cardinal value of any set. This understanding, obviously, is very powerful. In acquiring it, children have done what it took centuries for philosophers and mathematicians to formalize: they have come to understand the system of positive integers (Carey, 2009). It is the first symbolic mathematical knowledge that children acquire, and it is the foundation upon which all other formal math learning must be built (Dantzig, 1930).

In order to explain the developmental progression, researchers have drawn on two general stances: one nativist and the other empiricist. The 'principles-first' approach expounded by Rochel Gelman and colleagues (e.g., Gallistel & Gelman, 1992, 2000; Gelman, 1972, 1993; Gelman & Gallistel, 1978; Gelman & Greeno, 1989; Gelman & Meck, 1983; Leslie, Gallistel, & Gelman, 2007; Leslie, Gelman, & Gallistel, 2008) proposes that the principles that define and guide the counting routine ('the counting principles') are innately available as part of our intuitive number representations and serve both to help the child detect instances of counting in his or her environment and to guide the child's own attempts to count as he or she learns how to implement the process. The three core counting principles[1] are one-to-one correspondence, stable order, and cardinality, and they must be observed in order for the child to arrive at the correct result when enumerating a set. The *one-to-one correspondence principle* stipulates that each item may be counted once and only once. The *stable order principle* stipulates that the counting tags (i.e., count words) must always be applied in the same ordered sequence. *The cardinality principle* states that the last word in a count stands for the numerosity (i.e., cardinality) of the set. On this type of theory, children already 'know' (implicitly) the rules for the counting routine. Then, with experience enumerating sets of objects, they realize (perhaps implicitly) the correspondence between their internal representation of the counting routine (i.e., the counting principles) and the verbal counting routine, which obeys the same principles. If this characterization is accurate, then children undergo no revolution or conceptual change when they learn to count; rather, they merely have developed a verbal counting procedure that mirrors the nonverbal procedure they already possess.

The other traditional approach, the 'principles-after' theory (Briars & Siegler, 1984; Fuson, 1988; Fuson & Hall, 1983) posits that, initially, children merely imitate the counting routine based on what they have seen others (e.g., parents, teachers, peers) model for them. At this stage, reciting the count list is no different than reciting the alphabet—just a verbal script with no special meaning. Only after substantial experience witnessing counting in different contexts can children abstract the rules of counting that are

---

1 In fact, Gelman and Gallistel (1978) defined *five* counting principles in all. The three 'core' principles (one-to-one correspondence, stable order, and cardinality) are essential and any violations of these principles will result in an incorrect count. The other two principles (abstraction and order-irrelevance) are auxiliary, and provide a guide as to how generalizable, or abstract, the process is since you can count absolutely anything (e.g., the number of things in your bedroom that are blue and start with the letter d; abstraction), and it doesn't matter whether you count them in a sequence or skip around the group (e.g., counting right to left, or counting all the blue ones and then all the yellow ones; order-irrelevance), as long as you adhere to the essential principles, the count will be correct. Thus the auxiliary principles are important to the number concept, but are not themselves procedural rules for counting like the essential three.

essential (the principles of one-to-one correspondence, stable order, and cardinal principle), differentiating them from nonessential, but common features of counting, such as counting items from left to right, or counting them sequentially in a line. On these accounts, children have to construct an understanding of number through experience with concrete examples (Piaget, 1952), generalizing over many particular instances to abstract the notion that all sets named with a given cardinality are equivalent (i.e., they can be put in 1:1 correspondence; Mix & Sandhofer, 2007).

Although both nativist and empiricist approaches have their appeal, both also fall short of explaining the lengthy, stage-like growth seen in children's cardinal knowledge and their acquisition of the counting principles.

1.   Nativist Accounts of Cardinal Knowledge Development—Realizing the Principles

Nativist approaches have received a lot of attention of late, because we now have strong evidence that humans possess a nonverbal system—the *approximate number system* (ANS)—for representing and manipulating numbers and other quantities. Importantly, the ANS is functional at birth (Antell & Keating, 1983; Izard, Sann, Spelke, & Streri, 2009), which has led some researchers to argue that it is the source of the innately given counting principles (Gallistel & Gelman, 2000, 2005; Leslie et al., 2007; Leslie et al., 2008). After more than 30 years of research, we now have a basic theoretical understanding of the nature of the ANS (Brannon & Roitman, 2003; Dehaene & Brannon, 2011; Gallistel, 2011; Gallistel & Gelman, 2005; Spelke, 2000; Walsh, 2003) and which brain areas are involved in number processing (Ansari, 2008; Dehaene, Piazza, Pinel, & Cohen, 2003; Kersey & Cantlon, 2017; Neider, 2016), and we are beginning to understand how the representations change over the course of development (Libertus & Brannon, 2010; Starr, Libertus, & Brannon, 2013; for review, see Mou & vanMarle, 2014). However, the claim that the ANS provides an intuitive sense of the counting principles continues to be challenged (Carey, 2009; Le Corre & Carey, 2007; Sarnecka, Goldman, & Slusser, 2015; Sarnecka & Wright, 2013).

Current characterizations of the ANS suggest that it is present from birth and continues to support numerical reasoning into adulthood (Cordes, Gelman, Gallistel, & Whalen, 2001; Izard et al., 2009; Whalen, Gallistel, & Gelman, 1999). It represents both discrete (number) and continuous (area, length, duration, volume, etc.) quantities as imprecise analog magnitudes, similar to the way in which a line can be used to represent a quantity, with its length being proportional to the represented value. For example, if a 10cm line represents the quantity '1', then a line with twice the length, 20cm, would represent '2', and a line with four times the length (40cm) would represent '4', and so on. Importantly, the magnitude representations are imprecise, and the amount of variability (i.e., error) increases in proportion to the represented magnitude. Thus a representation of '40' is twice as fuzzy, or variable, than a representation of '20'. A consequence of this *scalar variability* is that the discriminability of any two numbers depends not on their absolute difference, but on their ratio, in accord with Weber's law. It is easier to discriminate 10 from 20 items than 90 from 100 items, even though the values differ by the same absolute difference in both cases (10 units), and such ratio-dependent performance is seen as a hallmark of the ANS and is evidenced throughout the lifespan (Feigenson, Dehaene, & Spelke, 2004; Gallistel & Gelman, 2000, 2005; Xu & Spelke, 2000).

In addition to being imprecise, another important feature of the ANS is that its representations are amodal. That is, the system represents numerosity independent of the sensory modality in which that information was received. Indeed, infants, children, and adults can represent numbers of visual items (Barth et al., 2006; Barth, La Mont, Lipton, & Spelke, 2005; Xu & Spelke, 2000), sounds (Lipton & Spelke, 2003; vanMarle & Wynn, 2006), and even tactile events (e.g., touches on one's hand; Plaisier, Tiest, & Kappers, 2010). Moreover, already at birth, humans can detect numerical correspondences across sensory modalities and will look longer at visual displays that match the number of sounds they are hearing, compared to visual displays that do not (Izard et al., 2009). Other studies have also demonstrated crossmodal matching in infants using various stimuli (Feigenson, 2011; Jordan & Brannon, 2006; Starkey, Spelke, & Gelman, 1983, 1990), including tactile-visual stimuli (Feron, Gentaz, & Streri, 2006). Findings such as these not only demonstrate that humans are innately prepared to represent the numerosities of sets they encounter out in the world but also the fact that even newborns can detect numerical correspondences across modalities suggests that ANS representations are inherently abstract.

With regard to the counting principles, Gallistel and Gelman (2000) have been one of the strongest proponents to argue that the operations and processes that comprise the ANS actually instantiate the three core counting principles. The nonverbal counting process they describe is iterative and operates by adding one 'cupful' into an accumulator for each item counted (imagine adding cupfuls of water into a container). Because the accumulator is always incremented by roughly the same amount, the fullness at the end of the count (total magnitude) represents the cardinal value of the set because it is proportional to the number of items counted. Furthermore, after the last item is counted, the contents of the accumulator are transferred to memory, where they become subject to scalar variability when read back out of memory as recalled magnitudes. The one-to-one correspondence principle is observed by way of the accumulator being incremented exactly once for each item counted. This discrete incrementing process in turn produces magnitudes with a discrete ordering, as required by the stable order principle. And just as the last number counted in the set stands for the cardinality of the set as stipulated by the cardinality principle, the fullness of the accumulator stands for the cardinality of the represented set. Refer to

Figure 7.1 for an illustration of the mechanism as envisioned by Gallistel and Gelman (2000).

Because of these structural properties, the ANS provides a ready explanation for the source of the counting principles. However, the mere fact that they are innately available poses a problem for the theory. If all children need to do is realize the correspondence between the verbal counting system and the nonverbal system, then it is unclear why it takes children so long to acquire the cardinal meanings of the count words and come to grasp the counting principles (Wynn, 1990, 1992a). In addition, the sudden shift that occurs when children acquire the meaning of 'four' is not easily accounted for on a 'principles-first' account, which would presumably predict smooth, continuous

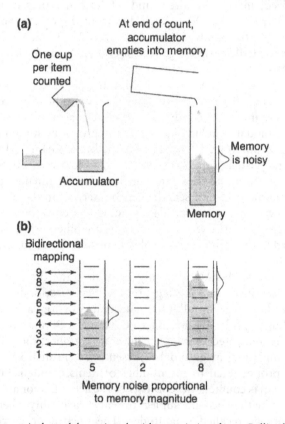

Figure 7.1 Theoretical model reprinted with permission from Gallistel and Gelman (2000) of ANS. In (a) we can see the iterative accumulation process that ensures one-to-one correspondence by incrementing the accumulator once, and only once, for each item counted. The contents are transferred to memory, where the retrieval process introduces scalar variability. According to this model, children learning the meanings of the number words map the verbal number labels onto representations from the ANS stored in memory.

change. Another problem for this view is the imprecise nature of ANS representations (Carey, 2009; Le Corre & Carey, 2007). Because the procedure for producing the 'next number'—the successor function—involves adding (exactly) one to the current set, it is not clear how it could provide an understanding of the notion 'add (exactly) one' (cf. vanMarle et al., 2016). Some theorists have solved this problem by taking an even more nativist approach and simply building in an understanding of 'exact integer' in the innately given number concept (e.g., Leslie et al., 2007; Leslie et al., 2008).

2. Empiricist Accounts of Cardinal Knowledge Development—Inducing the Principles

Empiricist and constructionist accounts of how children learn to count date back at least as far as Piaget (1952), who argued that children's knowledge of number follows a protracted development culminating around age 6 or 7 in their understanding of number as characterizing sets that is conserved over superficial changes in the spatial arrangement of the items. In fact, children show evidence of understanding conservation much earlier than Piaget suggested so long as the task used to measure their understanding is designed to minimize irrelevant task demands (Mehler & Bever, 1967). Nonetheless, modern contructivist accounts (e.g., Mix & Sandhofer, 2007) suggest children are doing something very different than merely noticing similarities between their intuitive sense of number and verbal number system. Instead, they are assumed to be literally building their understanding of number through repeated experiences seeing the creation of sets and instances of counting. After enough cumulated experience, children are thought to somehow infer the meaning of the number words not by mapping them onto preexisting representations (e.g., ANS representations), but rather by abstracting over the many instances of counting and labeling they have witnessed to notice the invariant property (i.e., their numerosity) of all sets named by a given number. Before making this induction, children have only rudimentary bits of a concept. For example, they may know initially that number words pick out the dimension of numerosity, but not know which number word goes with which numerosity. And when they do encounter a labeling event (e.g., "Look! Three cookies—one, two, three!"), they encode it in a context-specific way, generalizing a number word (e.g., 'two') only after having used it in its original context for some time (Mix, 2004, Mix, Sandhofer, & Baroody, 2005). In the end, however, children have undergone a genuine conceptual change, moving from a context-dependent, instance-specific rudimentary understanding of numbers and counting, to an abstract, principled way of thinking about numbers as invariant properties of sets. By predicting discontinuous development, empiricist accounts do provide a much better fit to the developmental pattern seen in the GiveN task. However, like nativist accounts, empiricist accounts have their own limitations.

Perhaps the biggest problem constructivists face is that without a mechanism for detecting and identifying the dimension of numerosity, and, furthermore, for representing the cardinal value of sets the child encounters, it is not logically possible to determine whether two sets seen on different occasions are numerically equivalent.[2] And if numerical equivalence cannot be established, then there are no grounds to license the child to make the necessary abstraction over those labeling instances. Indeed, it is question begging to assume that children learn the cardinal meanings of the number words by detecting numerical equivalence over distinct labeling instances, while simultaneously arguing that number words, themselves, are what guide children to the relevant dimension in the first place (cf. Sarnecka et al., 2015).

3. Finding Middle Ground—Bootstrapping and Language-Mediated Accounts

Many years of research and discussion in the literature strongly suggest that we are not born with a full, adult-like understanding of number (even if only implicitly), nor are we born a numerical 'blank slate' without even a foothold on the relevant dimension we need to attend to in order to find meaning in the verbal count list. Both extreme views fail, and, moreover, they both fail to account for the empirical data characterizing the growth of cardinal knowledge. In an attempt to overcome limitations of the traditional views, several researchers proposed bootstrapping accounts that incorporate both domain-general processes and domain-specific learning mechanisms to explain why children show the performance pattern they do in the GiveN task.

A prime example of this sort of account was given by Carey (2009) in her excellent book titled *The Origin of Concepts*. Further developing ideas stemming from empirical work done in collaboration with Mathieu Le Corre (Le Corre & Carey, 2007), Carey argued that although children do map their newly acquired verbal number words onto domain-specific, innately available core representations, it is not the ANS that provides these representations. Instead, Carey (and Le Corre & Carey, 2007) invoked a mechanism of visual attention—the object-tracking system (OTS)—that is, well studied in adults and appears to be functional in infants from the earliest months of life (e.g., Feigenson, Carey, & Hauser, 2002; Feigenson et al., 2004; Leslie, Xu, Tremoulet, & Scholl, 1998; Wynn, 1992b).

The OTS is often studied in the context of multiple object tracking, where a subject must visually track a subset of identical items through a period of dynamic movement. The task taps into mid-level mechanisms of visual

---

2 Of course, when two or more sets are simultaneously perceptible, numerical equivalence may be established by putting them into one-to-one correspondence. However, the constructivist claim is that the abstraction occurs over time, hearing the same label applied to different sets on different occasions.

attention that are believed to help in solving the visual correspondence problem (Kahneman, Treisman, & Gibbs, 1992) and to attentively track moving objects (Pylyshyn & Storm, 1988; see Scholl, 2001, for review). To accomplish this, the OTS assigns an index to each item to be tracked. Importantly, the indexes are 'sticky' and hang with the object, 'pointing' to its location as it moves through space and undergoes occlusion. An index can have one or more of the object's features linked to it (e.g., size, shape, color, kind), but spatio-temporal information (i.e., the object's location) is prioritized, allowing subjects to successfully track an object even when it undergoes featural changes while occluded (e.g., Flombaum et al., 2009; vanMarle & Scholl, 2003; Xu & Carey, 1996). In contrast to the ANS, which has no theoretical upper limit on the number of items it can represent, a defining feature of the OTS is its strict capacity limit; it can only track as many objects as it has indexes, which appears to be about three in infants (Feigenson et al., 2004) and about four in adults (Pylyshyn & Storm, 1988).

Strictly speaking, the OTS is not a *number* mechanism, because the model does not include a means to represent the cardinality of an indexed set (total number of active indexes). However, Carey (2009) argued that it nonetheless implicitly codes for numerosity because sets of indexes can be stored in long-term memory and later retrieved and compared to sets of active indexes in working memory. What is more, Carey also argued that rather than mapping their earliest number words ('one', 'two', and 'three') onto ANS representations, children in fact map them onto set representations derived from the OTS that can be stored in and retrieved from long-term memory. On this account, when a child is attending to a set of three objects, they set up a working memory model $\{i_a, i_b\}$ that they then store in long-term memory. Over time, the set representation becomes associated with the verbal label 'three' that often accompanies the retrieval of that remembered set. Thus, on this account, the slow rate and piecemeal progress that we see as children learn the meanings of 'one', 'two', and 'three' simply reflects the time it takes to form strong associations between the verbal labels children are experiencing and the set representations they have stored in long-term memory. Over this period of time, the child is also gathering the data that will eventually license the induction of *successor principle*—an understanding that moving to the *next word* in the verbal count list results in a set with exactly *one more individual* than the previous set. This realization then leads to conceptual change when children move from being subset-knowers ('one-', 'two-', or 'three-knowers') to CP-knowers. Thus children use what is provided by the OTS[3] to bootstrap their way into an understanding of number and counting. Although this account still appeals to a nonverbal

---

3 Set-based natural language quantifiers (e.g., singular, dual, and trial markers) also play a role in Carey's account, but the conceptual work is mostly carried by the OTS. See Carey (2004, 2009) and Le Corre and Carey (2007).

mechanism as the foundation of later number knowledge, it nonetheless requires learning and experience to drive what Carey (2009) argues is a genuine conceptual change as children transition from subset-knower to CP-knower. As such, this view offers a sort of middle ground between the nativist and empiricist views.

Spelke and Tsivkin (2001) and Spelke (2011) have argued for a similar, middle-of-the-road approach. They acknowledge the shortcomings of both the OTS and ANS, and propose that both core systems play important foundational roles in the development of the number concept. The ANS provides the counting principles (especially cardinality), and the OTS provides the notion of exact integer and the successor function. However, because these two systems are functionally distinct and independent of language, a third system—language—is required before their properties can be linked together into a sensible concept. For Spelke (2011), the verbal count list is this link. Recall that children know and can recite the verbal count list before they understand the meanings of the individual number words. With experience seeing and practicing counting, both the OTS and ANS representations become associated with the number words. For example, when faced with a set of three cookies and told, "Look, three cookies!" both systems generate representations of the set. The ANS generates a fuzzy magnitude that gets stored in memory (and possibly compared to previous instances of 'three'), while the OTS creates a set, $\{i_a, i_b, i_c\}$, that gets stored in long-term memory, and both of these representations become associated with the verbal label 'three'. Over time, through common activation with a given label, the child has access simultaneously to the principles afforded by the ANS and the properties of the OTS, allowing them to link them together to construct an understanding of number (Spelke, 2011; Spelke & Tsivkin, 2001).

4.   The Merge Model

Relative to the more extreme views, the merits of both Carey's (2009) and Spelke's (2011) proposals are numerous. However, they differ from each other substantially with respect to the role played by the ANS. For Spelke, both mechanisms provide critical components to the developing system of number knowledge. For Carey, however, despite being a dedicated system of number representation, the ANS plays no role in the acquisition of the meanings of the first few number words. Instead, the OTS in combination with the system of natural language quantifiers provides the foothold on the counting system. Recent data, however, from a large-scale longitudinal study (van-Marle et al., 2016) can help us arbitrate between Spelke's dual-mechanism model on the one hand, and Carey's OTS-only model on the other.

What role non-numerical processes (such as the OTS) may play in infants' numerical reasoning has been a point of serious contention among infant researchers for at least three decades. And the suggestion that small numbers may be processed differently than larger numbers goes back even further to

Taves (1941) and Kaufman, Lord, Reese, and Volkmann (1949) who found that adults were faster, more accurate, and more confident when estimating small numbers of items than larger numbers. Critically, Kaufaman et al. found a clear discontinuity in the accuracy and reaction time (RT) slopes for small (one to five) and large numbers (six plus), such that there was a linear increase in RT (and decrease in accuracy) for each additional item in the large sets, while the slopes remained almost flat for set sizes one to five. Therefore, early investigations suggesting that infants' numerical sensitivity was fragile and limited to only the smallest of set sizes (not above three; e.g., Antell & Keating, 1983; Starkey & Cooper, 1980; Strauss & Curtis, 1981; van Loosbroek & Smitsman, 1990) probably did not come as a surprise and were taken to bolster this two-system view. When researchers later started exploring the OTS as a possible "small number mechanism" (e.g., Feigenson & Carey, 2003; Feigenson et al., 2002; Scholl & Leslie, 1999; Simon, 1997, 1999), the two-system view gained further support. The appeal of the two-system view was purely data driven in early investigations due to the discontinuities observed in subjects' ability to discriminate small and large numbers. For developmental researchers, however, it also provided theoretical grounds for arguing against nativist views of the development of cardinal knowledge. Because the OTS is not itself a number system, its involvement in the acquisition of number knowledge is appealing to those who believe number knowledge is constructed or developed via bootstrapping from other concepts, in this case, set representations derived from the OTS. As a result of this controversy, we now have an impressive body of data collected on the matter, and the data suggest that both the OTS and the ANS are present and functional in infants and continue to operate into adulthood (for relevant reviews see Anderson & Cordes, 2013; Hyde, 2011; Mou & vanMarle, 2014; Trick, 2008).

So given the current broad agreement that both mechanisms are available to children as they begin to learn to count, the next step is to determine what role, if any, each mechanism plays in that process. The OTS-only (Carey, 2009) and dual-mechanism (Spelke, 2011) proposals both attempt to do this, but as we argue in recent work (vanMarle et al., 2016), like the older nativist and empiricist views, both proposals fall short. The major limitation of the OTS-only view is the lack of cardinal representations in the system. Sets of indexes held in working memory or long-term memory are still just sets of indexes. Unless there is another representation to indicate the total number of indexes (the set's cardinality), there seems to be no way for the learner to determine the cardinal meanings of the number words (Gallistel, 2007). Additionally, the successor function does not seem sensible in the absence of a cardinal value—*add one* would lack meaning both because the cardinal value of 'one' is undefined, and because the cardinal value of the set to which 'one' should be added is undefined (Leslie et al., 2007). Similarly, the major limitation of Spelke's (2011) dual-mechanism view is that it adopts the OTS 'as-is', and even though cardinal meaning is available in the ANS, it is unclear how a mere association between, say, the word 'two', a set of two indexes, and a magnitude representing 'about 2' somehow makes it possible for a child to *conceptually* link the representations.

Using data from a large-scale longitudinal study conducted by David Geary and me, we recently designed a study that would arbitrate between the OTS-only and dual-mechanism views, and also test our own view, the Merge model. The Merge model builds on Spelke's (2011) dual-mechanism model and is largely similar, save for two important deviations. First, our Merge model predicts that while both systems may contribute to the early stages of cardinal knowledge acquisition, due to its limited capacity, the influence of the OTS will wane over time while the ANS will continue to support cardinal knowledge gains. Spelke's model, in contrast, predicts that once linked, the OTS and the ANS should continue to both contribute to cardinal reasoning over time. Second, while Spelke's model suggests that the critical transition to CP-knower results from the induction of the successor and cardinal principles via the linking of the OTS and ANS, our Merge model predicts that it will be driven by the ANS and not the OTS. Our reasoning here depends on the fact that the OTS is capacity limited; even if children map their earliest acquired count words to set representations, they can only get so far before they run into sets that exceed the capacity of the OTS. When this happens, the role of the OTS may diminish, eliminating competition between the systems, thus making it easier for the child to realize the correspondence between the verbal counting system and the principles inherent in their ANS.

To test these three possibilities, we examined preschool children's early ANS and OTS abilities, and the emergence of cardinal knowledge on the GiveN task (vanMarle et al., 2016). We were particularly interested in whether one or both of the abilities would predict the critical transition from 'subset-knower' to 'CP-knower' to determine which mechanism might drive the conceptual development. Approximately 200 children completed the ANS task (the acuity, or precision, or their ANS representations), the OTS task (tracking small sets of objects through occlusion), and the cardinal knowledge task (GiveN) at the beginning of Year 1 and Year 2 of preschool (details on the various tasks and procedure can be found in vanMarle et al., 2016). We also collected demographic information (race/ethnicity, parental education) and measured children's domain-general abilities (IQ, executive function, and preliteracy). As one might expect, given the very different functions of the OTS and ANS, performance on the two tasks was not correlated at either time point, suggesting that object tracking and number representation are indeed distinct skills. Importantly, consistent with our Merge model, both traditional regression analyses and a novel Bayesian analysis indicated that while both the ANS and the OTS reliably predicted cardinal knowledge at the beginning of preschool, by the second year, only ANS continued to be related significantly to children's cardinal knowledge. And, critically, beginning of preschool ANS, but not OTS, was a reliable predictor of whether a child transitioned to CP-knower at Time 2.

Based on such findings, we argued that single-mechanism views, such as Carey's (2009) and Le Corre and Carey's (2007), cannot be right. Both the OTS and ANS were strong predictors of cardinal knowledge at the start of preschool, lending further support to the old idea that numbers are processed in multiple ways. Note that this also lends support to Carey's suggestion that the OTS plays a critical role as children try to learn the meanings of the first few number words. However, we also found strong support that the ANS plays an early and long-lasting role in children's cardinal knowledge acquisition, and, importantly, that it alone drove the critical transition to CP-knower. This suggests, of course, that the ANS may really be the source of children's understanding of the counting principles, undermining the notion of conceptual change. However, I would argue that although this shift may not represent a genuine novel conceptual change, it nonetheless indicates conceptual change in the sense that it signifies an explicit recognition of the inherent structure of the ANS.

Further studies are now underway in my lab and many others to determine whether training children's ANS abilities can promote faster change in cardinal knowledge. Ultimately, understanding which early skills best predict critical milestones in children's mathematical development can inform the design and development of interventions that can be targeted to children who begin preschool behind their peers in key skills in hopes of closing those gaps prior to school entry.

References

Anderson, U. S., & Cordes, S. (2013). 1< 2 and 2< 3: Non-linguistic appreciations of numerical order. *Frontiers in Psychology*, 4(5), 116–124.

Ansari, D. (2008). Effects of development and enculturation on number representation in the brain. *Nature Reviews: Neuroscience*, 9(4), 278–291.

Antell, S. E., & Keating, D. P. (1983). Perception on numerical invariance in neonates. *Child Development*, 54, 695–701.

Barth, H., La Mont, K., Lipton, J., Dehaene, S., Kanwisher, N., & Spelke, E. S. (2006). Non symbolic arithmetic in adults and young children. *Cognition*, 98(3), 199–222.

Barth, H., La Mont, K., Lipton, J., & Spelke, E. S. (2005). Abstract number and arithmetic in preschool children. *Proceedings of the National Academy of Sciences of the United States of America*, 102(39), 14116–14121.

Brannon, E. M., & Roitman, J. (2003). Nonverbal representations of time and number in non human animals and human infants. In W. Meck (Ed.), *Functional and neural mechanisms of interval timing* (pp. 143–182). New York: CRC Press.

Briars, D., & Siegler, R. S. (1984). A featural analysis of preschoolers' counting knowledge. *Developmental Psychology*, 20(4), 607–618.

Bynner, J. (1997). Basic skills in adolescents' occupational preparation. *Career Development Quarterly*, 45, 305–321.

Carey, S. (2004). Bootstrapping and the origins of concepts. *Daedalus*, 59–68.

Carey, S. (2009). *The origin of concepts*. New York: Oxford University Press.

Condry, K. F., & Spelke, E. S. (2008). The development of language and abstract concepts: The case of natural number. *Journal of Experimental Psychology: General*, 137, 22–38.

144   *Kristy vanMarle*

Cordes, S., Gelman, R., Gallistel, C. R., & Whalen, J. (2001). Variability signatures distinguish verbal from nonverbal counting for both large and small numbers. *Psychonomic Bulletin & Review, 8*, 698–707.

Dantzig, T. (1930). *Number: The language of science.* New York: The Macmillan Company.

Dehaene, S., & Brannon, E. M. (2011). *Space, time, and number in the brain: Searching for the foundations of mathematical thought.* Oxford: Elsevier.

Dehaene, S., Piazza, M., Pinel, P., & Cohen, L. (2003). Three parietal circuits for number processing. *Cognitive Neuropsychology, 20*(3–6), 487–506.

Duncan, G. J., Dowsett, C. J., Claessens, A., Magnuson, K., Huston, A. C., Klebanov, P. et al. (2007). School readiness and later achievement. *Developmental Psychology, 43*, 1428–1446.

Feigenson, L. (2011). Predicting sights from sounds: 6-month-old infants' intermodal numerical abilities. *Journal of Experimental Child Psychology, 110*(3), 347–361.

Feigenson, L., & Carey, S. (2003). Tracking individuals via object-files: Evidence from infants' manual search. *Developmental Science, 6*, 568–584.

Feigenson, L., Carey, S., & Hauser, M. (2002). The representations underlying infants' choice of more: Object-files versus analog magnitudes. *Psychological Science, 13*, 150–156.

Feigenson, L., Dehaene, S., & Spelke, E. S. (2004). Core systems of number. *Trends in Cognitive Sciences, 8*, 307–314.

Feron, J., Gentaz, E., & Streri, A. (2006). Evidence of amodal representation of small numbers across visuo-tactile modalities in 5-month-old infants. *Cognitive Development, 21*, 81–92.

Flombaum, J. I., Scholl, B. J., & Santos, L. R. (2009). Spatiotemporal priority as a fundamental principle of object persistence. In B. Hood & L. Santos (Eds.), *The origins of object knowledge* (pp. 135–164). Oxford, USA: Oxford University Press.

Fodor, J. A. (1983). *The modularity of mind.* Cambridge, MA: MIT Press.

Frege, G. (1980). *The foundations of arithmetic: A logico-mathematical enquiry into the concept of number* (J. L. Austin, Trans.). Evanston, IL: Northwestern University Press. (Original published 1884).

Fuson, K. C. (1988). *Children's counting and concepts of number.* New York: Springer-Verlag Publishing.

Fuson, K. C., & Hall, J. W. (1983). The acquisition of early number word meanings: A conceptual analysis and review. In H. P. Ginsburg (Ed.), *The development of mathematical thinking* (pp. 49–107). New York: Academic.

Gallistel, C. R. (2007). Commentary on Le Corre & Carey. *Cognition, 105*, 439–445.

Gallistel, C. R. (2011). Mental magnitudes. In S. Dehaene & E. M. Brannon (Eds.), *Space, time, and number in the brain: Searching for the foundations of mathematical thought* (pp. 3–12). New York: Elsevier.

Gallistel, C. R., & Gelman, R. (1992). Preverbal and verbal counting and computation. *Cognition, 44*, 43–74.

Gallistel, C. R., & Gelman, R. (2000). Non-verbal numerical cognition: From reals to integers. *Trends in Cognitive Sciences, 4*, 59–65.

Gallistel, C. R., & Gelman, R. (2005). Mathematical cognition. In K. Holyoak & R. Morrison (Eds.), *The Cambridge handbook of thinking and reasoning* (pp. 559–588). New York: Cambridge University Press.

Geary, D. C., Hoard, M. K., Nugent, L., & Bailey, H. D. (2013). Adolescents' functional numeracy is predicted by their school entry number system knowledge. *PLoS One, 8*(1), e54651. doi: 10.1371/journal.pone.0054651

Gelman, R. (1972). The nature and development of early number concepts. In H. W. Reese (Ed.), *Advances in child development* (Vol. 3, pp. 115–167). New York: Academic Press.

Gelman, R. (1993). A rational-constructivist account of early learning about numbers and objects. In D. Medin (Ed.), *Learning and motivation* (Vol. 30, pp. 61–96). New York: Academic Press.

Gelman, R., & Gallistel, C. R. (1978). *The child's understanding of number.* Cambridge, MA: Harvard University Press.

Gelman, R., & Greeno, J. G. (1989). On the nature of competence: Principles for understanding in a domain. In L. B. Resnick (Ed.), *Knowing and learning: Issues for a cognitive science of instruction* (pp. 125–186). Hillsdale, NJ: Erlbaum.

Gelman, R., & Meck, E. (1983). Preschoolers' counting: Principles before skill. *Cognition, 13,* 343–359.

Hyde, D. (2011). Two systems of non-symbolic numerical cognition. *Frontiers in Human Neuroscience, 5,* 150.

Izard, V., Sann, C., Spelke, E. S., & Streri, A. (2009). Newborn infants perceive abstract numbers. *Proceedings of the National Academy Sciences, USA, 106,* 10382–10385.

Jordan, K. E., & Brannon, E. M. (2006). The multisensory representation of number in infancy. *Proceedings of the National Academy of Sciences, USA, 103,* 3486–3489.

Kahneman, D., Treisman, A., & Gibbs, B. J. (1992). The reviewing of object files: Object specific integration of information. *Cognitive Psychology, 24,* 174–219.

Kaufman, E., Lord, M., Reese, T., & Volkmann, J. (1949). The discrimination of visual number. *The American Journal of Psychology, 62*(4), 498–525.

Kersey, A. J., & Cantlon, J. F. (2017). Neural tuning to numerosity relates to perceptual tuning in 3–6 year-old children. *The Journal of Neuroscience, 37*(3), 512–522.

Le Corre, M., & Carey, S. (2007). One, two, three, four, nothing more: An investigation of the conceptual sources of the verbal counting principles. *Cognition, 105,* 395–438.

Le Corre, M., Van de Walle, G., Brannon, E. M., & Carey, S. (2006). Re-visiting the competence/performance debate in the acquisition of the counting principles. *Cognitive Psychology, 52,* 130–169.

Lee, M. D., & Sarnecka, B. W. (2010). A model of knower-level behavior in number-concept development. *Cognitive Science, 67,* 34–51.

Lee, M. D., & Sarnecka, B. W. (2011). Number-knower levels in young children: Insights from Bayesian modeling. *Cognition, 120,* 391–402.

Leslie, A. M., Gallistel, C. R., & Gelman, R. (2007). Where integers come from. In P. Caruthers, S. Laurence, & S. Stich (Eds.), *The innate mind, vol. 3: Foundations and the future* (pp. 109–138). New York: Oxford University Press.

Leslie, A. M., Gelman, R., & Gallistel, C. R. (2008). The generative basis of natural number concepts. *Trends in Cognitive Sciences, 12*(6), 213–218.

Leslie, A. M., Xu, F., Tremoulet, P. D., & Scholl, B. J. (1998). Indexing and the object concept: Developing 'what' and 'where' systems. *Trends in Cognitive Sciences, 2*(1), 10–18.

Li, P., Le Corre, M., Shui, R., Jia, G., & Carey, S. (2003). *Effects of plural syntax on number word learning: A cross-linguistic study of the role of singular—plural in number word learning.* Paper presented at the 28th Annual Boston University Conference on Language Development, Boston, MA.

Libertus, M. E., & Brannon, E. M. (2010). Stable individual differences in number discrimination in infancy. *Developmental Science, 13*(6), 900–906.

Lipton, J. S., & Spelke, E. S. (2003). Origins of number sense: Large-number discrimination in human infants. *Psychological Science, 14*(5), 396–401.

Mehler, J., & Bever, T. G. (1967). Cognitive capacity of very young children. *Science, 158*(3797), 141–142.

Mix, K. S. (2004). *How spencer made number: First uses of the number words.* (Technical Report No. 255). Bloomington: Indiana University Cognitive Science Program.

Mix, K. S., & Sandhofer, C. M. (2007). Do we need a number sense? *Integrating the Mind*, 293–326.

Mix, K. S., Sandhofer, C. M., & Baroody, A. (2005). Number words and number concepts: The interplay of verbal and nonverbal processes in early quantitative development. In R. V. Kail (Ed.), *Advances in child development and behavior* (Vol. 33, pp. 305–346). New York: Academic Press.

Mou, Y., & vanMarle, K. (2014). Two core systems of numerical representation in infants. *Developmental Review*, 34(1), 1–25.

Neider, A. (2016). The neuronal code for number. *Nature Reviews: Neuroscience*, 17, 366–382.

Parsons, S., & Bynner, J. (1997). Numeracy and employment. *Education and Training*, 39(2), 43–51.

Piaget, J. (1952). *The child's conception of number.* London, England: Routledge & Kegan Paul.

Plaisier, M. A., Tiest, W. M. B., & Kappers, A. M. L. (2010). Range dependent processing of visual numerosity: Similarities across vision and haptics. *Experimental Brain Research*, 204, 525–537.

Pylyshyn, Z. W., & Storm, R. W. (1988). Tracking multiple independent targets: Evidence for a parallel tracking mechanism. *Spatial Vision*, 3, 179–197.

Ritchie, S. J., & Bates, T. C. (2013). Enduring links from childhood mathematics and reading achievement to adult socioeconomic status. *Psychological Science*, 24, 1301–1308.

Rivera-Batiz, F. (1992). Quantitative literacy and the likelihood of employment among young adults in the United States. *The Journal of Human Resources*, 27, 313–328.

Sarnecka, B. W., Goldman, M. C., & Slusser, E. B. (2015). How counting leads to children's first representations of exact, large numbers. In R. Cohen Kadosh & A. Dowker (Eds.), *Oxford handbook of numerical cognition* (pp. 291–309). New York: Oxford University Press.

Sarnecka, B. W., Kamenskaya, V. G., Yamana, Y., Ogura, T., & Yudovina, J. B. (2007). From grammatical number to exact numbers: Early meanings of 'one', 'two', and 'three' in English, Russian, and Japanese. *Cognitive Psychology*, 55, 136–168.

Sarnecka, B. W., & Wright, C. E. (2013). The idea of an exact number: Children's understanding of cardinality and equinumerosity. *Cognitive Science*, 37(8). doi: 10.1111/cogs.12043

Scholl, B. J. (2001). Objects and attention: The state of the art. *Cognition*, 80, 1–46.

Scholl, B. J., & Leslie, A. M. (1999). Explaining the infant's object concept: Beyond the perception/cognition dichotomy. In E. Lepore & Z. Pylyshyn (Eds.), *What is cognitive science?* (pp. 26–73). Oxford: Blackwell.

Simon, T. J. (1997). Reconceptualizing the origins of number knowledge: A 'non-numerical' account. *Cognitive Development*, 12, 349–372.

Simon, T. J. (1999). The foundations of numerical thinking in a brain without numbers. *Trends in Cognitive Sciences*, 3, 363–364.

Slusser, E., & Sarnecka, B. W. (2011). Find the picture of eight turtles: A link between children's counting and their knowledge of number-word semantics. *Journal of Experimental Child Psychology*, 110, 38–51.

Spelke, E. S. (2000). Core knowledge. *American Psychologist*, 55, 1233–1243.

Spelke, E. S. (2011). Natural number and natural geometry. In E. Brannon & S. Dehaene (Eds.), *Attention and performance, vol. 24: Space, time and number in the brain: Searching for the foundations of mathematical thought* (pp. 287–317). Oxford: Oxford University Press.

Spelke, E. S., & Tsivkin, S. (2001). Initial knowledge and conceptual change: Space and number. In M. Bowerman & S. Levinson (Eds.), *Language acquisition and conceptual development*. Cambridge, UK: Cambridge University Press.

Starkey, P., & Cooper, R. G., Jr. (1980). Perception of numbers by human infants. *Science, 210*, 1033–1035.

Starkey, P., Spelke, E. S., & Gelman, R. (1983). Detection of intermodal numerical correspondences by human infants. *Science, 222*, 179–181.

Starkey, P., Spelke, E. S., & Gelman, R. (1990). Numerical abstraction by human infants. *Cognition, 36*, 97–127.

Starr, A., Libertus, M., & Brannon, E. M. (2013). Number sense in infancy predicts mathematical abilities in childhood. *Proceedings of the National Academy of Science, 110*(45), 18116–18120. doi: 10.1073/pnas.1302751110

Strauss, M. S., & Curtis, L. E. (1981). Infant perception of numerosity. *Child Development, 52*, 1146–1152.

Taves, E. H. (1941). Two mechanisms for the perception of visual numerousness. *Archives of Psychology, 37*(265), 1–47.

Trick, L. M. (2008). More than superstition: Differential effects of featural heterogeneity and change on subitizing and counting. *Perception and Psychophysics, 70*(5), 743–760.

van Loosbroek, E., & Smitsman, A. W. (1990). Visual perception of numerosity in infants. *Developmental Psychology, 26*(6), 916–922.

vanMarle, K., Chu, F. W., Mou, Y., Seok, J. H., Rouder, J., & Geary, D. C. (2016). Attaching meaning to the number words: Contributions of the object tracking and approximate number systems. *Developmental Science*. doi: 10.1111/desc.12495

vanMarle, K., & Scholl, B. J. (2003). Attentive tracking of objects versus substances. *Psychological Science, 14*(5), 498–504.

vanMarle, K., & Wynn, K. (2006). Six-month-old infants use analog magnitudes to represent duration. *Developmental Science, 9*(5), F41–F49.

Walsh, V. (2003). A theory of magnitude: Common cortical metrics of time, space and quantity. *Trends of Cognitive Sciences, 7*, 483–488.

Whalen, J., Gallistel, C. R., & Gelman, R. (1999). Nonverbal counting in humans: The psychophysics of number representation. *Psychological Science, 10*, 130–137.

Wynn, K. (1990). Children's understanding of counting. *Cognition, 36*, 155–193.

Wynn, K. (1992a). Children's acquisition of the number words and the counting system. *Cognitive Psychology, 24*, 220–251.

Wynn, K. (1992b). Addition and subtraction by human infants. *Nature, 358*, 749–750.

Wynn, K. (1995). Origins of numerical knowledge. *Mathematical Cognition, 1*, 35–60.

Xu, F., & Carey, S. (1996). Infants' metaphysics: The case of numerical identity. *Cognitive Psychology, 30*, 111–153.

Xu, F., & Spelke, E. S. (2000). Large number discrimination in 6-month-old infants. *Cognition, 74*, B1–B11.

8   Seeing Numbers as Affordances

Max Jones

1. Introduction

Despite the widespread and well-founded belief that we lack perceptual access to number, research in the cognitive sciences "over the last few decades has firmly established that number is a perceptual attribute" (Anobile, Cicchini, & Burr, 2016, 24). Our capacity for numerical perception is hard to explain on the basis of either intuitive accounts of perception or classical approaches from the cognitive sciences. However, as suggested by Kitcher (1984), Gibson's ecological approach to perception can provide an explanation of numerical perception in terms of *affordances* (Gibson, 1979). We can perceive number by perceiving opportunities for engaging in enumerative actions. However, in order to explain this in a way that is compatible with contemporary evidence, it is necessary to refine Kitcher's approach, focusing on a more fine-grained notion of enumerative action.

The possibility of providing an ecological account of numerical perception has significant implications for issues in the philosophy of mathematics. At first sight, the existence of numerical perception seems to support a realist approach (Maddy, 1990; Franklin, 2014), which is capable of avoiding the problem of epistemic access (Benacerraf, 1973; Field, 1989). However, only a very weak form of realism is supported, which fails to guarantee mathematical truth. Moreover, the move from numerical perception to mathematical realism relies on assuming a naïve form of realism about the objects of perception, thereby overlooking a whole host of debates in the philosophy of perception.

Although the existence of numerical perception cannot decide between realism and antirealism, it exposes the fact that metaphysical assumptions in the philosophy of mathematics are closely tied to assumptions about the nature and objects of perception. If some mathematical entities are perceivable, then metaphysical issues in the philosophy of mathematics are not isolated from metaphysical issues in the philosophy of perception, and new theories of perception that call for radical alterations of the latter may have a significant impact on the former.

2. Against Numerical Perception

Most philosophers of mathematics hold that we lack perceptual access to number. On the one hand, realists tend to take a platonist approach according to which mathematical entities or structures are abstract entities. As a result, numbers are taken to be inaccessible to perception, since abstract entities lack spatiotemporal location and causal efficacy. If we can access numbers at all, we do so via abstraction, which is a *cognitive* rather than *perceptual* process. On the other hand, antirealists tend to see mathematical entities as mere mental constructions or useful fictions, which, likewise, are constructed or created on the basis of thought, rather than discovered through perception.

This widespread rejection of numerical perception can be seen to stem from Frege's infamous critique of Mill's Empiricism (Frege, 1950, 9–14, 27–32). Frege points out an important apparent difference between number and perceivable properties, such as color:

> It marks, therefore, an important difference between colour and number, that a colour such as blue belongs to a surface independently of any choice of ours . . . The number 1, on the other hand, or 100 or any other number, cannot be said to belong to the pile of playing cards in its own right, but must belong to it in the way that we have chosen to regard it . . . What we choose to call a complete pack is obviously an arbitrary decision, in which the pile of playing cards has no say.
>
> (1950, 29)

In other words, there is no characteristic manner in which we perceive the world as separated into parts (ibid., 30). Number can seemingly only be applied to objects in the world in virtue of a particular conceptualization. As such, apprehension of number requires concepts, and concepts are widely held to be cognitive rather than perceptual mental entities. To the extent that we can apply number to the world, this is a projection of the products of cognition rather than something that is directly accessed through perception.

This perspective is also supported by classical approaches to perception from the cognitive sciences. Classical approaches see the role of perception as constructing a detailed representation of the spatial layout of objects and their properties in the environment on the basis of the passive reception of incoming sensory signals. This representation is then passed on to a separate *cognitive* system in which processes such as conceptualization, reasoning, and decision-making take place. On some occasions, these processes will result in further signals being passed to a separate *action* system, which coordinates overt behavior. This characterization of the mind has been labeled the 'classical cognitive sandwich model' (Hurley, 1998, 401–402).

In line with Frege's critique, on this picture, it seems as though number is a product of cognition. Our perceptual systems must first construct a representation of the environment before the objects in the environment are conceptualized, and only once these conceptual representations have been delivered to the cognitive system can some form of enumeration operation be conducted so as to give rise to an apprehension of number. Of course, we do not only apprehend the number of entities in a collection 'in our heads'. We sometimes engage in more active counting, for example, by pointing to or attending to each object in a collection in sequence, while, at each stage, memorizing the corresponding number word or position on the number line. However, this too suggests that number is the product of thought rather than perception since each transition between perception and action is mediated by cognitive processes, such as memorizing the relevant numerical value and deciding to attend to the next object in the collection.

A further reason, on the classical picture, for seeing numerical apprehension as cognitive rather than perceptual arises from the multimodal nature of numerical properties. While most perceivable properties, such as color or pitch, can only be apprehended through single sensory modalities, such as vision or audition, number can seemingly be applied to various different sensory modalities. For example, we can enumerate dots, beeps, or touches, as well as stimuli from multiple modalities, such as enumerating beeps *and* flashes. On the classical picture, perceptual systems are divided into separate unimodal processing streams that transmit information from specific sensory receptors. Cognitive systems, on the other hand, are where information from multiple sensory modalities is integrated and, as such, would seem to be the locus of enumeration. For many, the distinction between unimodality and multimodality is definitive of the distinction between perception and cognition. The multimodality of numerical properties, thus, provides more reason, on the classical picture, for seeing number as a product of thought rather than a property that we directly perceive.

Both an intuitive understanding of the mind and the classical cognitive perspective suggest that our apprehension of number is a product of thought rather than something that we directly perceive, thus supporting the consensus in the philosophy of mathematics that number is not perceivable. However, in recent years, evidence from the scientific study of numerical cognition has emerged that threatens to undermine this consensus.

3.   Evidence for Numerical Perception

A range of recent evidence from the cognitive sciences suggests that we possess the capacity for *numerical perception* (Jones, 2016). It is now well established that a wide range of species possess the capacity for *subitizing*— that is, rapidly and automatically apprehending the number of entities in a perceived collection (Dehaene, 1998). This capacity is reliably accurate when dealing with collections of one to four entities, but the level of accuracy decreases logarithmically as the size of the target collection increases, roughly in line with the so-called Weber-Fechner law (Dehaene, 2003). This capacity is arguably supported by a dedicated approximate number system (ANS), which is an innate system located in the horizontal intraparietal sulcus (Dehaene, Molko, Cohen, & Wilson, 2004; Nieder & Dehaene, 2009).[1]

There are a number of reasons for seeing this capacity as a capacity of numerical *perception*. First, our apprehension of number is both rapid and automatic. We are able to apprehend the number of entities quicker than we are able to categorize said entities, and quicker than we can attend to each entity in sequence (Piazza, Izard, Pinel, Le Bihan, & Dehaene, 2004). Moreover, we are able to do so automatically without conscious awareness (Bahrami et al., 2010).

There is also evidence to suggest that our apprehension of number is direct, in the sense that it is not inferred from other perceivable properties (Burr & Ross, 2008; Ross & Burr, 2010; Anobile, Cicchini, & Burr, 2016). For example, the number of dots in a collection could, in principle, be approximated on the basis of the density and extent of the collection. However, systematically varying the properties from which number could potentially be inferred fails to impact on our capacity to apprehend number.

Our capacity to apprehend number also exhibits properties that are characteristic of perceptual systems. Variance in accuracy in accordance with the Weber-Fechner law was initially found to apply to perceptual properties (Dehaene, 2003), such as illumination or weight, and can be seen as a signature feature of perceptual apprehension. Moreover, apprehension of number is subject to adaptation effects, another signature feature of perceptual apprehension, whereby repeated exposure to larger collections can lead one to systematically underestimate subsequently presented smaller collections (Burr & Ross, 2008; Burr, Anobile, & Turi, 2011; Castaldi, Aagten-Murphy, Tosetti, Burr, & Morrone, 2016), similar to the way that things look bluer when one takes off a pair of yellow tinted glasses, having adapted to the yellow tint.

The ANS also arguably fulfills Fodor's criteria for perceptual input modules (Fodor, 1983; Jones, 2016). Although the issue of modularity is controversial, Fodor's criteria were initially put forward as a means for distinguishing perceptual input systems from cognitive systems. As such, the burden of proof lies with the opponent of numerical perception to explain why the ANS fits these criteria while failing to qualify as a perceptual system, without thereby disqualifying systems that are unequivocally agreed to be perceptual systems.

---

1 Some (Carey, 2009) have argued that our capacity for numerical apprehension is supported by two distinct systems: the object-tracking system (OTS), which is responsible for the reliably accurate apprehension of one to four objects, and the ANS, which is responsible for the approximate apprehension of more than four objects. However, this distinction will be set aside here since the arguments and evidence for our capacity for numerical apprehension being understood as a perceptual capacity apply equally to both systems, and there is evidence to suggest that the ANS is active alongside the OTS in detecting small numerosities (Burr et al., 2011).

At this stage, one might object that the ANS is not a system for numerical perception, since most "cognitive scientists distinguish between *numerosities*, the concrete, discrete magnitudes that animals represent, and *numbers*, the abstract entities that are studied by mathematicians and philosophers of mathematics" (De Cruz, 2016, 3). In other words, we perceive numerosities, not numbers. However, this distinction it is not tenable in the current context since assuming that there is a distinction between *concrete* numerosities and *abstract* numbers involves assuming the impossibility of numerical perception from the outset. It is without doubt the case that our perceptual access to number fails to specify fully the sophisticated number *concepts* studied by mathematicians and philosophers of mathematics. However, this is no different from the relationship between perception and concepts in other domains. Numerical perception provides approximate representations of number rather than representations of approximate number or numerosity, and our perceptual access only partially contributes to our fully fledged number concepts (Jones, 2016).

Given these considerations, there are good reasons to think that the ANS is a perceptual system for numerical perception. As such, we should take Dehaene seriously when he states,

> Number appears as one of the fundamental dimensions according to which our nervous system parses the external world. Just as we cannot avoid seeing objects in colour . . . and at definite locations in space . . . in the same way numerical quantities are imposed on us effortlessly.
>
> (1998, 71)

However, given the aforementioned incompatibility between the notion of numerical perception and classical and intuitive approaches to perception, it may be necessary to look to an alternative account of perception in order to explain this capacity.

4.   An Ecological Approach to Numerical Perception

Although the majority of philosophers of mathematics reject the notion of numerical perception, there have been some dissenters. Maddy, in her early work, argues that we can perceive number by perceiving sets (1990, 50–58), and Kitcher attempts to resurrect a Millian account, arguing that we can perceive arithmetical properties by perceiving opportunities to engage in collective activity (1984, 108). Interestingly, both make reference to Gibson's ecological theory of perception (Maddy, 1990, 50, fn 37; Kitcher, 1984, 11–12, 108).

Gibson's (1979) ecological approach to perception is a radical departure from the classical approach to perception and, thus, may offer a way to overcome the apparent incompatibility of classical approaches and numerical perception. Moreover, the potential applicability of this approach to numerical perception is particularly interesting given the recent resurgence in interest in and support for Gibson's ideas (Chemero, 2011; Anderson, 2014).

The ecological approach to perception challenges the traditional approach in a number of ways. Perception involves *active* exploration of the environment rather than *passive* reception of sensory stimuli. It involves continuous dynamic interaction with the environment rather than being the result of computations based on static image-like inputs. As a result of these features, organisms are able to glean far richer information directly from invariant feature of their interaction with the environment, negating the need to build up a rich and detailed inner representation of the external world.

Most importantly for current concerns, perception is, thus, action oriented, in the sense that organisms are able to perceive information directly that is relevant for action. Rather than representing objects and their properties before inferring their relevance for action, organisms perceive *affordances*, which are best understood as opportunities for action. For example, in perceiving a set of stairs, one perceives its climbability; in perceiving a mug, one perceives its graspability; in perceiving a chair, one perceives its sit-on-ability. An important consequence of the action-oriented nature of perception is that perception becomes organism-relative. For example, while humans perceive mugs as affording grasping, mice might perceive them as affording hiding in. Despite the organism-relativity of affordances, they need not be seen as subjective or mind dependent. There are objective facts of the matter regarding what a given organism can or cannot do, and, as such, affordances can be seen as objective features of reality. For instance, they can be seen as dispositional properties of objects (Turvey, 1992) or relational properties of organism – environment systems (Chemero, 2003).

Importantly for current concerns, the ecological approach to perception lacks some of the features that rendered traditional approaches as incompatible with the notion of numerical perception. Unlike traditional approaches, the ecological approach is not committed to a clear separation between perception, cognition, and action, where interactions between perception and action are mediated by cognition. It allows for the possibility of perceiving rich and complex features that are directly relevant for action, without the need for cognitive processing of inner representations. The ecological approach also dispenses with understanding perception in unimodal terms, characterizing the senses in terms of relational *perceptual systems* rather than sensory modalities (Gibson, 1966). Moreover, given the organism-relative nature of perception, it is compatible with there not being a single characteristic manner of ascribing number to collections, and with numerical perception being in some sense observer-relative, without number thereby becoming a mere subjective projection of thought. Number can be *perceived* in terms of "what *the world* will let us do *to it*" (Kitcher, 1984, 108) rather than being something that we project onto the world from within.

5.   Perceiving Numerical Affordances

Kitcher attempts to develop an empiricist account of the epistemology of arithmetic by providing an explicitly Gibsonian account of our ability to perceive numerical properties (1984, 11). He argues that children learn about number by "engaging in *activities* of collecting and segregating" objects into collections (1984, 107). We are thereby able to perceive number directly by perceiving affordances for engaging in such collective activities. For example, we can see that a collection consists of three objects by directly perceiving that it is possible to carry out a particular sequence of object manipulations whereby the objects are gathered together within a spatially bounded region. Importantly, the type of action sequence one carries out in collecting a three-membered collection will differ significantly from that which will be involved in doing so with a four- or five-membered collection (and so on), and, as such, each number will be associated with a specific type of action sequence whose affordance can be directly perceived (at least for relatively small numbers). In essence, we perceive numbers by perceiving possibilities for forming collections through object manipulation.

At face value, a problem for Kitcher's account is that it seems to undermine the objectivity and necessity of mathematical truth. If numbers are based on what organisms can do, then the truth of arithmetical claims is seemingly contingent upon the existence of organisms that are able to carry out certain actions. Thus one might worry that, in order to provide an account of numerical perception, one must give up on the notion that mathematical truths are necessary truths. This seems far too drastic a step to take to account merely for some recent evidence from the cognitive sciences. However, Kitcher responds to this worry by claiming that the affordances relevant to numerical perception are *"universal affordances"* that is, possibilities of interaction, "which are afforded by *any* environment" (1984, 12).

At first sight, it isn't clear that this addresses the worry at hand, as mathematical truths would still be contingent on the existence of organisms, such as ourselves, capable of carrying out relevant interactions with the environment. However, it is possible to avoid this worry by understanding affordances as dispositional properties. On such an approach, for example, a chair affords sitting on for humans, even if there are no humans in existence to sit on it and make this disposition manifest (Chemero, 2011, 149). Similarly, collections of objects may afford enumeration even if there are no humans to count them.

A further problem for Kitcher's account arises from the fact that it fails to apply to environments where "everything in the world is nailed down" (Frege, 1950, 9). It is simply untrue that any environment affords manipulating objects so as to form collections, as one can easily imagine environments in which object manipulation is not possible. Affordances for enumerative action, where enumerative action is understood in terms of manipulating macroscopic objects, are not universal affordances.

As well as highlighting a problem with Kitcher's attempt to preserve the universal nature of mathematical truth, Frege's criticism also points to more pressing problems with Kitcher's account of numerical perception. One need not conjure up fanciful imaginary scenarios of worlds where everything is nailed down to see problems with understanding numerical perception in terms of object manipulation. Evidence about the nature of numerical perception is already in conflict with this kind of approach.

There are two ways in which Kitcher's account of numerical perception is in conflict with the evidence available from the cognitive sciences. If we can perceive number at all, then it is clear that we can perceive the number of collections of objects that we cannot, in practice, rearrange into spatially bounded collections. For example, we can perceive the number of mountains on the horizon or clouds in the sky. Kitcher responds to these worries by suggesting that we, first, learn about number by moving objects around and, later, on this basis, we gain the ability to "collect the objects in thought without moving them around" (Kitcher, 1984, 111). Thus our capacity for perceiving the number of non-manipulable objects is taken to be ontogenetically dependent on our capacity for object manipulation.

This response, however, is problematic, as we are able to perceive number for collections of entities that are not even manipulable in principle. One of the hallmarks of numerical perception is its multimodal nature. We are capable of auditory and tactile numerical perception to the same extent as visual numerical perception (Xu & Spelke, 2000; Barth, La Mont, Lipton, & Spelke, 2005; Piazza, Mechelli, Price, & Butterworth, 2006). However, sounds and touches are not even manipulable in principle. Perhaps, Kitcher would explain this capacity by claiming that we can imagine sounds or touches *as if* they were manipulable objects; however, as well as being quite a stretch of the imagination, this response is incompatible with evidence about the development of our capacity for numerical perception.

Young infants, as well as a wide range of nonhuman species, have been shown to be capable of numerical perception (Wynn, 1992; Izard, Sann, Spelke, & Streri, 2009; De Cruz, 2006), despite lacking the manual dexterity or requisite morphology to be capable of manipulating objects into collections. This suggests that numerical perception cannot be dependent on the capacity for object manipulation. Moreover, infants are equally capable of perceiving the number of auditory stimuli as they are of perceiving the number of manipulable objects (Izard et al., 2009), thus undermining Kitcher's notion that perception of number for non-manipulable collections is ontogenetically dependent on the capacity for object manipulation. As such, the evidence from the cognitive sciences suggests that our capacity for numerical perception cannot be based on our ability to perceive affordances for manipulating objects into collections. In order to explain numerical perception from an ecological perspective, it is necessary to explain affordances for enumerative actions in terms of some other form of action.

6.   Fine-Grained Numerical Affordances

The capacity to perceive number is independent of and developmentally more basic than the capacity to pick up macroscopic objects and arrange them into spatially bounded collections. However, this need not be seen as fatal for understanding numerical perception in terms of affordances. The problems with Kitcher's account arise as a result of taking a far too coarse-grained form of action to be relevant for perceiving numerical affordances.

Elucidations of ecological psychology tend to focus on affordances for relatively coarse-grained, deliberative actions, such as graspability or climbability. However, this should not be seen as a reflection of the details of the theory. Rather, these are simply affordances that are easier (although still somewhat cumbersome) to articulate. Yet this should not obscure the fact that most of the affordances we perceive will be more fine-grained, not necessarily associated with any conscious deliberation, and much harder to articulate, lacking any corresponding action verb. For example, one can perceive affordances for adjusting one's bodily position in a specific manner so as to fit through a gap, or for adjusting the position of one's eyes in a particular way so as to bring a different region of space into view.

The question thus arises as to which more fine-grained form of action is relevant for numerical perception, in the sense of being a common feature of all enumerative actions. A prime candidate for this role is *sequential attention*. Counting the objects in a collection requires attending to each object in sequence.[2] Importantly, this is not only the case for collections of macroscopic objects but also applies to non-manipulable collections, including collections of nonvisual stimuli such as beeps or touches. Moreover, unlike object manipulation, the capacity for sequential attention is something that we share with young infants and a wide range of other species. Kitcher was correct to focus on the collective activity of manipulating objects, not because this is the basis of our capacity for numerical perception, but because all instances of this behavior involve the more fundamental action of sequential attention. Whenever we perceive an affordance for bringing objects together into a collection, we perceive an affordance for sequentially attending to each object, but not vice versa.

Importantly, as with collective activity, the type of attention sequence one carries out in sequentially attending to the items in a three-membered

---

2 It is important to note here that *counting* must be distinguished from *numerical perception*. Counting involves sequentially attending to each object in a collection, while registering a position in a count list (either in thought or through external vocalizations or inscriptions). Numerical perception, on the other hand, has been shown to be too rapid, automatic, and direct to be dependent on sequential attention (see Section 2).

collection will differ significantly from that which will be involved in doing so with a four- or five-membered collection (and so on), and, as such, each number will be associated with a specific type of attention sequence whose affordance can be directly perceived. We perceive numbers by perceiving affordances for enumerative action, where enumerative action is understood in terms of sequential attention. We see numbers by seeing that we can count.

The idea that numerical perception is closely associated with our capacity for sequential attention is supported by neurological evidence. The intraparietal sulcus, the area in which the ANS is held to be located, is a system that is widely held to be involved in the coordination of spatial attention and oculomotor control (Grefkes & Fink, 2005; Goldberg, Bisley, Powell, & Gottlieb, 2006; Gillebert et al., 2011). This makes sense, since in order to coordinate attention to a number of objects in sequence, one must already have a rough sense of how many objects there are. This is further supported by evidence suggesting that the ANS plays a role both in perceiving the number of entities in a collection and in determining the number of self-produced actions (Anobile et al., 2016).

At this stage, a potential worry arises regarding whether the notion of affordances for sequential attention represents too much of a departure from the orthodox ecological approach. The ecological approach suggests affordances are opportunities for action; however, it's not immediately clear that attention qualifies as a form of action. In most ecologically valid scenarios attending to a stimulus involves orienting the eyes in the direction of the stimulus (Findlay & Gilchrist, 2003), even in the case of some nonvisual stimuli. However, it is well established that we can also covertly shift attention without engaging in any overt orientation behavior (Posner, 1980). As such, some might argue that attention is a purely cognitive process rather than an action, and that it is not the kind of thing that we perceive opportunities for. However, there are good reasons to resist this move. There is a wealth of behavioral and neurological evidence to suggest a close relationship between overt and covert attention (Rizzolatti & Craighero, 2010), even in the case of nonvisual attention (Driver & Spence, 1998). Moreover, overt movement need not be seen as an essential feature of action, particularly in the context of the ecological approach. Attention involves organisms actively altering their relationship with the environment and, as such, it is plausible that they are able to perceive opportunities to engage in such activity.

The ecological approach to perception, unlike traditional theories of perception, can offer an explanation of numerical perception. We perceive number by directly perceiving opportunities to engage in enumerative action, and one possible way of fleshing out this notion of enumerative action is in terms of sequential attention. Given the widespread consensus in the philosophy of mathematics that numerical perception is not possible, it is important to assess the potential ramifications of overturning this consensus.

7.   Numerical Perception and Mathematical Realism

At first sight, it is tempting to see the possibility of numerical perception as providing support for some form of realist position in the philosophy of mathematics (e.g., Maddy, 1990; Franklin, 2014). If perception of affordances is relatively widespread and numerical perception is perception of affordances, then numbers can be seen as on a par with a wide range of other objects of perception. Furthermore, it is natural to adopt a realist perspective with regard to the objects of perception, so, for the sake of parity, one should also adopt a realist perspective with respect to at least some mathematical entities, such as those numbers that we can perceive. The fact that numerical perception seems to be an evolved capacity, present in infants and a wide range of other animals, arguably provides further support for a realist approach (De Cruz, 2016).

Such a move is also tempting, as the possibility of numerical perception seemingly provides an easy way of sidestepping the epistemic access problem for at least some mathematical entities (Benacerraf, 1973; Field, 1989). The access problem arises due to most realists' commitment to the abstract nature of mathematical entities. However, if we directly perceive some mathematical entities then our access to these is seemingly no more mysterious than our perception of other objects.

Particular features of the ecological approach to perception can also be seen to support a realist approach. The ecological approach is often explained as a form of direct realism, and affordances are understood to be objective features of reality (Turvey, 1992; Chemero, 2003, 2011). Thus if some numbers are nothing more than affordances, they too might be understood as objective features of this kind. Moreover, if affordances are understood as dispositions, then numbers could even be seen to exist in the absence of any agents (Turvey, 1992).

However, this move from numerical perception to mathematical realism is problematic in a number of ways. First, the type of realism that is supported by the existence of numerical perception is extremely weak, to the extent of being almost insignificant. The capacity for numerical perception is only reliably accurate when dealing with collections of one to four entities, and, even if one accepts that approximate numerical perception of larger collections still provides a basis for realism, this capacity has clear limitations. Thus numerical perception may only provide us with perceptual access to an insignificant finite portion of the entities of arithmetic.

Given the grounding of numerical perception in terms of affordances, this problem is exacerbated. In order for affordances to be objective, they must be related to the *actual* capacities of organisms that perceive them. However, it is abundantly clear that every organism we know of is only capable of carrying out finite enumerative actions in a finite lifetime. As such, perception of numerical affordances alone can only ever provide reason to believe in some arbitrary initial portion of the natural numbers, which is organism-relative and, thus, determined by biological factors that are irrelevant to mathematical practice.

The possible existence of numerical perception may provide some motivation to be ontologically committed to an insignificantly small portion of the natural numbers, but this is obviously not sufficient to provide the basis for arithmetical truth. Ontological commitment to only *some* mathematical entities fails to address Benacerraf's challenge to provide a unified account of mathematical truth (1973, 661) and, thus, fails to support semantic realism. Our perceptual access to number is, on its own, woefully inadequate for explaining the sophisticated *concept* of number employed by practicing mathematicians (Rips, Bloomfield, & Asmuth, 2008). This does not mean that numerical perception is irrelevant. As with any other concept, our number concepts go beyond our immediate perceptual access.[3] However, it does suggest that the existence of numerical perception fails to support a realist approach.

Kitcher responds to these worries by arguing, "Mathematics is an idealized science of particular universal affordances" (1984, 12). Our arithmetical knowledge is an idealization based on our perceptual access to number, achieved by imagining the possibility of an ideal agent able to engage in indefinite enumerative actions. Thus numbers and other mathematical entities need not be seen as any more controversial than the idealizations of science, such as ideal gases or frictionless planes (1984, 116–117). However, this response fails to support a realist approach to mathematical entities, merely shifting the debate to the equally controversial debate in the philosophy of science about whether we should be realists about scientific idealizations; as such, Kitcher's approach seems equally compatible with an instrumentalist approach to mathematics.

A further problem with the move from the possibility of numerical perception to mathematical realism is that it assumes a highly contentious naïve form of realism in the philosophy of perception. Although perceptual access to a certain kind of entity provides *prima facie* evidence for the existence of those entities, it is clear that considerations of perceptual access alone can settle the debate between realists and antirealists in any domain. One only has to look to the intransigent nature of the debate about the existence of color, something that we incontrovertibly have perceptual access to, in order to see that numerical perception alone, however it is understood, fails to support realism about number.

This problem is further exacerbated by the controversial nature of affordances. There is much disagreement about the metaphysical status of affordances, even among proponents of an ecological approach (Chemero, 2011, 135–154). Gibson himself was unclear about the status of affordances

---

3 As such, evidence for numerical perception may be able to contribute to arguments for realism based on its interaction with the many other, often culturally augmented, aspects of our capacity for numerical *cognition* (e.g. De Cruz, 2016).

(Chemero, 2003), stating, "An affordance is neither an objective nor a subjective property; or it is both if you like . . . It is both physical and psychical, yet neither" (Gibson, 1979, 129). On some interpretations, such as Turvey's dispositional account (1992), affordances can be seen as objective properties that exist independently of the perceiver. Whereas, on other accounts, it isn't clear that affordances qualify as objective mind-dependent entities, and certainly not the kind that could help to support a realist approach in the philosophy of mathematics. As such, whether an ecological account of numerical perception can support some form of mathematical realism depends on, far from settled, issues about the nature and existence of affordances.

An interesting potential upshot of understanding numerical perception in terms of affordances is that it may lend some support to Putnam's suggestion that debates in the philosophy of mathematics need not be framed in terms of foundational metaphysical debates between realism and antirealism (Putnam, 1967). Putnam and, later, Hellman (1989), argue that mathematics can be understood in *modal* terms without altering mathematical content. Affordances are opportunities for *possible* action in the environment and, as such, one could argue that perception of affordances gives us access to *modal* content (Nanay, 2011). Thus if one conceives of numerical perception in terms of affordances, this might provide the basis for an account of our basic epistemic access to modal mathematical content, thereby supporting modalist approaches (Putnam, 1967; Hellman, 1989). After all, Putnam's notion of 'objects – modalities dualities' (1967, 11) could be seen as a clearer characterization of affordances than is offered by many of the proponents of the ecological approach.

Even if one buys into the whole story presented earlier, and accepts that we perceive number by perceiving affordances for sequential attention, this is not sufficient to decide between realist and antirealist approaches in the philosophy of mathematics. However, this does not mean that the possible existence of numerical perception has no consequences for metaphysical and epistemological debates in the philosophy of mathematics. The existence of numerical perception would place significant constraints on how one addresses these issues. As the preceding discussion demonstrates, the mere possibility that we perceive number entails that debates about the existence of mathematical entities cannot be conducted in isolation from emerging evidence from the study of mathematical cognition or from broader developments in the philosophy of perception.

8. Conclusion

The ecological approach to perception offers a potential explanation of numerical perception. However, whether the ecological approach is the best approach and whether or not the evidence from the cognitive sciences are best understood in terms of numerical perception remain contentious issues, which can only be decided through further empirical investigation and debate. Despite this, the mere possibility of numerical perception, while failing to support a realist or antirealist approach, still has important implications for metaphysical and epistemological issues in the philosophy of mathematics. The mere suggestion that some mathematical entities might be on a par with some everyday objects of perception is enough to ensure that debates about the status of the former cannot be conducted independently from debates about the status of the latter.

As highlighted by the other contributions to this volume, it is important that philosophers of mathematics pay attention to evidence from the cognitive sciences regarding the nature of logical and mathematical cognition. However, the possibility of explaining numerical perception in terms of the ecological approach suggests that this narrow focus alone may not be sufficient. Radical theories of perception, such as Gibson's (1979), entail an equally radical revision of the ontology of perceptual experience (Chemero, 2003), thereby changing the metaphysical landscape in which mathematical entities either do or do not fit. Philosophers of mathematics must, thus, also be sensitive to broader developments in the cognitive sciences and the philosophy of perception since background assumptions about the nature of perception shape our intuitions about the existence and nature of mathematical entities, as well as our epistemic access to them.

References

Anderson, M. (2014). *After phrenology*. Cambridge, MA: MIT Press.

Anobile, G., Arrighi, R., Togoli, I., & Burr, D. (2016). A shared numerical representation for action and perception. *eLife, 5*, e16161.

Anobile, G., Cicchini, G. M., & Burr, D. (2016). Number as a primary perceptual attribute: A review. *Perception, 45*(1–2), 5–31.

Bahrami, B., Vetter, P., Spolaore, E., Pagano, S., Butterworth, B., & Rees, G. (2010). Unconscious numerical priming despite interocular suppression. *Psychological Science, 21*(2), 224–233.

Barth, H., La Mont, K., Lipton, J., & Spelke, E. (2005). Abstract number and arithmetic in preschool children. *Proceedings of the National Academy of Sciences of the United States of America, 102*(39), 14116–14121.

Benacerraf, P. (1973). Mathematical truth. *The Journal of Philosophy, 70*(19), 661–679.

Burr, D., Anobile, G., & Turi, M. (2011). Adaptation affects both high and low (subitized) numbers under conditions of high attentional load. *Seeing and Perceiving, 24*(2), 141–150.

Burr, D., & Ross, J. (2008). A visual sense of number. *Current Biology, 18*(6), 425–428.

Carey, S. (2009). *The origin of concepts*. New York: Oxford University Press.

Castaldi, E., Aagten-Murphy, D., Tosetti, M., Burr, D., & Morrone, M. (2016). Effects of adaptation on numerosity decoding in the human brain. *NeuroImage, 143*, 364–377.

Chemero, A. (2003). An outline of a theory of affordances. *Ecological Psychology, 15*(2), 181–195.

Chemero, A. (2011). *Radical embodied cognitive science*. Cambridge, MA: MIT Press.

De Cruz, H. (2016). Numerical cognition and mathematical realism. *Philosophers' Imprint*, 16(16), 1–13.

De Cruz, H. (2006). Towards a Darwinian approach to mathematics. *Foundations of Science*, 11(1–2), 157–196.

Dehaene, S. (2003). The neural basis of the Weber-Fechner law: A logarithmic mental number line. *Trends in Cognitive Sciences*, 7(4), 145–147.

Dehaene, S. (1998). *The number sense*. Oxford: Oxford University Press.

Dehaene, S., Molko, N., Cohen, L., & Wilson, A. (2004). Arithmetic and the brain. *Current Opinions in Neurobiology*, 14(2), 218–224.

Driver, J., & Spence, C. (1998). Crossmodal attention. *Current Opinion in Neurobiology*, 8(2), 245–253.

Field, H. (1989). *Realism, mathematics and modality*. Oxford: Oxford University Press.

Findlay, J., & Gilchrist, I. (2003). *Active vision: The psychology of looking and seeing*. Oxford: Oxford University Press.

Fodor, J. (1983). *The modularity of mind*. Cambridge, MA: MIT Press.

Franklin, J. (2014). *An Aristotelian realist philosophy of mathematics*. New York: Palgrave Macmillan.

Frege, G. (1950). *Foundations of arithmetic* (J. L. Austin, Trans.). Oxford: Blackwell.

Gibson, J. (1979). *The ecological approach to visual perception*. Boston, MA: Houghton, Mifflin.

Gibson, J. (1966). *The senses considered as perceptual systems*. Oxford: Houghton, Mifflin.

Gillebert, C., Mantini, D., Thijs, V., Sunaert, S., Dupont, P., & Vandenberghe, R. (2011). Lesion evidence for the critical role of the intraparietal sulcus in spatial attention. *Brain: A Journal of Neurology*, 134(6), 1694.

Goldberg, M., Bisley, J., Powell, K., & Gottlieb, J. (2006). Saccades, salience and attention: The role of the lateral intraparietal area in visual behavior. *Progress in Brain Research*, 155, 157–175.

Grefkes, C., & Fink, G. (2005). The functional organization of the intraparietal sulcus in humans and monkeys. *Journal of Anatomy*, 207(1), 3–17.

Hellman, G. (1989). *Mathematics without numbers: Towards a modal-structural interpretation*. Oxford: Clarendon Press.

Hurley, S. (1998). *Consciousness in action*. Cambridge, MA: Harvard University Press.

Izard, V., Sann, C., Spelke, E., & Streri, A. (2009). Newborn infants perceive abstract numbers. *Proceedings of the National Academy of Sciences*, 106(25), 10382–10385.

Jones, M. (2016). Number concepts for the concept empiricist. *Philosophical Psychology*, 29(3), 334–348.

Kitcher, P. (1984). *The nature of mathematical knowledge*. New York: Oxford University Press.

Maddy, P. (1990). *Realism in mathematics*. New York: Oxford University Press.

Nanay, B. (2011). Do we sense modalities with our sense modalities? *Ratio*, 24(3), 299–310.

Nieder, A., & Dehaene, S. (2009). Representation of number in the brain. *Annual Review of Neuroscience*, 32, 185–208.

Piazza, M., Izard, V., Pinel, P., Le Bihan, D., & Dehaene, S. (2004). Tuning curves for approximate numerosity in the human intraparietal sulcus. *Neuron*, 44(3), 547–555.

Piazza, M., Mechelli, A., Price, C., & Butterworth, B. (2006). Exact and approximate judgements of visual and auditory numerosity: An fMRI study. *Brain Research*, 1106(1), 177–188.

Posner, M. (1980). Orienting of attention. *Quarterly Journal of Experimental Psychology*, *32*(1), 3–25.

Putnam, H. (1967). Mathematics without foundations. *The Journal of Philosophy*, *64*(1), 5–22.

Rips, L., Bloomfield, A., & Asmuth, J. (2008). From numerical concepts to concepts of number. *Behavioral and Brain Sciences*, *31*(6), 623–642.

Rizzolatti, G., & Craighero, L. (2010). Premotor theory of attention. *Scholarpedia*, *5*(1), 6311.

Ross, J., & Burr, D. (2010). Vision senses number directly. *Journal of Vision*, *10*(2), 1–8.

Turvey, M. (1992). Affordances and prospective control: An outline of the ontology. *Ecological Psychology*, *4*(3), 173–187.

Wynn, K. (1992). Addition and subtraction by human infants. *Nature*, *358*(6389), 749–750.

Xu, F., & Spelke, E. (2000). Large number discrimination in 6-month-old infants. *Cognition*, *74*(1), B1–B11.

9 Testimony and Children's Acquisition of Number Concepts

Helen De Cruz

1. The Puzzle of Number Acquisition

At around 3.5 to 4 years of age, children in Western and other numerate cultures experience a profound shift in their understanding of numbers: they come to understand how counting works. They can use number words to denote the cardinality of collections of items in a precise fashion by placing each item to be counted into a one-to-one correspondence with elements of a counting list and using the last item to denote the cardinality of the set (see e.g., Sarnecka, in press; Le Corre, 2014). Children's acquisition of number concepts is often conceptualized in terms of individual discovery and personal reconstruction. For example, Carey (2009, 302) writes that children learn to individuate three items "*before figuring out* how the numeral list represents natural number". Davidson, Eng, and Barner (2012, 163) put it this way: "Sometime between the ages of 3-and-a-half and 4, children *discover* that counting can be used to generate sets of the correct size for any word in their count list" (emphasis added in both).

   While young children's understanding of natural numbers is remarkable, it is misleading to suggest that each child figures this out individually. Young children grow up in an environment with numerate adults and older children who use counting systems that are transmitted over many generations. Across cultures, humans use a variety of ways to count, including tallying, body-part counting, counting rods, and abacuses. Children are born in these rich cultural environments and adopt the counting systems of their parents, further extending them and building on them. Our counting systems are in line with other cultural inventions, where successive generations build on what earlier generations have already achieved. Tomasello, Carpenter, Call, Behne, and Moll (2005, 688) describe this process of learning as the ratchet effect: children are born in an environment of collective artifacts (e.g., abacuses) and social practices (e.g., body-part counting), which structure their cognitive ontogenies, and allow them to build on the cultural achievements of previous generations. It is not clear when the earliest counting systems arose, but archaeological evidence suggests that numeracy arose substantially earlier than writing. Bone, antlers, or ochre sticks with regular incisions appear in the archaeological record from about 77,000 years ago, but it is unclear whether these notches are purely decorative or might have numerical meaning (Cain, 2006). From about 25,000 years ago, artifacts appear that have groupings of markings, indicating that quantities may have been meaningful. Two bones from Ishango, Congo (25,000 years BP), have notches that are grouped in interesting ways. For instance, one of these bones has two of its three sides covered with notches that add up to 60 (Pletser & Huylebrouck, 1999).

In this chapter, I will argue that testimonial transmission plays a crucial role in how children learn about numbers. This involves both transmission of propositional knowledge (knowledge-that) and the transmission of skills (knowledge-how). The chapter is structured as follows: in Section 2, I review children's acquisition of the cardinal principle (CP). Section 3 describes how this cognitive change is usually framed in terms of individual discovery. This view—which I term the child as a lone mathematician—fits in a broader picture of children as individual learners and discoverers. Section 4 presents an alternative account of children's learning, the child as a social learner who acquires knowledge through testimony. While traditionally (particularly in philosophy), testimonial transmission has been framed in terms of propositional knowledge, I argue for a broader view where testimonial transmission also involves demonstration and transmission of knowledge-how. In Section 5, I outline how this approach is fruitful in explaining how children acquire number concepts.

2.   Children's Acquisition of the Cardinal Principle

Children typically learn to count through a stable developmental sequence. Toddlers between 18 months and 2 years can recite counting lists (e.g., 'one, two, three') as a meaningless string of words; they cannot use the number words to determine the cardinality of collections. At around 2 years of age, they become subset-knowers (Wynn, 1990). The first stage is to be a one-knower: a child can correctly give one item when requested—e.g., "Can you give me one toy"?—but performs at chance level for higher quantities (e.g., the child gives three toys when two are requested). The next stage is to become a two-knower: children can correctly give one or two toys but perform at chance level for three or more; they then become three-knowers and, occasionally, four-knowers—always in that order.

One would expect that the next step is becoming a four- or five-knower, but after three or four, a crucial cognitive change takes place. Children become *CP* knowers: they understand that counting can be used to denote the correct cardinality of any given collection of items. The CP states that if n is followed by m in the counting sequence, a set with cardinality n will have cardinality m if an item is added. While this seems straightforward (indeed, we use this principle without conscious reflection when we count), no other animal seems to understand it, in spite of extensive efforts by experimenters to train animals in numeracy. For example, Ai, a female chimpanzee, learned to use Arabic digits one to nine to denote quantities, but had to learn each separate number by brute association. She did not make the shift to CP at the numbers four or five (Biro & Matsuzawa, 2001).

CP builds on phylogenetically older numerical abilities, which we share with many other animals. Even insects, such as honeybees, can recognize small quantities up to three or four precisely—a capacity that is termed 'subitizing' (e.g., Dacke & Srinivasan, 2008). Humans also subitize: they are more confident, less error-prone, and much quicker when handling small sets up to four items, compared to larger sets (Revkin, Piazza, Izard, Cohen, & Dehaene, 2008). Human newborns can distinguish between sets up to three or four items (Antell & Keating, 1983). Numerical cognition for larger quantities is approximate: for instance, we can see that putting 16 + 16 together is smaller than 40, but without counting, we don't know how much smaller (Pica, Lemer, Izard, & Dehaene, 2004). Young chicks (Rugani, Vallortigara, & Regolin, 2013) and guppies (Bisazza, Piffer, Serena, & Agrillo, 2010) can estimate and compare different quantities, such as four versus eight without any prior training. However, animals are limited in their capacity to distinguish between numerosities larger than three or four exactly. A wide range of studies indicates that animals can distinguish between two and three, but fail to tell the difference between five and six (e.g., Petrazzini, Lucon-Xiccato, Agrillo, & Bisazza, 2015). Animal numerical cognition above three or four is approximate and becomes increasingly less precise as cardinalities increase (see e.g., Scarf, Hayne, & Colombo, 2011 for a comparative study with pigeons and primates).

Since only humans are able to represent cardinalities above three or four precisely, cognitive scientists and philosophers have speculated how children acquire number concepts. Experimental studies on children's numerical cognition suggest that it is a mixture of knowing-that (propositional knowledge) and knowing-how (procedural knowledge). Examples of propositional knowledge are knowing that numbers can be extended indefinitely, or that number words denote precise cardinalities rather than approximate ones. Already before they are able to count, young children realize that if you add items to a collection of 'five', the resulting cardinality cannot be 'five' anymore, and that 'six plus more' is no longer 'six', but that 'a lot plus more' is still 'a lot' (Sarnecka & Gelman, 2004). As children grow older and have more experience with formal mathematics, they learn additional properties of numbers, such as that two is prime, or that three is both prime and odd. Knowledge-how involves elements of the counting routine, such as being able to recite number words from memory in the correct order. It also involves the ability to keep track of what is already counted when counting a collection of items. In many cultures, numerical competence includes being able to use correct procedures, for instance, to make calculations on an abacus or counting board, being able to carry numerals when multiplying, or using the correct procedure for a long division. Such skills require extensive practice, often in a formal school context.

3. The Child as a Lone Mathematician

How do we explain young children's acquisition of knowledge about numbers, in particular the acquisition of CP? Standard accounts tend to conceptualize young children as lone mathematicians who come up with the requisite concepts independently. This is in line with the popular view of the child as a little scientist, introduced by the developmental psychologist Piaget (1929), but going back as early as Rousseau (1762/1999). Piaget was impressed by how quickly children pick up information about their environment, and he was intrigued by the errors they made in the process. These errors suggested to him that children are not passive receptacles of knowledge, but active learners. Children learn by actively engaging with the world and changing their ideas about it as a result of their experience, a bit like scientists who conduct experiments and formulate hypotheses on the basis of them. This view has contemporary defenders, for example, Gopnik, Meltzoff, and Kuhl (2001), who draw an explicit comparison between babies and scientists engaged in individual experimentation. Carey (2009, 20) develops the analogy between scientists and children further, conceptualizing developmental changes in individual children's minds in terms of mini-scientific revolutions, with processes in individual minds being similar to "those described in the literature on history and philosophy of science" (Carey, 2009, 20).

Carey (2004, 2009) regards children's acquisition of the CP as a cognitive revolution. She uses a Quinean bootstrapping account to explicate it. Bootstrapping is the process whereby a child acquires a new concept by first using a system of external symbols (counting words), initially without knowing what they mean. Through analogical and inductive inference, children come to realize what the symbols stand for. First, children learn the counting list as meaningless lexical items. During the next phase, they realize that 'one' corresponds to one item in a collection (e.g., one biscuit), and they do the same with 'two' and 'three'. Children probably rely on their subitizing capacity to associate the number words 'one', 'two', 'three' (and sometimes 'four') with the corresponding quantities. However, subitizing ends at three or four, so for larger quantities, this system does not work anymore, which may explain why children become one-knowers, two-knowers, three-knowers, and, occasionally four-knowers, but never five-knowers. According to the bootstrapping account, it is at this point that a crucial induction takes place: the child notices an analogy between the next in the numeral list and next in series of objects to be counted. She realizes (for small numerosities within the subitizing range) if 'x' is followed by 'y' in the counting sequence, adding an individual to a set with cardinal value x results in a set with cardinal value y. She generalizes this for quantities > 4. In this way, the successor function is established. The successor function plays a crucial role in the Dedekind-Peano axioms of arithmetic—a 19th-century formalization of arithmetic. The successor function is a primitive function, which takes that if a given n is a natural number, so is its successor—i.e., $S(1) = 2$, $S(2) = 3$, and so on. The successor function obeys several of the Dedekind-Peano axioms of arithmetic—for instance, that the successor of a natural number is also a natural number and that there is no natural number whose successor is zero. It is remarkable, given that this function was spelled out only in the 19th century, that 3- and 4-year-old children around the world routinely hit upon this principle through induction.

The bootstrapping account is not without controversy. Rips, Asmuth, and Bloomfield (2006), for example, have argued that it is insufficient to explain how children actually learn the natural numbers. They present a rather contrived case of 3-year-old twins (Fran and Jan) to argue their case. Fran learns the standard natural numbers, while Jan learns a peculiar, modular system in which the count sequence cycles back around from 'nine' to 'none' and then goes back to 'one', 'two', etc., again. Clearly, Jan ends up with a badly underdetermined system since 'two' can refer to collections of 2, 12, 22, . . . items. Although both children have used induction, only Fran learns the natural numbers. Thus, according to Rips et al. (2006), induction by itself does not suffice to establish natural numbers. Rips, Bloomfield, and Asmuth (2008) instead propose that children use abstract mathematical schemas, such as commutativity $(a + b = b + a)$, to understand natural numbers. How do children learn such abstract schemas? According to Rips et al., they do so through physical experiences and manipulating the environment:

> Suppose that children initially notice that two similar sets of objects—for example, two sets of three toy cars—can be matched one-to-one. At a later stage, they may extend this matching to successively less similar objects—three toy cars matched to three toy drivers—and eventually to one-to-one matching for any two sets of three items. This could yield the general concept of sets that can be matched one-to-one to a target set of three objects—a possible representation for three itself.
>
> (Rips et al., 2008, 637)

Again, children are conceptualized as individual discoverers who actively experiment and acquire sophisticated mathematical knowledge through this.

The picture of the preschooler who learns about numbers, including sophisticated concepts such as the successor function and commutativity, is that of the lone mathematician—akin to Srinivasa Ramanujan (1887–1920) who, seemingly on his own, made substantial contributions to several areas of mathematics. The tradition of the child as lone scientist/mathematician is not only descriptive but also normative: children *ought* to learn in this way. Educators, such as Maria Montessori, have devised education systems that encourage children to learn through individual discovery. The idea behind this is that individual discovery is a form of learning that is superior to social learning, which is regarded as mere parroting that precludes a genuine understanding of what is being learned.

4. Learning Through Testimony

An alternative developmental picture sees young children primarily as social learners who benefit from the vast knowledge and experience of adults and older children. This tradition goes back to Thomas Reid (1764) who regarded testimony as one of the most important channels through which we learn; like perception or memory, testimony is a basic source of knowledge. Although we may occasionally be led astray by liars or people who are genuinely mistaken, testimony is an important source of information for facts outside of our immediate experience. Indeed, without testimony, we would not even know the place and date of our own birth. Reid (1764) formulated several arguments for why testimony is a crucial and basic source of knowledge: humans have a natural inclination to believe what they learn from others, and they only become skeptical when they have reasons to be doubtful (e.g., the testifier is a known liar). Young children are more gullible than adults, showing that it is an innate tendency. As he wrote,

> It is evident, that in the matter of testimony, the balance of human judgment is by nature inclined to the side of belief; and turns to that side of itself when there is nothing put in the opposite scale . . . if credulity were the effect of reasoning and experience, it must grow up and gather strength, in the same proportion as reason and experience do. But if it is the gift of nature, it will be strongest in childhood, and limited and restrained by experience; and the most superficial view of human life shows, that the last is really the case, and not the first.
>
> (Reid, 1764, 478–479)

Reid's account of testimony is both descriptive and normative: humans acquire vast stores of knowledge through testimony, and it is proper to do so. For testimony to work, two elements need to be in place: trustfulness and truthfulness, or as Reid calls them, the principle of credulity and the principle of veracity—"two principles that tally with each other" (Reid, 1764, 474). Children need to be willing to learn counterintuitive, strange ideas, such as that the earth is round even though it is in their own experience flat, if they want to learn the correct scientific view of the earth. Parents and teachers must be 'benevolent communicators' (Csibra & Gergely, 2009, 152) who are willing to teach and not to deceive. The strange scenario Rips et al. (2006) offer of one of the twins being taught a modular counting method (by a 'diabolic parent', as they put it) is unlikely to occur in real life.

This Reidian picture of children as trustful social learners is gaining increasing popularity in cognitive science (see e.g., Harris, 2012, for a comprehensive overview). Csibra and Gergely (2009) have argued that humans are natural pedagogues and that communication has evolved to facilitate the transmission of knowledge. Applying this model of the child as social learner to numerical cognition, we come to the view of children as socially embedded mathematicians. Likening children to adult scientists or mathematicians is inherently problematic; for one thing, children do not really conduct experiments. However, even if one were to insist upon such a close analogy, it is important to point out that mathematicians rely on testimony and other forms of socially transmitted information. Thurston (2006) observes,

> Mathematical knowledge can be transmitted amazingly fast within a subfield. When a significant theorem is proved, it often (but not always) happens that the solution can be communicated in a matter of minutes from one person to another within the subfield. The same proof would be communicated and generally understood in an hour talk to members of the subfield . . . When people are doing mathematics, the flow of ideas and the social standard of validity is much more reliable than formal documents. People are usually not very good in checking formal correctness of proofs, but they are quite good at detecting potential weaknesses or flaws in proofs.
>
> (Thurston, 2006, 43, 46)

Geist, Löwe, and Van Kerkhove (2010) argue that mathematicians frequently rely on testimony to decide matters such as the soundness of a mathematical proof. For instance, many proofs rely on theorems that have been proved before, and individual mathematicians using these theorems do not meticulously check for themselves whether these earlier proofs are correct. Moreover, reviewers for papers for mathematical journals do not check all proofs in detail for themselves, but rely on the author's own scrutiny, only checking results in detail when it seems warranted to do so.

Given that children's early knowledge of numerical concepts is largely procedural (correctly using the counting routine to denote cardinalities of sets), one may ask whether testimony is a good model for the acquisition of number concepts. When we think of testimony, we tend to think of propositional knowledge. Indeed, Poston (2016) argues that knowing-how is not transmitted as easily as knowing-that. When I tell my daughter that Jimi Hendrix was a famous guitarist, she thereby comes to know that Hendrix was a famous guitarist. However, when someone demonstrates a guitar riff from Hendrix's repertoire to her, she thereby does not come to know that riff. It would likely take her many weeks to learn the riff, and if her guitar skills are not up to it, she may never learn how to perform it correctly. Through testimony, we can learn fairly complex pieces of propositional knowledge, which does not seem to be the case for knowledge-how. This disanalogy between the transmission of knowing-that (which you can do by simply stating that p) and knowing-how (which you cannot do by simply demonstrating φ) leads Poston (2016) to conclude that knowing-how is not reducible to knowing-that.

While Poston (2016) is correct that learning a skill requires additional resources from the learner (e.g., practice), there are nonetheless striking parallels between verbal testimony and demonstration of knowing-how. Buckwalter and Turri (2014) argue that demonstration has a close analogue to the knowledge norm of assertion. According to the knowledge norm of assertion, when you assert that *p*, you should know that *p* (Williamson, 2000). There are several motivations for this—e.g., it would be paradoxical to say *p*, but I don't know that *p*, and people will often challenge an assertor by asking, "How do you know that?" Similarly, in showing a skill (demonstration), you need to know how to perform that skill. It would be paradoxical to say, "I will show you how to φ, but I don't know how to φ". You can also challenge a demonstrator—e.g., one can ask a bungling ski instructor, "Do you actually know how to ski?"

If we look at testimony in the broader context of social learning, it is clear that demonstrating skills preceded verbal testimony, perhaps by over one million years. Evidence for explicit teaching and demonstration appears as early as 2.5 million years ago. Early human stone tool technology, such as Oldowan (ca. 2.5 million years ago) and Acheulean (ca. 1.5 million years ago), require some form of social learning that almost certainly involved demonstration and very likely also explicit teaching—e.g., how to correctly strike the core (demonstrating the angle with which you have to hit the core so as to obtain a razor-sharp flake that can be used to cut meat), or which pebbles to flake from. Demonstration and trusting demonstration may be a uniquely human capacity. While chimpanzees are sophisticated social learners, they do not faithfully follow demonstrators. For example, Horner and Whiten (2005) showed children and juvenile chimpanzees a complex way to open a puzzle box in order to retrieve a candy. Both chimps and children opened the box in the way that was demonstrated. However, in a second try, a transparent box was used that made it obvious that part of the actions were superfluous. The children continued using the way that was demonstrated to them (this tendency of sticking to a learned suboptimal routine is called overimitation). The chimpanzees went for a more direct way to obtain the candy that deviated from the demonstration. The human default position is one of trust. Children show a high degree of trust when an adult demonstrates for them the 'proper way' to do something: they assume, by default, that demonstration presupposes knowing-how. Overimitation has not only been found in Western children (who might be influenced by schooling) but also in children who live in groups with little formal education, such as in South-African and Australian hunter-gatherer societies (Nielsen, Mushin, Tomaselli, & Whiten, 2014).

Humans are the only species that have cumulative cultural transmission in a variety of domains: they build on the accomplishments of earlier generations and create more complex cultural elements, including artifacts (e.g., the *atl atl*, a small device that attaches to a spear to make it go much further and faster when thrown), ways of preparing food (to make it less toxic, tastier, or easier to digest), and stories (including very long and complex epic tales). Cumulative culture is not altogether absent in nonhuman animals. For example, New Caledonian crows (*Corvus moneduloides*), which use tools made of the edges of *Pandanus* leaves to help them catch invertebrates, use design innovations by conspecifics to improve their own tools (Hunt & Gray, 2003). However, their cumulative culture is limited to the domain of tools. By contrast, human learning takes place in an environment populated with potential teachers and artifacts, and concerns a wide range of domains. According to Sterelny's (2012) apprenticeship model, humans grow up in environments that are 'seeded' with artifacts that help and support learning novel skills and information. Children learn by combining information from the social world with information from the physical world. For example, in a workshop, tools tend to be laid out in an order that facilitates completing specific tasks, such as making or repairing furniture. Through observing accomplished woodworkers or through active teaching, or likely a combination of both, young apprentices learn both the requisite propositional knowledge (e.g., distinguishing between the top rail, splat, and back rail of a chair) and knowledge-how (e.g., how to use a hole-saw properly). In this way, the testimonial transmission of many complex domains of culture requires both a transmission of propositional knowledge and a transmission of skills through demonstration.

5.   The Testimonial Transmission of Number Concepts

We are now in a position to examine how the testimonial transmission of number concepts might work. When children learn to count at about 3.5–4 years of age, they do not yet have a fully fledged concept of natural numbers. For example, children of this age tend to fail the Piagetian conservation task (Piaget, 1941): when one lays out a row of items, they tend to think that its number changes if the spatial configuration of the row changes (e.g., by widening the gaps between the items). Although Piaget's theoretical rationale for why young children fail this task, and only succeed in it about age 5–7, is no longer generally accepted, failure of toddlers to pass the conservation task is robust and has often been replicated. Young CP-knowers do not realize that number words further in the counting list are larger than numbers earlier in the counting list (the later-greater principle), even though it seems like this could be easily inferred from the CP and its applications (Le Corre, 2014).

Moving to more sophisticated concepts, the development of children's concept of zero follows a different trajectory compared to other natural numbers. It seems to proceed in three phases (Wellman & Miller, 1986). In the earliest phase, preschoolers recognize the Arabic numeral 0 and the noun 'zero', but they do not know what it means. In the second phase, children understand zero to refer to nothing, but they do not understand its relationship to other small numbers (e.g., they are equally likely to say that zero is smaller than three, or vice versa, whereas they understand that one is smaller than three). The third phase is when children realize zero is the smallest natural number, and they can make accurate comparisons between zero and other small numbers.

Children's understanding of infinity also appears several years after they learn to count. Many counting systems, be they number words or other ways to denote numbers (e.g., pointing to body parts, where each body part symbolizes a natural number), are quite limited, often not exceeding 20 or 30. In these cultures, people can come up with higher numbers on an ad hoc basis (e.g., counting fingers and toes of several people present). Children's understanding of infinity only arises at about 8 years of age, many years after their ability to count collections has developed. Eight-year-olds have some understanding of potential and actual infinity, but they still do not appreciate that an infinite set is immeasurably bigger than a finite set (for example, they mistakenly think that a very large set, such as the number of grains of sand on a beach, is almost infinite); indeed, even many numerate adults fail to grasp this unbridgeable gap between very large sets and infinite sets (Falk, 2010). It is worth pointing out that the concepts of zero and infinity are lacking in many cultures that use natural numbers (e.g., Roman numerals do not have a symbol for zero). Thus knowledge of numbers is not all-or-nothing. It has many components, such as understanding how counting works, what cardinalities are, and how to perform calculations, as well as more complex, culturally restricted concepts, such as zero and infinity.

How do children learn about natural numbers in a social context? Children tend to grow up in environments that are seeded with artifacts and ideas that help them acquire number concepts. These include not only special learning tools, such as abacuses for children or foam numbers for the bathtub, but also the symbolic representations of numbers that adults use in their day-to-day lives. In this environment, children acquire knowledge of numbers through their parents, older siblings, and more formal learning situations (e.g., preschool and elementary school).

Learning the counting list by rote memory is a crucial step toward learning about natural numbers. Counting is ostensive and deliberate: across the world, there are counting songs and routines to help children memorize the first few counting words. In English, this is usually the first five ("Five Little Monkeys", "Five Little Ducks") or the first ten counting words ("Ten in a Bed", "This Old Man"). During these early encounters, children do not yet grasp the semantic content of these number words, but the songs do help them familiarize themselves with counting routines. Interestingly, such counting routines and songs are absent in languages without exact number words and in homesigners (deaf children and adults who live with hearing parents and carers without using a fully developed sign language) (Spaepen, Coppola, Spelke, Carey, & Goldin-Meadow, 2011).

Levine, Suriyakham, Rowe, Huttenlocher, and Gunderson (2010) recorded 7.5 hours of natural conversations (in several sessions) between parents and toddlers aged 14 to 30 months to understand how talking about numbers ('number talk') influences numerical cognition. Parents were informed that this was a study on language development, but they did not learn that it was specifically about number words, so as not to influence them to use more number words than they would do under ordinary circumstances. The authors found that parents used an average of 90.8 number words over the 7.5 hours recorded. Remarkably, only an average of 6.3 instances of this number talk consisted of prompts where children had to give cardinal values—e.g., "How many cars do you see?" Most of the time, number talk was testimonial. About 50% of number talk by parents consisted of stating cardinal values ("Look, three fish!"); 32% consisted of counting songs, rhymes, and routines. When the toddlers engaged in number talk, about 62% consisted of counting routines and 28% of cardinal values. These utterances demonstrate the importance of transmitting knowing-how to count in children's early acquisition of number concepts. Both the toddlers and their parents included counting routines in their number talk. Levine et al. (2010) found that children whose parents engaged more in number talk performed better at a cardinality test at 46 months. Given that high socioeconomic status (SES) parents tend to talk more to their children, they controlled for the amount of talking and for SES. They found that parents who talked more about number, taking into account SES, had children with more knowledge of the cardinal meanings of number words at 46 months.

Number talk not only helps children to master the counting routine but also helps them to become aware of the fact that number words and symbols refer to discrete quantities. This has been demonstrated in the blocks and water task (Slusser, Ditta, & Sarnecka, 2013) where young children are presented with two bowls, one containing countable items (blocks of different colors) and another containing colored water. The experimenter asks the toddlers either a number question, "Which one has *five*?", or a question that does not involve numbers ("Which one has *orange*?"). N-knowers, who can only enumerate collections up to three or four, do well within their range but perform less consistently at higher numbers (five, six). However, when experimenters start with small numbers and work their way up, these younger children tend to choose the bowl with the blocks when asked, "Which one has five/six", indicating that they have some understanding that these cardinalities refer to discrete entities, not to uncountable quantities. Number talk, where parents, other carers, and older siblings refer to discrete sets, may help young children to understand this property of number words.

Next to linguistic input, artifacts can help support children's developing numerical cognition. Board games, such as the Game of Goose, which often involve numerical properties (e.g., throwing a die that shows n requires the child to move her piece n spaces forward), have a positive impact on children's later mathematical achievements. As Siegler and Booth (2004, 441) point out, this may be because "board games provide children with strongly correlated spatial, temporal, kinaesthetic, and verbal or auditory cues to numerical magnitude". Siegler and Ramani (2009) found that especially designed games that were linear, rather than having numbers in a spiraling or circular pattern, were particularly effective in helping young children from low SES backgrounds to close the numeracy gap with their high SES peers. Under the apprenticeship model of human learning, the environment is seeded with artifacts that help to support the development of numeracy. However, the lack of knowledge of natural numbers in homesigners indicates that an environment seeded with cues for numbers is not sufficient. There needs to be active testimonial transmission as well, which is lacking in these transmission situations where there is no adequate sign language to transmit number concepts.

If the acquisition of number concepts is the result of testimony to skills and propositional knowledge, we can understand why not all human cultures have natural numbers. Some cultures, such as the Pirahã (e.g., Frank, Everett, Fedorenko, & Gibson, 2008) and the Mundurucú (e.g., Pica et al., 2004), lack words to denote exact cardinalities. In a video[1] recorded by the anthropologist Pierre Pica, an old man and the medicine man of a Mundurucú village attempt to count 10 and 13 seeds, with limited success. As can be seen, these men are not used to count—e.g., they do not separate what has already been counted and what still has to be counted, something young children in numerate cultures learn early on. Mundurucú do not have counting routines, although, intriguingly, their language does have a count/mass noun distinction, indicating that they see the difference between discrete and continuous quantities (Pica & Lecomte, 2008). While this distinction is vital for children to understand natural numbers, it is clearly not enough to help them count.

The testimonial model of knowledge transmission can also explain why nonhuman animals fail to understand CP. Learning CP, as we have seen, depends on an understanding of the counting routine. Children learn the counting routine when they are young toddlers, far too early to grasp its meaning. Humans are willing to copy actions even if they do not understand them, which helps them to acquire the correct sequence of counting words. This is a big difference with the chimpanzee Ai who was taught to use Arabic digits in Biro and Matsuzawa's (2001) studies. Ai did not learn the counting routine, but instead relied on estimation to enumerate collections of items and assign them to the correct numeral. As Biro and Matsuzawa (2001) observe, while Ai might have been similar to children in the fact that she learned symbols for numbers, this is just one aspect of learning how to count. She was not raised in a supportive, rich environment filled with numerical cues.

---

1 Pierre Pica, *Mundurucu Seed Counting* (recorded in 2012). www.youtube.com/watch?v=9iXh8wte3gM

6. Conclusions

The question of how children learn number concepts is unresolved. In this chapter, I have argued that testimony plays an important role in the transmission of number concepts, both to knowing-how (the skills involved in counting) and to knowing-that (propositional knowledge about number words, such as that they refer to discrete quantities). Children grow up in a world seeded with artifacts, counting songs, and other cultural features that help them to count. I have pointed out the shortcomings of the model of children as lone mathematicians who discover sophisticated mathematical principles by themselves. In order to understand numerical cognition, we should not only pay attention to what may go on in the minds of young children but also to what goes on in their broader learning environment.

Acknowledgments

Many thanks to Sorin Bangu, Max Jones, Dirk Schlimm, Johan De Smedt, and audiences in Bergen, Norway, and Bristol, UK for their helpful comments on an earlier version of this chapter.

References

Antell, S. E., & Keating, D. P. (1983). Perception of numerical invariance in neonates. *Child Development, 54*, 695–701.

Biro, D., & Matsuzawa, T. (2001). Chimpanzee numerical competence: Cardinal and ordinal skills. In T. Matsuzawa (Ed.), *Primate origins of human cognition and behavior* (pp. 199–225). Tokyo: Springer.

Bisazza, A., Piffer, L., Serena, G., & Agrillo, C. (2010). Ontogeny of numerical abilities in fish. *PLoS One, 5*, e15516.

Buckwalter, W., & Turri, J. (2014). Telling, showing and knowing: A unified theory of pedagogical norms. *Analysis, 74*, 16–20.

Cain, C. R. (2006). Implications of the marked artifacts of the Middle Stone Age of Africa. *Current Anthropology, 47*, 675–681.

Carey, S. (2004). Bootstrapping and the origin of concepts. *Daedalus, 133*, 59–68.

Carey, S. (2009). *The origin of concepts*. Oxford: Oxford University Press.

Csibra, G., & Gergely, G. (2009). Natural pedagogy. *Trends in Cognitive Sciences, 13*, 148–153.

Dacke, M., & Srinivasan, M. V. (2008). Evidence for counting in insects. *Animal Cognition, 11*, 683–689.

Davidson, K., Eng, K., & Barner, D. (2012). Does learning to count involve a semantic induction? *Cognition, 123*, 162–173.

Falk, R. (2010). The infinite challenge: Levels of conceiving the endlessness of numbers. *Cognition and Instruction, 28*, 1–38.

Frank, M. C., Everett, D. L., Fedorenko, E., & Gibson, E. (2008). Number as a cognitive technology: Evidence from Pirahã language and cognition. *Cognition, 108*, 819–824.

Geist, C., Löwe, B., & Van Kerkhove, B. (2010). Peer review and knowledge by testimony in mathematics. In B. Löwe & T. Müller (Eds.), *PhiMSAMP: Philosophy of mathematics: Sociological aspects and mathematical practice* (pp. 155–178). London: College Publications.

Gopnik, A., Meltzoff, A. N., & Kuhl, P. (2001). *The scientist in the crib: What early learning tells us about the mind.* New York: Perennial.

Harris, P. L. (2012). *Trusting what you're told: How children learn from others.* Cambridge, MA: Harvard University Press.

Horner, V., & Whiten, A. (2005). Causal knowledge and imitation/emulation switching in chimpanzees (*Pan troglodytes*) and children (*Homo sapiens*). *Animal Cognition, 8,* 164–181.

Hunt, G. R., & Gray, R. D. (2003). Diversification and cumulative evolution in New Caledonian crow tool manufacture. *Proceedings of the Royal Society of London B: Biological Sciences, 270,* 867–874.

Le Corre, M. (2014). Children acquire the later-greater principle after the cardinal principle. *British Journal of Developmental Psychology, 32,* 163–177.

Levine, S. C., Suriyakham, L. W., Rowe, M. L., Huttenlocher, J., & Gunderson, E. A. (2010). What counts in the development of young children's number knowledge? *Developmental Psychology, 46,* 1309–1319.

Nielsen, M., Mushin, I., Tomaselli, K., & Whiten, A. (2014). Where culture takes hold: "Overimitation" and its flexible deployment in Western, Aboriginal, and Bushmen children. *Child Development, 85,* 2169–2184.

Petrazzini, M. E. M., Lucon-Xiccato, T., Agrillo, C., & Bisazza, A. (2015). Use of ordinal information by fish. *Scientific Reports, 5,* 15497.

Piaget, J. (1929). *The child's conception of the world* (J. Tomlinson & A. Tomlinson, Trans.). London: Routledge and Kegan Paul.

Piaget, J. (1941/1952). *The child's conception of number* (C. Gattegno & F. M. Hodgson, Trans.). London and New York: Routledge.

Pica, P., & Lecomte, A. (2008). Theoretical implications of the study of numbers and numerals in Mundurucú. *Philosophical Psychology, 21,* 507–522.

Pica, P., Lemer, C., Izard, V., & Dehaene, S. (2004). Exact and approximate arithmetic in an Amazonian indigene group. *Science, 306,* 499–503.

Pletser, V., & Huylebrouck, D. (1999). The Ishango artefact: The missing base 12 link. *Forma, 14,* 339–346.

Poston, T. (2016). Know how to transmit knowledge? *Noûs, 50,* 865–878.

Reid, T. (1764). *An inquiry into the human mind, on the principles of common sense.* Edinburgh: Millar, Kincaid & Bell.

Revkin, S. K., Piazza, M., Izard, V., Cohen, L., & Dehaene, S. (2008). Does subitizing reflect numerical estimation? *Psychological Science, 19,* 607–614.

Rips, L. J., Asmuth, J., & Bloomfield, A. (2006). Giving the boot to the bootstrap: How not to learn the natural numbers. *Cognition, 101,* B51–B60.

Rips, L. J., Bloomfield, A., & Asmuth, J. (2008). From numerical concepts to concepts of number. *Behavioral and Brain Sciences, 31,* 623–642.

Rousseau, J.-J. (1762/1999). Emile. In *Oeuvres Complètes* (Vol. 4). Paris: Pléiade.

Rugani, R., Vallortigara, G., & Regolin, L. (2013). Numerical abstraction in young domestic chicks (*Gallus gallus*). *PloS One, 8,* e65262.

Sarnecka, B. W. (in press). Learning to represent exact numbers. *Synthese.*

Sarnecka, B. W., & Gelman, S. A. (2004). Six does not just mean a lot: Preschoolers see number words as specific. *Cognition, 92,* 329–352.

Scarf, D., Hayne, H., & Colombo, M. (2011). Pigeons on par with primates in numerical competence. *Science, 334,* 1664–1664.

Siegler, R. S., & Booth, J. L. (2004). Development of numerical estimation in young children. *Child Development, 75,* 428–444.

Siegler, R. S., & Ramani, G. B. (2009). Playing linear number board games—but not circular ones—improves low-income preschoolers' numerical understanding. *Journal of Educational Psychology, 101,* 545–560.

Slusser, E., Ditta, A., & Sarnecka, B. (2013). Connecting numbers to discrete quantification: A step in the child's construction of integer concepts. *Cognition, 129,* 31–41.

Spaepen, E., Coppola, M., Spelke, E. S., Carey, S. E., & Goldin-Meadow, S. (2011). Number without a language model. *Proceedings of the National Academy of Sciences, 108,* 3163–3168.

Sterelny, K. (2012). *The evolved apprentice: How evolution made humans unique.* Cambridge, MA: MIT Press.

Thurston, W. (2006). On proof and progress in mathematics. In R. Hersh (Ed.), *18 unconventional essays on the nature of mathematics* (pp. 37–55). New York: Springer.

Tomasello, M., Carpenter, M., Call, J., Behne, T., & Moll, H. (2005). Understanding and sharing intentions: The origins of cultural cognition. *Behavioral and Brain Sciences, 28,* 675–691.

Wellman, H. M., & Miller, K. F. (1986). Thinking about nothing: Development of concepts of zero. *British Journal of Developmental Psychology, 4,* 31–42.

Williamson, T. (2000). *Knowledge and its limits.* Oxford: Oxford University Press.

Wynn, K. (1990). Children's understanding of counting. *Cognition, 36,* 155–193.

10 Which Came First, the Number or the Numeral?

Jean-Charles Pelland

Introduction

Despite the tremendous progress made by research into the origins of numerical cognition since the late 20th century, an important developmental question remains unanswered: given the limitations of our innate cognitive machinery, how did the content of advanced number concepts emerge? While many have attempted to address this issue (e.g., Gelman & Gallistel, 1978; Gallistel, Gelman, & Cordes, 2006; Hurford, 1987; Dehaene, 1997/2011; Butterworth, 1999; Lakoff & Núñez, 2000; Wiese, 2004; De Cruz, 2008; Rips, Asmuth, & Bloomfield, 2008; Carey, 2009; Coolidge & Overmann, 2012; Menary, 2015; Malafouris, 2010), most proposed solutions rely on the presence of numerical symbols in the environment, often in linguistic format, to bridge the gap between the output of our innate representational systems and the content of our advanced number concepts. In this chapter, I will argue that this externalist approach needs to be replaced by one that focuses more on internal cognitive processes, since any appeal to external symbols for numbers must come *after* we have explained the emergence of numerical cognition *internally*, given that external symbols for numbers depend on the construction of internal representations with numerical content for their existence. My main point is methodological: I claim that relying on external symbols for numbers in explaining what makes advanced numerical cognition possible leads us to an incomplete—or, worse, incoherent—account of the origins of our ability to think numerically, by taking for granted the very thing we are trying to explain.

In the first part of the chapter, I briefly discuss relevant data concerning numerically tuned neural systems and describe how their limitations imply a gap with the content of our more advanced number concepts. In the second part, I summarize what I consider to be some of the main theories that have attempted to explain how to bridge this gap—namely, Stanislas Dehaene's number line (1997/2011), Susan Carey's Core Cognition (2009), and Helen De Cruz's Darwinian approach (2006, 2008)—and highlight their reliance on external symbols for numbers. I argue in Section 3 that relying on such symbols to explain the development of number concepts puts the cart before the horse and leaves out important details concerning where numerical content comes from. Finally, Section 4 discusses potential replies and shows that none of them allows externalism to give us a complete account of the origins of numerical cognition.

1.   What Are We Trying to Explain?

In recent years, research in a wide variety of fields has produced vast amounts of data strongly supporting the existence of (at least) two neural systems that are in some way dedicated to presenting information about quantities of objects in our environment (Feigenson, Dehaene, & Spelke, 2004).[1] Evidence suggests that human infants, adults, and nonhuman animals all have an innate representational system designed to give them a rough idea of how many things there are in the part of their environment they are paying attention to, regardless of modality. An important feature of this system is that the precision of the representations it delivers decreases as cardinality of the attended stimulus increases, and its performance follows Weber's law, displaying size and distance effects. This system is often labeled the approximate number system (ANS) to reflect this limited precision. Another representational system that plays an important role in many accounts of how we develop number concepts is the object-file system (OFS). Though it seems more likely that this system is dedicated to maintaining online representations of small collections of objects rather than to representing their cardinality itself (Simon, 1997; Carey, 2009, 151), considerable data nevertheless suggests that it allows subjects (whether infant, adult, or nonhuman) to track the number of stimuli to which they are attending, if this number is kept below four (Kahneman, Treisman, & Gibbs, 1992; Hauser, Carey, & Hauser, 2000).

While both the ANS and the OFS are involved in representing information about quantities of things in our environment, their obvious size and precision limitations suggest that a major change must occur in order to allow us to develop the fully fledged number concepts used in arithmetic, whose precision and infinite extension cannot be accommodated by either innate system. Without such a major change, it is mysterious how we could even come to think about the number six, say, given that neither system can explicitly represent precise numbers larger than four. There is thus a significant gap between the content generated by the neural systems we are born with and the content we use in precise arithmetical thinking. In the next section, I summarize three well-received proposals that have tried to explain how we bridge this gap.

---

1 For a full review, see Carey, 2009, ch. 4 and Dehaene, 1997/2011, section 1. For a critical review that questions whether the ANS reacts to approximate numbers of objects instead of other continuous magnitudes like size or length, see Leibovich, Katzin, Harel, and Henik (2017). For a discussion of the format of the representations of the ANS, see Ball (2016).

2. Review of Recent Accounts of the Origin of Numerical Cognition

2.1 *Dehaene's Number Line*

Stanislas Dehaene's (1997/2011) work on the cerebral bases of number arguably laid the groundwork for much of the current research on the biological origins of number. To explain the ontogenetic emergence of numerical cognition, Dehaene holds that when children learn to count, numerals such as those in a count list are mapped onto representations of the ANS. Learning the meaning of the number words is then a matter of mapping these to a precise location on an innately given number line.[2] If Dehaene is right, without precise symbols, there would be no advanced number concepts:

> Certain structures of the human brain that are still far from understood enable us to use any arbitrary symbol, be it a spoken word, a gesture, or a shape on paper, as a vehicle for a mental representation. Linguistic symbols parse the world into discrete categories. Hence, they allow us to refer to precise numbers and to separate them categorically from their closest neighbours. Without symbols, we might not discriminate 8 from 9.
>
> (Dehaene, 1997/2011, 79)

There are indeed good reasons to think that language is involved in the development of arithmetic (Donlan, Cowan, Newton, & Lloyd, 2007; Sarnecka, Kamenskaya, Ogura, Yamana, & Yudovina, 2007). For example, there is evidence of a strong dissociation between the parts of the brain involved in approximate number representations and those used for precise arithmetic, which appears to involve cognitive resources often associated with language (Dehaene, Spelke, Pinel, Stanescu, & Tsivkin, 1999).[3]

As for the historical development of our numerical abilities, Dehaene describes a few key steps that suggest how numerals came to invade our environment (Dehaene, 1997/2011, 80–81). At first, we may only have had words for the first few numbers (1, 2, 3), since these are associated with easily perceived aspects of the environment:

> When our species first began to speak, it may have been able to name only the numbers 1, 2, and perhaps 3. Oneness, twoness, and threeness

---

2 Dehaene presents data suggesting that the ANS takes the form of an analog number line with logarithmic structure (Dehaene, 2003; Izard & Dehaene, 2007). Núñez (2011) and Núñez et al. (2012) have challenged the universality of this number line and its logarithmic format.

3 This being said, the role of language in advanced numerical cognition remains a controversial subject, given that there is also evidence that some aspects of mathematical reasoning are independent from language (e.g. Brannon, 2005; Amalric & Dehaene, 2016). For a discussion of why language might not be necessary for the development of number concepts, see Overmann (2015) and Malafouris (2010).

are perceptual qualities that our brain computes effortlessly without counting. Hence, giving them a name was probably no more difficult than naming any other sensory attribute, such as red, big, or warm.

(Dehaene, 1997/2011, 80)

There is indeed evidence that number words did not come out all at once, as can be seen in persisting discontinuities in many languages between the first few number words and the rest of the count list (Hurford, 1987).[4]

At a later stage, we mapped quantities in the environment to parts of our body, to which we pointed when we wanted to designate a number in particular. Using our body parts was easy to imitate and would have come naturally: "All children spontaneously discover that their fingers can be put into one-to-one correspondence with any set of items" (1997/2011, 81). Later, we started using words for these body parts, allowing us free use of our hands when communicating numerical content. Finally, since the nonrecursive nature of words linked to body parts meant limited efficiency in commercial exchanges, we developed a more generative numerical syntax. Many languages still display relics of this era, with number words containing references to body parts in their etymology (Ifrah, 1998; Dehaene, 1997/2011; Hurford, 1987).[5]

2.2   Carey's Core Cognition

In her 2009 opus *The Origin of Concepts*, Susan Carey goes through an impressive body of empirical data, with the aim of explaining how it is possible to develop concepts whose expressive power exceeds that of the resources with which they were built. According to Carey, we are born with innate perceptual input analyzers that take data from the senses to yield representations with conceptual content. Among the core concepts whose origin falls into this category are AGENT, OBJECT, and NUMBER.[6]

As Carey sees it, to bridge the gap and explain how to develop concepts whose expressive power is incommensurate[7] with that of its building blocks, we must appeal to what she calls 'Quinian bootstrapping'. A key element in Quinian bootstrapping is the learning of lists of symbols, which, at first, have no meaning apart from their place in the list. In the case of numbers, this is the count list (1, 2, 3, 4 . . .), which children learn to recite long before they have any developed number concepts. Once children know this list by heart, there is a slow piecemeal process in which they learn the meaning of the first few words in the list. It's only many months after they

---

4 See also Overmann (2014).
5 For example, the word 'digit' refers to both numerals and fingers.
6 Names of concepts are in capital letters.
7 A conceptual system is incommensurate with another if it is impossible to express its content using the vocabulary of the other.

have learned this list that children start correctly applying the word 'one'. Months later, they form the ability to use 'two' and, after another prolonged period, 'three' (Wynn, 1992). From this partial understanding of their count list, children eventually come to use all symbols in their count list correctly and realize that the next element in the list represents the next number.

To explain how children generalize from partial meanings to the meaning of the list as a whole, Carey appeals to non-deductive modeling processes that are used in many problem-solving mechanisms, including induction.[8] According to Carey, it is language—or, rather, the singular/plural distinction and other quantifiers in natural languages—that allows us to notice analogies between our count lists and the sets of objects being tracked by our OFS. The idea here is that children would learn that certain regularities in how words are used are analogous to regularities in quantities of objects tracked by their OFS, which would allow them to generalize that going to the next element in the list of number words means adding one object. Natural language quantifiers help make the numerical properties of elements within the count list more salient by highlighting the analogy between adding one item to the object file and going to the next element in the list.

Initially, the singular/plural distinction, more prevalent in our linguistic environment, allows us to notice similarities between words being used and attending to one object. Other quantifiers such as 'both' and 'a pair' also come to solidify this analogical learning until the key induction to the successor principle takes place. Note that, unlike Dehaene, Carey holds that the ANS plays no part in this process. Only the OFS, linguistic set-based quantification, and 1 – 1 correspondence are needed to explain ontogenetic development of number concepts on Carey's model.

2.3 De Cruz and the Extended Mind

Helen De Cruz's (2006, 2008) Darwinian approach to numerical cognition rests on the assumption that the advanced numerical cognition involved in doing arithmetic is constituted by the interaction between humans and external numerical symbols. While the idea that mathematical practice depends on external objects is not new, it is important to realize that on this framework, external artifacts are considered *constitutive* parts of cognitive processes rather than mere triggers for internal processes or essential mnemonic devices.[9]

8 Carey (2009, 307). Carey notes that there is evidence that children use semantic and syntactic bootstrapping in many other cases of conceptual development. More details on Carey's account of the bootstrapping of number concepts can be found in Carey (2009, 325–329) and Carey (2011), 120. See also Shea (2011).

9 See Macbeth (2013) for a discussion of the constitutive role of notation in mathematical practice, including Kant's and Rotman's view that notation is constitutive of this activity. For more on what form this constitutivity may take, see Dutilh Novaes (2013).

The inspiration for this approach is taken from Clark and Chalmers' (1998) proposal that the mind can extend beyond the barriers of our skull to include parts of our environment. Consider Otto, one of the many intuition pumps that Clark and Chalmers introduce to motivate their controversial idea. Otto suffers from Alzheimer's disease and carries a notebook in which he has written useful information to compensate for his memory loss. According to Clark and Chalmers, Otto's consulting of his notebook to retrieve information is functionally comparable to that of someone looking up the information in his or her biological memory, and they take this to mean cognition can include things outside our heads.

Without entering the debate about whether or not this view makes sense (for this, see Menary, 2010), we can see how De Cruz wants to apply it to numerical cognition. Here the idea is that external representations of numbers—whether in the form of body parts, tally systems, or numerals—are constitutive parts of extended cognition systems in which numerical symbols play an essential part:

> External symbolic representations of natural numbers are not merely converted into an inner code; they remain an important and irreducible part of our numerical cognition . . . During cognitive development, the structure of the brain is adapted to the external media that represent natural numbers in the culture where one is raised. In this way, the interaction between internal cognitive resources and external media is not a one-way traffic but an intricate bidirectional process: we do not just endow external media with numerical meaning, without them we would not be able to represent cardinalities exactly.
>
> (De Cruz, 2008, 487)[10]

As we grow up, our brain adapts to the dependable presence of external numerical symbols in our environment and forms couplings with these. In a very real sense, then, our brains are wired in a way that is reflective of the numeral-enriched environment in which we grew up.

Evidence for this view can be seen in the fact that cultural variability in external artifacts affects individuals' performance in various mathematical tasks (De Cruz, Neth, & Schlimm, 2010) as well as which parts of the brain are used in certain operations. For example, Tang et al. (2006) obtained data that suggest mental calculation recruits different networks in subjects from communities that learned how to calculate using abacuses when compared to subjects from communities that learned how to calculate using multiplication tables. The difference in activation patterns likely

---

10 Malafouris (2010), Coolidge and Overmann (2012), and Menary (2015) also apply variants of extended cognition to their accounts of numerical cognition.

reflects the difference in external artifacts used when learning how to calculate, showing that our brain has integrated external object manipulation in its circuitry.

3. Putting the Cart Before the Horse: Why Symbols Can't Bridge the Gap

While they may disagree on which cognitive systems play what part in the development of number concepts, the three theories we just looked at all rely on the presence of external symbols with numerical content at some point in their explanation of how mature number concepts emerge. In this section, I discuss why such externalism is problematic. The problem I want to talk about is this: how, in each of these theories, can we rely on external symbols for numbers in our explanation of the development of numerical content when the existence of such symbols in turn depends on the existence of number concepts?

For starters, consider the role played by integer lists in Carey's ontogenetic account. The problem here is, how could such numeral lists possibly emerge without someone first having had some kind of number concepts? The fact that someone (or, much more likely, many people, in a gradual process of personal innovation and cultural transmission) had to come up with this counting routine means that it was possible to think about numbers (or perhaps, more basically, about precise quantities) without relying on words. This in turn suggests that there can be individual-level development of number concepts without external support, which would seem to imply that some other artifact-free cognitive processes are involved in the origin of basic number concepts. But if we must appeal to such internalist processes to describe the origin of number concepts, then how can externalism ever hope to offer a complete picture of the origin of numerical cognition?

My claim is that, if we are trying to explain the ontogeny of number concepts, our theory should apply to *everyone* capable of thinking about numbers. But since some people seem to have been able to think about numbers without external aids in the (distant) past, any account that depends on such support will not apply to every case of numerical cognition.[11] At best, such externalist accounts could describe how numerical cognition emerges in a numeral-enriched environment. Even so, the fact that it is possible to

---

11 While Coolidge and Overmann (2012), Overmann (2015), and Malafouris (2010) offer similar lines of reasoning concerning the emergence of numerical cognition in numeral-impoverished environments, their criticism is mostly aimed at the role of language in this process. Importantly, these accounts also rely on external symbols—either in the form of fingers or clay tokens—to explain how we transcended the limitation of our innate cognitive systems. The point I am making here goes further and applies to *any* external representation with numerical content, including numerals and number words, but also clay tokens, body parts, and other 'enactive signs' (Malafouris, 2010).

develop some basic number concepts without external support seems to sug-gest that cases that do involve external support might somehow appeal to a more fundamental process, which the externalist framework is leaving out. So while it may seem unfair to Carey to criticize her for not taking into consideration historical development, given that her theory is aimed at the individual, ontogenetic level, there is reason to do so: the ontogeny of number concepts in a world where symbols for numbers abound cannot be completely separated from past cases of numeral-free ontogeny since the former depends on the latter in important ways.

This line of thinking was perhaps behind Overmann, Wynn, and Coolidge's (2011) comment on Carey's *précis* of her 2009 book when they ask, "In the absence of a numeral list, how could a concept of natural num-ber ever have arisen in the first place?" (Overmann et al., 2011, 142) Carey appeals to extended cognition in a cultural setting to answer this question:

> Understanding the invention of tally systems would involve understand-ing how people came to the insight that beads could serve this symbolic function, rather than decorative uses, or as markers of wealth, or myr-iad others. That is, the availability of an artifact that could serve as the medium of a tally system doesn't explain how it came to be one. Now that we are in the realm of speculation, I believe, contra Overmann et al., that body counting systems could well also play an extended cogni-tion role in the cultural construction of integer representations.
> (Carey, 2011, 159)

Unfortunately, it is difficult to see how this can tell us the whole story, since this simply raises the question of how these body-counting systems emerge, if not by some kind of purely blood-n-bones intracranial cognitive process. Assuming that our first number symbols were parts of the body, how did we come to point to these and intend to communicate numerical content without first having developed this numerical content by some other cogni-tive process that did not rely on external symbols for number? For example, how could a person point to their right knee or left thumb to communicate 'six' without first having come to some understanding of SIX? Whatever process allowed such symbols to emerge, shouldn't *that* be where we look to find the key to the mystery of the development of numerical content?

Embracing an extended cognition framework does not help here, since accepting external objects as constitutive parts of cognitive systems does not tell us where the content associated with these objects come from. Coming back to Otto, it is important to notice that everything he wrote in his notebook got there because it was in his head beforehand. Applying this analogy to the case of (extended) numerical cognition seems to suggest that we must look in our heads to find out where numerical symbols and artifacts come from.

Even if we wish to deny that the content involved in using body parts is fully numerical—it seems likely, after all, that these body parts only required mastery of the concept PRECISE QUANTITY, for example, without the formal properties associated with more advanced number concepts—it is still mysterious how such content could arise from any of the innate systems mentioned earlier. Dehaene's claim that perceptual qualities such as oneness and twoness can be computed effortlessly may indeed be right, but that does not explain how we could get concepts such as NUMBER or PRECISE QUANTITY from these. The fact that some innumerate cultures such as the Pirahã and the Mundurukú have words for only the first few numbers and yet no developed number concepts (e.g., Gordon, 2004; Pica, Lemer, Izard, & Dehaene, 2004) tells us it is not enough to have hands, tools, and even language to come up with concepts such as NUMBER.[12]

4. Potential Replies

If I am right to be worried about the problem outlined in the previous section, it looks like externalist theories of numerical cognition have swept a crucial historical question under the explanatory rug: how did the first NUMBER emerge? This elicits another question: what is the relationship of this primitive number concept to the one we learn today? In this section, I propose two potential externalist answers to these questions and argue that they cannot explain the origin of numerical content. The first answer tries to deny that the only source of symbols with numerical content is in our head by providing potential cases of novel content without internal support. My rebuttal involves showing that none of the cases considered involve the right kind of content novelty. The second externalist response is that we can separate pre- and post-numeral cases of ontogeny, and that we can explain the difference between these via mechanisms of cultural evolution. Against this, I maintain that there is ontogenetic continuity in the historical development of numerical cognition and that appealing to cultural evolution brings us back to individual-level construction of novel content without external support.

4.1 Symbols and Minds

One way to show how externalism could explain the origin of numerical content is to deny that the only way a person can use an external object as a symbol for number is by first having an internal representation of its content. This way, externalists can appeal to symbols for number without having to explain their origin in terms of intracranial processes. It seems true that there are, after all, many cases of novel symbols arising externally. For example, we need not stretch our imagination very far to conceive

---

12 Similarly, Roberson (2009, 172) describes cultures with words for certain colors, but no word for the concept COLOR itself, suggesting that generalizations and category abstractions don't come cheap.

of a cat stepping on a keyboard and accidentally typing a numeral that has never before been thought of by a human. No one has ever thought about such content, by hypothesis, and yet the symbol that expresses it has come to be, so this might look like a case of novel content without internal representation.

However, there is reason to doubt that this example involves the right type of content novelty. After all, the novel content in our feline case is only possible to the extent that it is part of an already existing system capable of generating endless meaningful symbols. But as we saw in our discussion of Carey's bootstrap, what we are looking for is something that can explain discontinuities between conceptual *systems*. Keeping this in mind, such examples can't cause much harm since the underlying representational system is already in place, and any 'novel' content seems very much within the expressive power of the system that generates it. On the other hand, given their limitations, there is no way the ANS or the OFS could accidentally generate anything as precise as the concept EIGHT, so that symbols for this concept could not have been the result of some form of psychological 'accident'.

Perhaps a better example of this possibility would be the symbolic estrangement proposed by Lyons, Ansari, and Beilock (2012). Here the idea is that symbols for numerical quantities gradually become estranged from the mappings to the ANS as we start to consider very large numbers (e.g., 1,000,000 and 1,000,001), to the point where the meaning associated with these symbols are not quantities in our heads, but relations to other external symbols. However, in this case, too, even if symbols for large numbers come to refer to other symbols, the system in which this change in meaning takes place is grounded in the smaller representations that are mapped to things in our heads, so we cannot consider this to be an example of a symbol whose meaning was determined independently of (internal) mental content, nor as a case of conceptual discontinuity.

While granting that accidental or estranged content cannot help the externalist, some may be tempted to interpret certain historical shifts in mathematical notation as examples of symbols being used before having any mental content associated with them. Dutilh Novaes, for example, writes, "There seem to be a number of examples in the history of mathematics where specific notations were adopted even before it became clear which concept(s), if any, they singled out" (2013, 56). As examples, she cites the fact that zero was only thought of as a number hundreds of years after it had been used as a notational placeholder in calculations and that the ambiguous ontological status of infinitesimals did not prevent people from using the associated notation. However, in both cases, we are dealing not with symbols without any previously constructed mental content but rather symbols associated with mental content that either changes over time or is not precisely determined with respect to the system in which they have been introduced.

4.2 You Say 'Number', I Say 'Precise Quantity'

Another option to save externalism would be to propose that how number concepts arose in individuals in the past is different from present-day ontogenesis. While this is certainly appealing, it is difficult to see what reasons would motivate us to tell two separate developmental stories, given that it looks like we are talking about the same cognitive ability—namely, the ability to think about numbers. Even considering the cultural variation described earlier between cultures who learn to calculate via abacuses and those who use multiplication tables, there is compelling evidence suggesting that the same brain regions are involved in more elementary tasks, such as number comparisons (e.g., Feigenson et al., 2004; Dehaene, 1997/2011), which suggests that cultural variation may affect which brain areas are used for certain number-related tasks (e.g., calculation) rather than which processes underlie their ontogeny.[13] Similar considerations applied to the domain of color further weaken the appeal of this reply: despite living in different environments from those in which people lived thousands of years ago, we wouldn't want to say that the process of developing concepts for individual colors, say, was radically different in the past from how it works now. The same sort of considerations would seem to block divorcing present from past ontogeny for the case of numerical cognition.

This reply still holds some promise, though: the output of the process that allowed us to develop concepts such as PRECISE QUANTITY is different from that which allowed us to get NUMBER, so what reason would we have to think the same process applies in both cases? While there is little doubt that the historical emergence of numerical cognition proceeded in successive steps over great many generations, the continuity of underlying ontogenetic processes that I am proposing can accommodate such distinct outputs at different historical stages of complexity, including one person's realization that there is such a thing as precise quantities, while another later coming to realize that such precise quantities can increase endlessly, say.

To see how this is a legitimate possibility, consider again the case of perception: presumably, the same internal process that yields RED is also involved in producing other color outputs, such as PINK and BLUE. While this process is (presumably) not the same as the one that underlies the formation of COLOR, it seems plausible that the same process involved in the emergence of COLOR would also underlie DARK COLOR, for example. Similarly, the same process could be behind the concepts NUMBER and PRECISE QUANTITY, despite the fact that the content is not the same. Thus, even if, historically, there was a stepwise progression, I speculate that

13 Recent evidence (Amalric & Dehaene, 2016) also suggests that expert mathematicians working in a wide variety of fields recruit the same neural systems involved in basic number comparison tasks.

increasing demands for attention to quantitative stimuli could have been inputs to the same cognitive process at every level.

Even assuming we could come up with different ontogenetic stories, say, by appealing to environmental differences and their influence on ontogeny, another problem with this divorce is that it would force us to explain how early versions of number concepts evolved into more advanced ones and how both have equal claim to being number concepts. To describe how the different stages of numerical cognition are related to one another, externalists could perhaps appeal to mechanisms of cultural evolution. The next subsection considers this possibility and shows that appeals to processes of cultural evolution cannot help the externalist give a full account of the different stages of evolution of the concept of number, as would be required by the proposed divorce of past and present ontogeny.

4.3   Cultural Evolution and Numerical Cognition

Talk of cumulative culture taps into an important intuition concerning how number concepts emerged: it is patently false to claim that these appeared fully formed, complete with all their formal properties, in a single individual's head.[14] Rather, it seems more appropriate to describe the history of mathematics as one of individuals reflecting upon historically constructed ideas and adding their bit to an increasingly large body of knowledge. At no point does the emergence of novel mathematical content require a person to reinvent the whole body of mathematics, and the same would seem to apply to the simpler concepts involved in numerical cognition.

Given that mechanisms of cultural evolution can explain how content changes and evolves over generations, the externalist could appeal to these to explain how number concepts evolved, and we could thereby accept two (or more) ontogenetic stories to accommodate various levels of mathematical knowledge. Questions about the increasing role of numerical symbols in numerical cognition could be answered by mechanisms of cultural evolution, transmission, and inheritance (e.g., Dawkins, 1976/2006; Aunger, 2001; Richerson & Boyd, 2005; Sperber, 1996). Such mechanisms could perhaps explain how symbols in the environment could have acquired novel numerical content. For example, symbols for approximate quantities could have gradually spread and evolved due to their increasing usefulness in advancing societies.

In considering this option, however, we should keep in mind that, for the most part, theories of cultural evolution focus on mechanisms of transmission and inheritance at the *population* level (Richerson & Boyd, 2005),

---

14 This piecemeal historical process is perhaps what Carey is referring to when she mentions "the *cultural* construction of integer representations" (Carey, 2011, 159, emphasis mine) in the passage quoted earlier. For more on mechanisms of cumulative culture, see Charbonneau (2015).

often neglecting the mental states of the individual (Kirkpatrick, 2009; Sperber, 2006) and the mechanisms responsible for the generation of novel content (Charbonneau, 2015, 2016). And yet, without reference to an individual-level psychological process, it is difficult to see how a purely population-level description of the evolution of a concept could explain how its content changes. Much like genetic change in a species is explained in terms of genetic mutations in individuals, mechanisms of cultural evolution rely on individual-level psychological processes in their explanation of where cultural innovation and change come from. One person innovates, and, if others understand and value the innovation, it can spread via various mechanisms of cultural transmission.[15] The innovation itself, however, originates at the individual, psychological level. So if we want to understand how numerical cognition evolved over generations, we must first understand how it could have arisen through psychological processes in individuals. For this, population-level mechanisms of cultural evolution cannot help.

An externalist could try and explain the evolution of numerical content by appealing to transmission or imitation errors, or perhaps to cultural mutation, but this would not change the fact that these would have taken place in someone's head—initially, at least, in a numeral-free environment—and that any change in content must be explained at the psychological, individual level rather than in reference to mechanisms of cultural evolution. To see why this is true, it is important to note that while some innovations can be the result of imitation error, such modification by error does not seem to apply to the spread of numerical content. This is because there is good reason to think that the spread of conceptual content occurs via internal reconstruction, not mere imitation: "While the propagation of word sound may be seen as based on copying, that of word meaning cannot: it is re-productive, in the sense that it necessarily involves the triggering of constructive processes" (Claidière, Scott-Phillips, & Sperber, 2014, 3). These considerations show that an appeal to cultural evolution cannot help motivate the externalist's two-stories answer to the origin of numerical content since it cannot explain the innovation responsible for the emergence of numerical content in a world without numerical symbols, nor its spread via reconstruction, thus leaving out important details about the origin of numerical cognition.

5. Concluding Remarks

In the previous section, I explored potential replies to my charge that externalism about numerical cognition leaves out important details about the origin of numerical content. I argued that these replies were unsatisfactory by relying on claims about the relation between mental content

---

15 See Richerson and Boyd (2005) for great examples that illustrate the role of the individual in cultural evolution.

and symbols, the continuity of ontogenetic processes through the historical development of numerical cognition, and the role of individual-level psychological processes in mechanisms of cultural evolution. Even if these claims are mistaken, the externalist replies I considered could only help rid us of the apparent methodological circularity involved in taking a manifestation of numerical cognition to be one of its causes, and thus externalism would still not have a story about the origin of numerical content. Thus a different internalist approach to the origin of number concepts seems warranted, if only to explain the original development of numerical content that externalism leaves out.

If I am right, however, the internalist approach would apply to all cases of ontogenetic development of numerical content. Note that this would not mean that externalism needs to be abandoned. On the contrary, examining the causal role played by external symbols in numerical cognition can help us understand the internal processes involved in the development of number concepts, since knowledge of this causal process may help identify internal analogues that play the same role as external symbols. So while relying on external symbols might not give us the whole story, it can move us in the right direction.

Acknowledgments

Research for this chapter was supported by grant 163129 *from the Fonds de recherche du Québec—Société et culture (FRQSC)*. Thanks to Ophelia Deroy and the Institute of Philosophy (London) for their support. Many thanks to Mathieu Marion and Brian Ball for comments on an earlier version of this chapter.

References

Amalric, M., & Dehaene, S. (2016). Origins of the brain networks for advanced mathematics in expert mathematicians. *Proceedings of the National Academy of Sciences of the United States of America, 113*, 4909–4917. doi: 10.1073/pnas.1603205113

Aunger, R. (Ed.). (2001). *Darwinizing culture: The status of memetics as a science.* Oxford: Oxford University Press.

Ball, B. (2016). On representational content and format in core numerical cognition. *Philosophical Psychology*, 1–21. doi: 10.1080/09515089.2016.1263988

Brannon, E. M. (2005). The independence of language and mathematical reasoning. *Proceedings of the National Academy of Sciences of the United States of America, 102*(9), 3177–3178.

Butterworth, B. (1999). *The mathematical brain.* London: Macmillan.

Carey, S. (2009). *The origin of concepts.* New York: Oxford University Press.

Carey, S. (2011). Précis of the origin of concepts. *Behavioral and Brain Sciences, 34*(3), 113–124. doi: 10.1017/S0140525X10000919

Charbonneau, M. (2015). All innovations are equal, but some more than others: (Re)integrating modification processes to the origins of cumulative culture. *Biological Theory, 10*(4), 322–335.

Charbonneau, M. (2016). Modularity and recombination in technological evolution. *Philosophy & Technology, 29*, 373–392.

Claidière, N., Scott-Phillips, T. C., & Sperber, D. (2014). How Darwinian is cultural evolution? *Philosophical Transactions of the Royal Society B: Biological Sciences, 369*(1642). doi: 10.1098/rstb.2013.0368

Clark, A., & Chalmers, D. (1998). The extended mind. *Analysis, 58*, 7–19.

Coolidge, F. L., & Overmann, K. A. (2012). Numerosity, abstraction, and the emergence of symbolic thinking. *Current Anthropology, 53*(2), 204–225. doi: 10.1086/664818

Dawkins, R. (1976/2006). *The selfish gene*. Oxford: Oxford University Press.

De Cruz, H. (2006). Why are some numerical concepts more successful than others? An evolutionary perspective on the history of number concepts. *Evolution and Human Behavior, 27*, 306–323.

De Cruz, H. (2008). An extended mind perspective on natural number representation. *Philosophical Psychology, 21*(4), 475–490.

De Cruz, H., Neth, H. & Schlimm, D. (2010). The cognitive basis of arithmetic. In B. Löwe & T. Müller (Eds.), *Philosophy of mathematics: Sociological aspects and mathematical practice* (pp. 59–106). London: College Publications, London. Texts in Philosophy 11.

Dehaene, S. (1997/2011). *The number sense: How the mind creates mathematics*. New York: Oxford University Press.

Dehaene, S. (2003). The neural basis of the Weber-Fechner law: A logarithmic mental number line. *Trends in Cognitive Sciences, 7*, 145–147.

Dehaene, S., Spelke, E., Pinel, P., Stanescu, R., & Tsivkin, S. (1999). Sources of mathematical thinking: Behavioral and brain-imaging evidence. *Science, 284*, 970–974.

Donlan, C., Cowan, R., Newton, E., & Lloyd, D. (2007). The role of language in mathematical development: Evidence from children with specific language impairments. *Cognition, 103*(1), 23–33.

Dutilh Novaes, C. (2013). Mathematical reasoning and external symbolic systems. *Logique & Analyse, 56*(21), 45–65.

Feigenson, L., Dehaene, S., & Spelke, E. (2004). Core systems of number. *Trends in Cognitive Sciences, 8*, 307–314.

Gallistel, C. R., Gelman, R., & Cordes, S. (2006). The cultural and evolutionary history of the real numbers. In S. C. Levinson & P. Jaisson (Eds.), *Evolution and culture* (pp. 247–274). Cambridge, MA: MIT Press.

Gelman, R., & Gallistel, C. R. (1978). *The child's understanding of number*. Cambridge: Harvard University Press/MIT Press.

Gordon, P. (2004). Numerical cognition without words: Evidence from Amazonia. *Science, 306*, 496–499.

Hauser, M. D., Carey, S., & Hauser, L. B. (2000). Spontaneous number representations in semi-free-ranging rhesus monkeys. *Proceedings of the Royal Society of London B, 267*, 829–833.

Hurford, J. R. (1987). *Language and number*. Oxford: Basil Blackwell.

Ifrah, G. (1998). *The universal history of numbers*. London: The Harvil Press.

Izard, V., & Dehaene, S. (2007). Calibrating the mental number line. *Cognition, 106*(3), 1221–1247. doi: 10.1016/j.cognition.2007.06.004

Izard, V., Pica, P., Spelke, E. S., & Dehaene, S. (2008). Exact equality and successor function: Two key concepts on the path towards understanding exact numbers. *Philosophical Psychology, 21*(4), 491–505. doi: 10.1080/09515080802285354

Kahneman, D., Treisman, A., & Gibbs, B. J. (1992). The reviewing of object files: Object specific integration of information. *Cognitive Psychology, 24*, 175–219.

Kirkpatrick, L. A. (2009). Between evolution and culture: Psychology at the nexus. In M. Schaller, S. J. Heine, A. Norenzayan, T. Yamagishi, & T. Kameda (Eds.), *Evolution, culture, and the human mind* (pp. 71–79). Mahwah, NJ: Lawrence Erlbaum.

Lakoff, G., & Núñez, R. E. (2000). *Where mathematics comes from: How the embodied mind brings mathematics into being*. New York: Basic Books.

Leibovich, T., Katzin, N., Harel, M., & Henik, A. (2017). From "sense of number" to "sense of magnitude": The role of continuous magnitudes in numerical cognition. *Behavioral and Brain Sciences, 40*, 1–62. doi: 10.1017/S0140525X16000960

Lyons, I. M., Ansari, D., & Beilock, S. L. (2012). Symbolic estrangement: Evidence against a strong association between numerical symbols and the quantities they represent. *Journal of Experimental Psychology: General, 141*(4), 635–641. doi: 10.1037/a027248

Macbeth, D. (2013). Writing reason. *Logique & Analyse, 221*, 25–44.

Malafouris, L. (2010). Grasping the concept of number: How did the sapient mind move beyond approximation? In C. Renfrew & I. Morley (Eds.), *The archaeology of measurement: Comprehending heaven, earth and time in ancient societies* (pp. 35–42). Cambridge: Cambridge University Press.

Menary, R. (Ed.). (2010). *The extended mind*. Cambridge: MIT Press.

Menary, R. (2015). Mathematical cognition: A case of enculturation. In T. Metzinger & J. M. Windt (Eds.), *Open MIND*. Frankfurt a. M., GER: MIND Group.

Núñez, R. E. (2011). No innate number line in the human brain. *Journal of Cross Cultural Psychology, 42*, 651. doi: 10.1177/0022022111406097

Núñez, R. E., Cooperrider, K., & Wassmann, J. (2012). Number concepts without number lines in an indigenous group of Papua New Guinea. *PLoS ONE, 7*(4), e35662. doi: 10.1371/journal.pone.0035662

Overmann, K. A. (2014). Finger-counting in the Upper Palaeolithic. *Rock Art Research, 31*(1), 63–80.

Overmann, K. A. (2015). Numerosity structures the expression of quantity in lexical numbers and grammatical number. *Current Anthropology, 56*(5), 638–653.

Overmann, K. A., Wynn, T., & Coolidge, F. L. (2011). The prehistory of number concept. *Behavioral and Brain Sciences, 34*(3), 142–144.

Pica, P., Lemer, C., Izard, V., & Dehaene, S. (2004). Exact and approximate arithmetic in an Amazonian indigene group. *Science, 306*(5695), 499–503.

Richerson, P., & Boyd, R. (2005). *Not by genes alone: How culture transformed human evolution*. Chicago: University of Chicago Press.

Rips, L. J., Bloomfield, A., & Asmuth, J. (2008). From numerical concepts to concepts of number. *Behavioral and Brain Sciences, 31*, 623–642.

Roberson, D. (2009). Color in mind, culture and language. In M. Schaller, A. Norenzayan, S. H. Heine, T. Yamagishi, & T. Kameda (Eds.), *Evolution, culture and the human mind* (pp. 167–184). Hove: Psychology Press.

Sarnecka, B. W., Kamenskaya, V. G., Ogura, T., Yamana, Y., & Yudovina, J. B. (2007). From grammatical number to exact numbers: Early meanings of "one", "two", and "three", in English, Russian, and Japanese. *Cognitive Psychology, 55*(2), 136–168.

Shea, N. (2011). New concepts can be learned: Review of Susan Carey: The origins of concepts. *Biology and Philosophy, 26*, 129–139.

Simon, T. J. (1997). Reconceptualizing the origins of number knowledge: A "non-numerical" account. *Cognitive Development, 12*, 349–372.

Sperber, D. (1996). *Explaining culture: A naturalistic approach*. Oxford, UK: Blackwell.

Sperber, D. (2006). Why a deep understanding of cultural evolution is incompatible with shallow psychology. In N. Enfield & S. Levinson (Eds.), *Roots of human sociality* (pp. 431–449). London: Berg.

Tang, Y., Zhang, W., Chen, K., Feng, S., Ji, Y., Shen, J., Reiman, E., & Liu, Y. (2006). Arithmetic processing in the brain shaped by cultures. *Proceedings of the National Academy of Sciences of the United States of America, 103*, 10775–10780.

Wiese, H. (2004). *Numbers, language, and the human mind*. Cambridge, NY: Cambridge University Press.

Wynn, K. (1992). Children's acquisition of the number words and the counting system. *Cognitive Psychology, 24*, 220–251.

11 Numbers Through Numerals
The Constitutive Role of External Representations

Dirk Schlimm

1. Introduction

There are many levels of mathematical cognition: The ability to distinguish collections of different sizes, identifying collections of given sizes, counting the objects in a collection, realizing the potential infinity of the number sequence, adding numbers and performing other arithmetical calculations, recognizing arithmetical principles such as commutativity, understanding mathematical proofs, and so on. My primary interest in this chapter is the particular contribution of numerals to the development of our understanding of the natural numbers (although similar considerations also apply to other external representations, such as an abacus).

Empirical work on numerical cognition has furnished us with an impressive body of results in the past few decades, and while there is still disagreement on the details of most issues, there have also been a number of widely accepted stable findings (Dehaene, 1997; Butterworth, 1999). One reason why empirical and theoretical research on numerical cognition is especially important for philosophy of mathematics is that educated Westerners typically spend many years learning the Indo-Arabic decimal place-value notation and encounter numbers on a daily basis in society, commerce, etc. This exposure leads to a great familiarity with numbers and their representations, which in turn can lead to biases and misconceptions when it comes to their investigation. For example, it is frequently, but falsely, believed that finger counting is a necessary step in the development of numerical abilities, that numeral systems always developed from a more object-specific system to a more abstract one, and that it is impossible to calculate with Roman numerals.[1]

The relevance of mathematical notations has been stressed for their role in the evolution of numerical cognition (Overmann, 2016), as vehicles of ontological innovations (Muntersbjorn, 2003), as 'epistemic symbols' (De Cruz and De Smedt, 2010), and as constituents of mathematical practices (Ferreirós,

---

1 See Crollen et al. (2011), Beller and Bender (2008), and Schlimm and Neth (2008) for arguments against these claims.

2015). Dutilh Novaes (2013, 63) has argued that "both from an ontogenetic and a phylogenetic point of view, the development of mathematical abilities is intimately related to acquiring mastery of such systems" as notations, number words, and abaci. In the present chapter, I intend to push this line of argument further by focusing on notational systems for natural numbers and investigate the ways in which they contribute to our understanding of numbers. To do so, I will first look at contemporary accounts of the development of mathematical abilities from basic cognition to more advanced mathematics (Section 2). These accounts typically end with the ability of counting the objects in a given collection and of using the technique of counting-on to perform simple additions. However, I will argue that this knowledge is not sufficient for arriving at an advanced conception of number that involves the potential infinity of numbers and being able to perform more advanced arithmetic, such as adding numbers greater than 100. For this next step in the development of mathematical cognition, we also need to understand the structural features of an external representation like the Indo-Arabic numerals (Section 3). On the one hand, this understanding is developed in part by learning how to perform computations with numerals, but on the other hand, the structural features of the external representations find their way into our mental representations. I conclude that an advanced understanding of arithmetic, which includes performing number comparisons and mental arithmetic, is mediated by symbolic reasoning with external representations. A philosophy of mathematics that does not take into consideration the role of external representations thus fails to include a crucial ingredient of mathematical thought.

2. From Basic Cognition to Simple Addition

2.1 *Numbers and Their Representations*

What do we mean by 'natural numbers'? Mathematicians can give a rigorous definition of the natural number structure in terms of the Dedekind-Peano axioms, but I take it that most humans have never heard of these axioms and still have some conception of (natural) numbers. Infants can distinguish between collections of one and two objects, and have some sense of permanency of the objects, but in what way this amounts to having a concept of one and two depends on one's understanding of what the numbers one and two are.[2] To avoid such ambiguities, I shall presently introduce the terminology used in this text (Table 11.1).

I shall refer to the infinite sequence of positive integers 1, 2, 3, . . . , as *natural numbers* or simply *numbers* since other kinds of numbers, such as rational,

---

2 See, for example, the discussion in (Sarnecka, 2008) and her comment that "no one (certainly not Rips or Carey) suggests that 4-year-olds talk explicitly about 'natural numbers', the way that philosophers and mathematicians do" (Sarnecka, 2015, 16).

Table 11.1 Terminology for systems of representations of numbers

External					Internal		
Symbolic				Iconic	Mind		Brain
verbal		formal		collections	approximate	exact	physical
grammatical	lexical	numerals	artifacts	('∴', '...')	(ANS)	(OTS)	(neurons)
('dogs')	('three')	('3, 'III')	(abacus)				

real, etc., won't be addressed in this chapter.[3] Numbers are abstract, and their structure is characterized mathematically by the second-order Dedekind-Peano axioms, but a conceptual understanding of them is richer than knowledge of their abstract structure. Our understanding of numbers and their properties is mediated by representations of them, which can be *internal*—i.e., located in the brain—or *external*—i.e., located outside human bodies. Among the external representations, we distinguish, following Peirce, between *symbolic* and *iconic* representations. Iconic representations, which are frequently just referred to as 'non-symbolic' in the literature, are collections of objects, such as arrays of dots, whose numerical meanings can be determined from the representation itself by counting. Such collections are also referred to as 'numerosities'.[4] For Peirce, iconic signs are characterized by a relation of resemblance to what they represent, but of course, there is no reason to think that an array of three dots resembles in any way the abstract number three, although it can be put in 1 – 1 correspondence with any collection of three objects. For symbolic representations, the link between the representation and what is represented is established by an arbitrary convention, like for the Indo-Arabic numerals or English number words. *Verbal* representations are part of a (spoken) language, whereas *formal* representations are not tied to a particular language. For example, an abacus or the Roman numerals can be used regardless of whether one speaks Latin or English; 'three', on the other hand, is a particular lexical term of English. *Lexical* representations are number words, whereas *grammatical* are those features of a language that indicate cardinality, such as quantifiers or plurals (e.g., the ending 's' in 'dogs' indicates a plural noun).

---

3 In beginning the natural numbers with one, I follow the historical tradition that includes Dedekind (1888). Since I will not address any issues related to the concept of zero, the choice of beginning with one seems to be appropriate.

4 For example, Gelman and Gallistel (1986, 51) introduce this term for a "set of one or more objects". However, instead of referring the collection itself, Butterworth (2005, 3) speaks of 'numerosity' to refer to "the number of objects in a set", and Coolidge and Overmann (2012, 204) consider 'numerosity' to be neither a collection nor a property of collections, but "the ability to appreciate and understand numbers". To avoid these ambiguities, I shall refrain from using the term 'numerosity' altogether and also talk about 'collections' instead of using the mathematical notion of set, which is abstract.

Although some of the instances that fall under the earlier distinctions are fairly straightforward, the distinctions themselves are subtle and more problematic than one might think at first. Compare, for example, the first three Roman numerals 'I', 'II', 'III', with the following first three arrays of dots '·', '· ·', '· · ·', or with a representation of the first three numbers on an abacus. They all are formed from one, two, and three basic elements, but have nevertheless been classified above as different kinds of representations. This is because the classification is one of representational *systems*, and as such, the Roman numerals are quite different from arrays of dots or an abacus, despite the fact that some individual elements can share salient structural features. However, what this entails is that in terms of required cognitive skills and effort, there might well be some overlap between different representational systems.

Internal representations of numbers can be at the level of *physical* locations in the brain (i.e., neurons or brain regions), which are identified by PET scans or fMRI studies during the performance of numerical tasks, or at a more conceptual, mental level.[5] The conceptual representations are inferred indirectly from stable patterns of behavior in reaction to specific stimuli (e.g., size effects and distance effects).[6] With regard to mental representations, we can distinguish between *homogeneous* models in which a single fundamental representation is assumed (like a mental number line) and *distributed* ones in which different types of representations are postulated. McCloskey (1992) has put forward a homogeneous abstract representation in which arithmetical facts are stored and computational procedures performed. According to this model, in a mental calculation, the external representations (verbal, formal, and iconic) of the problem formulation are first transcoded into the abstract internal representation. This is manipulated in order to obtain the result from which the output is finally produced in the required external representation.[7] In recent years, however, two different *core systems* have received particular attention. While animals, infants, children, and adults can reliably distinguish collections of sizes one to four (or up to three in some cases), comparisons of larger collections are more error-prone. Identifying the size of these very small collections can be done in constant time with a process called 'subitizing', whereas from about four onwards, the time increases with the size of the collection. The wealth of consistent experimental results of such findings led researchers to postulate a system that represents numerical magnitudes up to four in an exact manner (also referred to as 'parallel individuation system' or 'object-tracking system (OTS)') and an

5 Note, that this is the psychologists' understanding of concepts as mental representations, not philosophers' understanding, where concepts are considered to be abstract entities that are not necessarily physically realized.
6 See De Cruz, Neth, and Schlimm (2010) and Gaber and Schlimm (2015) for general overviews.
7 We shall return to this account in Section 3.3.

approximate number system (ANS) for the internal representation of greater numbers.[8] Because these systems are presumably at work also in fish, rats, and birds, they are considered to be evolutionarily ancient innate systems.[9]

2.2 On the Origin of Numerical Concepts

As an example of a contemporary account of the development of numerical cognition, I shall summarize briefly the view proposed by Susan Carey (2009), one of the leading researchers in this area. Her account is based on three innate core systems: first, a system that allows for the parallel individuation of small numerosities (up to four elements). Second, a system that represents analog magnitudes, which allows for the estimation and comparison of larger numerosities, but represents their cardinalities inexactly, and, third, a system that underlies quantification in natural language, which allows us to distinguish the meanings of 'one' and 'some' (or 'dog' and 'dogs'). Although empirical evidence for the existence of these systems can be found in infants and nonhuman beings, they are by themselves not enough for an understanding of natural numbers, according to Carey. To achieve such an understanding, she argues that the child has to generate inductively an 'external verbal numeral list' representation of natural numbers, which happens in several steps. First, the child has to memorize an ordered 'count list' of meaningless words, such as 'eeny, meeny, miny, mo' or 'one, two, three, four'. Then the meaning of each element of this list and the relevance of the list for the representation of numbers has to be learned. The latter includes knowing how to infer the meaning of new elements in the count list from their position in that list. Somewhat surprisingly, this knowledge is not identical from the ability of applying a counting routine (i.e., reciting the count list and establishing a one-to-one correspondence to a collection of objects), as empirical evidence shows (Carey, 2009, 316). At this point, usually by the age of 3–4 years, the child is able to count a given collection of objects, can determine a specific number of objects from a collection, and knows the relation between the counting list and the sizes of collections.

Carey gives a very detailed account of the cognitive and linguistic ingredients that she deems necessary for learning and understanding the sequence of natural numbers and their relations to the cardinalities of collections of objects. She also notes that the bootstrapping process described earlier only yields a finite representation that is limited by the length of the learned count list. While the further steps involved in learning basic arithmetic are beyond the scope of her account, she presents the following sketch:

> Explicitly asked about the existence of a highest number, 5-year-olds say no, and they explain that for any candidate highest number, someone could always add one to it. [. . .] Children also build models of addition

---

8 But see Gebuis, Cohen Kadosh, and Gevers (2016) for a challenge to this assumption.
9 See the references in footnote 6 for general overviews.

and subtraction based on the successor function and 1–1 correspondence, they conceptualize multiplication as repeated addition, and they begin to explicitly understand base 10 notation. By ages 6 through 8, children's arithmetical understanding is very rich and very firmly built on the concept of number as positive integer.

(Carey, 2009, 469–470)[10]

With regard to the development of addition, Butterworth (2005, 9) provides a more detailed account, distinguishing between three stages in which a problem such as adding three and five is tackled: (i) counting all, where the child forms a collection of three elements by counting 'one, two, three' and a collection of five elements by counting 'one, two, three, four, five' and then brings the collections together and counts all elements beginning from 'one'. (ii) When the child begins considering the sequence of number words as a 'breakable chain' (Fuson, 1988, 45), it can recite it from any starting point, not just from 'one'. This allows for *counting-on from first* such that the addition problem can be solved by starting with 'three' and counting-on the second collection of objects (or using one's fingers): 'four, five, six, seven, eight'. (iii) Finally, the child realizes that the result of counting is the same regardless of which of the addends is used first and so adopts the more efficient strategy of *counting-on from larger*, i.e., starting with 'five' and continuing 'six, seven, eight'. Abstractly, this amounts to exploiting the commutativity of addition.

The scope of addition based on counting-on is in practice quite limited. Even problems such as 72 + 35 would be difficult to solve with great confidence in a reasonable amount of time. Moreover, for being able to reason about larger numbers, say above 100, the internal structure of the external systems must be understood, which is not always an easy task (see Section 3.1.2). In the following, we shall take a look at how such a more advanced conception of numbers can be arrived at.

3.   Moving Beyond Words

3.1   Learning the Structure of Numbers From Verbal and Formal Representations

In the literature on developmental psychology, verbal representations have received more attention than formal ones. We have seen that Carey's account is based on what she calls an 'external verbal numeral list' or a 'numeral list representation of natural number', and although she allows for written and

10 An informative summary of the developmental milestones with regard to arithmetical abilities, beginning with the discrimination of small collections of objects according to their size by infants and proceeding to the retrieval of some arithmetical facts from memory by the age of 7, is presented in (Butterworth, 2005, 20, Table 1).

spoken languages as well as systems of notation to form a symbol system for bootstrapping (Carey, 2009, 306), her discussion of 'numeral lists' is mainly about lists of number words (i.e., verbal lexical representations and not formal numerals, according to the terminology shown in Table 11.1). Similarly, in Sarnecka's most recent account, which is based on Carey's and where a placeholder structure is gradually imbued with meaning to form a conceptual structure, the counting list consists of number words (Sarnecka, 2015, 1).

Perhaps because they are both symbolic, systems of number words and numerals are often treated as being informationally equivalent—i.e., of being able to express the same information—and differing only with regard to their computational properties (Simon, 1978; Larkin & Simon, 1987). Hurford expresses this sentiment quite clearly, when he writes that the Indo-Arabic notation "is just another linguistic system, more useful for doing sums with than orthographic representations of ordinary spoken language" (Hurford, 1987, 144). However, the Indo-Arabic system of numerals differs in essential ways from verbal systems, which casts doubt on their informational and computational equivalence. Let us therefore consider some pertinent differences between verbal and formal number representations.

3.1.1 Infinity

Greenberg (1978, 253) formulated as his first generalization of lexical systems that "every language has a numeral system of finite scope".[11] In other words, every verbal system can only represent finitely many numbers. The use of such a system as a counting list or placeholder structure thus gives us only representations for natural numbers up to a certain limit, but not for all natural numbers.[12] Some numeral systems (e.g., the Indo-Arabic decimal place-value system), on the other hand, get by with only a finite number of basic symbols to represent any natural number, with space restrictions imposing the only limitation. While one might be tempted to generalize this observation to all place-value notations, this is not correct with regard to historical notations, since "some positional systems are not infinitely extendable" (Chrisomalis, 2010, 371). As Chrisomalis also shows, the reliance on a finite set of basic symbols is not limited to place-value systems, but can also be found in some additive systems that use composite power-signs as multiplicands. Also artifacts, such as an abacus, can often be extended in a straightforward way (e.g., by lining up multiple abaci in a row).

---

11 Greenberg lists 'decillion' (10^{33}) as the lexical item with the highest value in American English. Kasner and Newman (1949, 23) famously introduced the invented terms 'googol' (10^{100}) and 'googolplex' (10^{googol}). To express a very large, but indefinite number, English has also terms such as 'gazillion' (Chrisomalis, 2016). However, the latter terms are usually not connected in a systematic way to the existing system.

12 Also Giaquinto's argument for a naturalistic account of our knowledge of cardinal numbers that "without positing powers that could not be countenanced in cognitive science" is restricted mainly to finite cardinals (Giaquinto, 2001, 16).

Given that the verbal representations are finite, how does a child learn that there are infinitely many natural numbers? Rips, Asmuth, and Bloomffeld (2006) made the case for the need of some general principles like the Dedekind-Peano axioms to establish a unique natural number structure. Their argument is that simple extrapolation from a finite set of samples does not guarantee, for example, that the natural numbers do not form a loop at some point. In addition, given that all languages have an upper limit for number words, it would also seem plausible to assume that there is also a largest natural number. In sum, "once you get past the numerals you have memorized, you do not automatically know how to continue" (Rips et al., 2006, B57). I consider these objections to be serious ones, because the move from the finite domain to an infinite one is not well understood psychologically and because it is well-known that such a move can lead to mathematical and logical difficulties. Some paradoxes of set theory result from such a move (e.g., that the set of all sets does not exist) and also Brouwer's philosophy of intuitionism can be seen as a reaction to an uncritical embracing of infinite sets. The solution proposed by Rips and his colleagues is to postulate knowledge of the Dedekind-Peano principles.

One difficulty with this is that these principles involve the successor function $s(n)$, which applies to all natural numbers n, and that in order to understand such a function, one also needs to have an infinite set at one's disposal, which amounts to knowing an axiom of infinity. Instead of beginning with such powerful assumptions, I suggest that one could also extrapolate the Dedekind-Peano principles from knowledge about the workings of external notations for numbers. Once the recursive structure of the sequence of Indo-Arabic numerals has been understood, one *does* know how to continue counting, even though one might not know verbal names of all the numerals. Because the construction principles of the sequence of numerals guarantee that elements never get shorter and are always distinct from each other, the possibility of loops and of a largest element are ruled out.[13]

3.1.2   Transparency

It is well-known that verbal systems of numeration, in particular in Western languages, often lack *transparency*, which means that their structure is not as regular as that of systems of numerals, which makes it more difficult to learn the sequence and to systematically extract their numerical values. Just a few examples: The English term 'twelve' is a single lexical item, just like 'eight', but its value is 12, which is represented by a two-digit numeral in the decimal system. Most number words in German for numbers over ten have

---

13 See also (Ferreirós, 2015, 97 and 184–189) for an argument that the Dedekind-Peano axioms are recognized as being true of the counting numbers. The 19th-century geometer Moritz Pasch also argued along similar lines for his empiricist philosophy of mathematics (Schlimm, 2018).

the *inversion property*, which means that the order of the words is inverted compared to the decimal numeral (e.g., 'dreiundzwanzig', literally three-and-twenty, stands for 23). French has the curious property of showing remnants of a base-20 system in number words above 60 (e.g., 'soixante-treize', literally sixty-thirteen, for 73, and 'quatre-vingts-dix-huit', literally four-twenty-ten-eight, for 98). On the other hand, East Asian languages, such as Japanese, Korean, and Chinese tend to have a much more regular structure where the number words more or less correspond to the reading of the decimal numerals from left to right (e.g., the Japanese term for 73 is 'nana ju san' and literally means seven-ten-three; see also (Bender, Schlimm, & Beller, 2015) for some Oceanic languages). While it has been shown that Chinese children learn their numerical vocabulary faster than their English-speaking peers, and that German children make many more transcoding errors (i.e., when translating between number words and numerals) than their Japanese peers, the exact contribution of language, as opposed to educational and cultural differences, is not entirely clear (see, e.g., Fayol & Seron, 2005; Miller, Kelly, & Zhou, 2005; Dowker, Bala, & Lloyd, 2008; Klein et al., 2013).

The importance of the Indo-Arabic numerals for magnitude comparisons and for understanding the structure of the verbal counting list has been highlighted by Fuson (1988). Like the accounts of Carey and Sarnecka, Fuson's account is based on a verbal counting list. In addition, however, Fuson (1988, 389) emphasizes that children have to learn that the list "has a decade structure between twenty and one hundred". She maintains that for the production of number words above 100, the structural features of the already learned counting list are essential, but that this is tightly connected to mastery of the system of numerals:

> [. . .] learning of this later part of the conventional list [i.e., number words above one hundred] typically occurs at the school age and is related to the learning of the base-ten place-value system of numeration rather than to counting, that is, number words above one hundred are primarily used to say (read) number symbols rather than to count.
>
> (Fuson, 1988, 389)

Also magnitude comparisons of larger numbers are based on knowledge of the formal numeral system, according to Fuson:

> For very large numbers, base-ten numbers and base-ten conceptual knowledge [. . .] are used to decide the cardinal relation between numbers such as 8,397 and 8,541. It is not at present clear for what size of numbers children move from using sequence relations to using relations derived from base-ten knowledge in order to decide cardinal order relations.
>
> (Fuson, 1988, 355–358)[14]

---

14 Of course, 'base-ten numbers' are not numbers, but numerals.

Thus, when a regular system of numerals is available, it seems to contribute in essential ways to the learning of the verbal number system and even to thinking about larger numbers.

The aforementioned considerations show that, at least for the cases of Western languages and the Indo-Arabic numerals, the verbal and the formal systems are neither informationally equivalent, because the verbal systems are finite, but the formal systems are indefinitely extendable, nor computationally equivalent, because the information about the structure of the system can be obtained much more easily from the formal system than from the verbal one. The label 'surface form' can be found in the literature for the distinction between verbal and formal systems, but in light of the aforementioned, this is misleading, because the differences lie much deeper than at the surface; they concern the internal structure of the systems in question.[15]

3.2   Learning the Structure of Notations

In this section, I discuss some evidence for the claim that being able to name, read, and write numerals and to use them for counting small collections is not sufficient for having an understanding of the internal structure of the numeral system. Thus, learning this structure is a developmental step that is beyond the accounts presented in Section 2.

3.2.1   Evidence for a Lack of Understanding of Place-Value Notation

Even if children can represent numbers in the decimal place-value notation, this does not mean that they actually understand the structure of the notation itself. In the insightful BBC documentary *Twice Five Plus the Wings of a Bird* (1986, 24 min 02 sec), we can witness a young girl who is able to perform multiplications with paper and pencil that involve three-digit numerals who is asked to add 53 and 4, which she does correctly. But when she is asked why she wrote the '4' under the '3' and not under the '5', she replies, "Um, that's the way the teacher does it" and is not able to give any further explanation. The documentary continues with Herbert Ginsburg summarizing:

> In school, the child encounters a very strange world with its own rules that are not explained to him very well.

$$
\begin{array}{cc}
12 & 12 \\
+\ 3 & +\ 3 \\
\hline
42 & 15
\end{array}
$$

[A paper in front of him shows: .]

---

15 To make this claim more precise, one would have to distinguish the systems of representation not only according to whether they are formed from terms in a language or are constituted by an extra-linguistic set of symbols but also according to their internal structure. Initial steps in this direction can be found in the works of Zhang and Norman (1995); Chrisomalis (2010); and Widom and Schlimm (2012), but much more needs to be done.

If you ask the child which of these two is correct, he might insist that this one is [pointing at the left calculation], because he says well, two plus zero is two, one plus three is four, that's 42. And so both those answers are just as good. Now this is a child, who if given 12 pieces of candy and another three, would be able to figure out perfectly well that there are 15 pieces of candy, and there would be no ambiguity involved whatsoever. It's only when he's introduced to the world of paper and pencil that some of the confusion begins.

The many difficulties that children have with understanding the place-value notation are well-known in mathematics education. Ashlock's *Error patterns in computation: using error patterns to improve instruction* (1998) discusses many systematic errors that elementary school children make in written arithmetic. These errors are often difficult to correct, because they result from the children learning incorrect algorithms (via over-generalization or faulty repair strategies) that work in certain cases, but not in others. Thus, although they apply the same algorithm, whether the result is correct or not remains a mystery to them.[16] Some simple but telling examples presented by Ashlock (1998) are shown in Table 11.2.

What is happening in these cases? In A1, each digit is processed individually to calculate the sum. Thus, here '43' is not understood as forty-three, but as four and three. According to Bergeron and Herscovics (1989, 106), this is often the first stage toward a gradual understanding of the Indo-Arabic numerals.[17] In A2, the child computes the result (15) correctly, splits it up into five units and one ten, but writes '10' instead of just '1' in the left column. The direction of operation in A3 is from left to right. The correct handling of carries poses a major difficulty for many children: in A4, they are simply disregarded, and in A5, the whole individual sums for each column

Table 11.2 Examples of common addition errors: Columns, direction, operation

A1:	A2:	A3:	A4:	A5:	A6:	A7:
	9	32				1
43	8	618	48	8 8	26	98
+ 26	+ 7	+ 782	+ 37	+ 3 9	+ 3	+ 3
15	105	1112	75	1117	11	131

16 See Lengnink and Schlimm (2010) for a more detailed discussion of these and other examples, and for a comparison with calculations using Roman numerals.

17 "[T]he understanding of positional notation grows gradually: from mere juxtaposition (numerals are written next to each other without regard to relative position), through a chronological stage (the order of production prevails over the relative position), to a final conventional level" (Bergeron & Herscovics, 1989, 106).

are written in the respective columns. Another common source of confusion is the presence of empty places in a column. The child in A6 can perform additions of two two-digit numerals correctly, but when faced with an empty place reverts to just adding the individual digits (as in A1). The problem of what to add if a place is empty is solved in A7 by reusing '3', the digit in the corresponding unit column. Note that the algorithms employed are not completely irrational, as they work well for simpler problems that the child has encountered before: A1 works if only single-digit numerals are to be added; A2, A3, A4, and A5 give the correct result if no carries are needed; and the children in A6 and A7 get the right answer if no empty places occur.

What all of the earlier systematic errors show is that the fundamental idea of place-value notation has not yet been understood by the children who make these mistakes. They can read and write individual numbers, as well as perform simple additions, but the internal structure of the numerals is nevertheless opaque to them.

3.2.2   Reflections on Conceptual Versus Procedural Understanding

We have seen earlier that the abilities of counting and counting-on, together with a solid grasp of reading and writing Indo-Arabic numerals do not yet amount to having an understanding of the place-value structure of the system of numerals. Otherwise, the frequent and systematic errors made by children in elementary school when learning paper and pencil arithmetic could have been caught and corrected by the children themselves. The nature of the relations between symbolic representations and mathematics instruction, and between concepts and procedures, have been mentioned as issues of current interest in a recent survey article by LeFevre (2016, 9). Nevertheless, the issue about the priority of conceptual and procedural knowledge (Hiebert & LeFevre, 1986) has been hotly debated for the past 30 years. On the one hand, conceptual understanding can support the development of correct arithmetical procedures. This is mentioned, for example, in a summary of research results relevant for mathematics education, complied by the National Research Council (USA):

> A good conceptual understanding of place value in the base-10 system supports the development of fluency in multi-digit computation. Such understanding also supports simplified but accurate mental arithmetic and more flexible ways of dealing with numbers than many students ultimately achieve.
>
> (Kilpatrick, Swafford, & Findell, 2001, 121)

On the other hand, the correct application of algorithmic procedures can in turn lead to a better conceptual understanding. Quoting from the same document as before:

> The process of developing fluency with arithmetic algorithms in elementary school can contribute to progress in developing the other strands of proficiency if time is spent examining why algorithms work and comparing

their advantages and disadvantages. Such analyses can boost conceptual understanding by revealing much about the structure of the number system itself and can facilitate understanding of place-value representations.
(Kilpatrick et al., 2001, 196)

That the influence between conceptual and procedural understanding goes both ways has also been argued for by Rittle-Johnson and Sigler (1998, 106) for the case of multi-digit addition and subtraction. Thus, instead of attempting to answer this chicken-and-egg question, Rittle-Johnson, Sigler, and Alibali (2001) have argued for an alternating and iterative process of the development of conceptual understanding and procedural fluency.

An argument one can frequently find against focusing on procedural knowledge in the teaching of arithmetic is that it is possible to learn and apply paper-and-pencil algorithms without a proper understanding of the underlying principles (e.g., Plunkett, 1979; Greeno, 1991, 198). This is, however, a general feature of formal systems—namely, that their meanings can be completely disregarded as long as one adheres to their formal rules of manipulation—and it has been discussed under the name of 'de-semantification' (Krämer, 2003; Dutilh Novaes, 2012). A new perspective on this discussion opened up in the previous paragraph: before we can remove the meanings from a formal system, it must first acquire them. Accordingly, we should speak of a non-trivial process of *semantification* of formal systems. As the earlier discussion of the learning of the structure of the Indo-Arabic numerals has shown, this process is much more complex than just stipulating the meanings of individual symbols and depends on the internal structure of the systems in question.

3.3 Numerals in the Mind

According to the classification of number representations presented in Table 11.1, there is a clear-cut separation between external and internal representations. We have seen in the previous section that formal representations are a vehicle for enhancing our mathematical conceptions. Now I would like to address the question of whether and how external and internal representations interact.

Given the empirical evidence for the existence of an internal ANS (see Section 2.1), it has been popular to consider the aim of learning an external representation of numbers as establishing a mapping between that representation and the ANS. In the literature, this is commonly referred to as the 'symbol-grounding problem' (Harnad, 1990).[18] Any differences in response

---

18 See Leibovich and Ansari (2016) for an overview. Analogous to the different understanding of 'concept' between philosophers and psychologists (see footnote 5) is also a different understanding of 'meaning'. In the cognitive science literature, to render an external symbol meaningful is often understood as relating it to an internal representation. For example, the meaning of 'twelve' is thus understood as the representation of 12 in the mind/brain.

times in tasks that involve iconic, verbal, and formal representations are then considered merely effects of transcoding the input to the internal format and again back to the respective output format. According to this view, once the transcoding of the input is completed, the structure of the external representation should have no effect on the processing (in terms of response times, errors, or locations of brain activity). In the remainder of this section, I will present some evidence against the view just described and I will argue that external representations are not just some arbitrary way of expressing thoughts, but they are constitutive of our thoughts themselves.

3.3.1   On the Relation Between External and Internal Representations

In a recent comparison task where participants had to decide which of two stimuli represents the greater quantity, Lyons, Ansari, and Beilock (2012) found that the performance was faster if both stimuli were presented as numerals than in a non-symbolic form, i.e., by patterns of dots. Moreover, mixing symbolic and non-symbolic representations—e.g., one stimulus consisting of dots and the other a numeral—resulted in a substantive increase of response times, whereas mixing two symbolic representations (numerals and number words) did not affect longer response times in general. Because one experiment also included intervals between the stimuli, the differences could not be simply attributed to differences in transcoding times. The authors conclude that "a numeral does not provide direct access to an approximate sense of the quantity it represents" and that "numerical symbols operate primarily as an associative system in which relations between symbols come to overshadow those between symbols and their quantity referents and may even become devoid of a strong sense of non-symbolic quantity per se". As a consequence, they contend that "at least for adults, the mature endpoint involves two relatively distinct representation systems for symbolic and non-symbolic numbers" (Lyons et al., 2012, 639–640). In sum, Lyons et al. argue for an 'estrangement', where either an initially tight connection between symbolic representations and the ANS becomes weakened over time or where such a connection never existed in the first place. An argument for the independence of exact calculations from the ANS is given by Frank and Barner (2012), who show that even drastic improvements in exact computation abilities are not correlated with improved performance in tasks that employ the ANS. Lyons and Ansari (2015) have critically reviewed the literature on the relation between symbolic processing and the ANS to address the question of "how young children first come to understand the numerical meaning of number symbols". They found that some studies argue for a relation between symbolic and non-symbolic number processing in kindergarteners or younger children, other studies argue against such a relation, still others argue for symbolic processing as being an intermediate step, and a recent study showed that symbolic processing predicts non-symbolic processing at a later stage.[19] Thus, they conclude that these empirical studies "have not provided robust evidence in support of a strong link between the non-symbolic, approximate representation of numerical magnitude and number symbols" (Lyons & Ansari, 2015, 110). In sum, the aforementioned studies cast serious doubts on the assumption that the meaningfulness of symbolic representations of numbers is established by connecting them to the ANS, and they suggest that symbolic and non-symbolic external representations might also have different mental representations.

3.3.2 *Representational Effects in Mental Arithmetic*

Another area of research that is relevant for my argument and that has received quite a bit of attention lately concerns the cognitive processes involved in mental arithmetic.[20] What makes it particularly difficult for cognitive scientists to pinpoint the underlying mechanisms is that adults use a wide variety of techniques for performing mental arithmetic, such as retrieval of memorized facts, application of general rules, and counting-on. Adults seem to rely only on retrieval of numerical facts when performing simple additions and multiplications of single-digit operands (e.g., Seitz & Schumann-Hengsteler, 2000).

There is increasing evidence that the structure of the verbal number system affects the performance in counting, number reading and comparison, and mental arithmetic (Dowker et al., 2008). In particular, the more irregularly the number words are formed, the more difficult the tasks become. Conversely, if a regular, or transparent (see Section 3.1.2), system is used, responses are generally faster and less prone to errors. As an example from mental arithmetic, even if addition problems are presented with numerals, Göbel, Moeller, Pixner, Kaufmann, and Nuerk (2014) found that the 'carry effect' (i.e., that operations involving carries take longer) of German-speaking children was more pronounced than that of Italian-speaking children, whose language does not have any inversions. Apparently, the number word structure of German adds inconsistent positional information that renders the task of identifying and keeping track of positions more difficult. The authors interpret this result as showing that language modulates mental arithmetic, even if the problems themselves are not explicitly linguistic.

If numbers were represented internally independently of external representations—e.g., on an analog mental number line or on the basis of a uniform placeholder structure built up from an initial element by repeated applications of a successor operator (1, s1, ss1, sss1, . . .)—then the native language of the subjects, or the form in which the stimuli are presented, should make no difference to the way in which they are mentally processed, e.g., in case of a comparison of two numeric values or of mental arithmetic.

19 See also Feigenson, Libertus, and Halberda (2013) for a similar overview.
20 See DeStefano and LeFevre (2004) for an overview.

210 *Dirk Schlimm*

The 'distance effect', according to which the comparison of two numbers is faster if their numerical distance is greater (Moyer & Landauer, 1967), is a well-established phenomenon. However, Verguts and de Moor (2005) were able to reproduce such a distance effect only if the two numbers were in the same decade (e.g., 61 and 67), but not if they were in different decades (e.g., 69 and 75). On the basis of this finding, the authors argue against a holistic mental processing of two-digit numbers and in favor of a decomposition of the numbers into units and decades followed by a single-digit comparison.[21] The 'unit-decade compatibility effect', according to which two numbers are compared faster and with more accuracy if the comparison of both units and decades yield the same decision (e.g., in 42 vs. 57, both 4 < 5 and 2 < 7) than if they yield opposite decisions (e.g., 47 vs. 62), also points to an internal representation of numbers that relies on more than the overall magnitude (Nuerk, Wegner, & Willmes, 2001). Decade effects for mental addition were also shown by Neth (2004), who asked subjects to add up sequences of single-digit numbers and found systematic differences in the processing times for each individual addition: the fastest additions where those that resulted in a round sum (e.g., 16 + 4), then those that did not require a carry (e.g., 16 + 3), and the slowest were those that required a carry (e.g., 16 + 5). These results are independent of the problem size and clearly reflect structural features of the system of numerals.

Leaving aside speculations about the exact cognitive processes that underlie Verguts and de Moor's, Nuerk et al.'s, and Neth's findings, what matters for the present chapter is that these results about mental number comparison and arithmetic depend crucially on specific features of the external representations of numbers. The effects of the format in which arithmetic tasks are presented also provide evidence against an abstract internal representation, as Campbell explains,

> One consistent result is that simple arithmetic with visual number words (seven × five) is much more difficult than with Arabic numerals (7 × 5). Arithmetic with written words can be as much as 30% slower and 30% more error-prone compared with Arabic problems (Campbell, 1994). [. . .] This makes it unlikely that word format costs can be attributed simply to encoding differences. [. . .] the problem-size effect is larger for problems in written-word format (e.g., three + eight) than digit format (3 + 8) [. . .] The finding that word-format costs increase with problem difficulty suggests that format directly affects calculation.
>
> (Campbell, 2015, 144–145)

21 The terminology of 'two-digit numbers' and 'multi-digit numbers' that is used here following the literature, is somewhat misleading. The fact that a number is represented with two digits is not a property of the number itself (like being even or a prime), but of its decimal place-value representation. For example, while 37 has two digits in the Indo-Arabic representation, it has only one digit in a sexagesimal place-value notation (where '10' stands for 60) and six digits in the Roman notation ('XXXVII').

The importance of fluency with numerals for the development of one's math skills is highlighted by the fact that symbolic skills appear to be good predictors for later mathematical performance. A study by Moeller, Pixner, Zuber, Kaufmann, and Nuerk (2011) indicated that early place-value understanding is a significant predictor of later addition performance and the carry effect. Similarly, Kolkman, Kroesbergen, and Leseman (2013, 95) concluded that "symbolic and mapping skills" were important predictors for math performance and Sasanguie, Göbel, Moll, Smets, and Reynvoet (2013, 418) emphasize "the importance of learning experiences with symbols for later math abilities".

Studies that involve different systems of numerals are rare, but Krajcsi and Szabó (2012) have shown that sign-value and place-value systems that involve a comparable number of symbols yield different response times and accuracies in comparison tasks and simple additions. On the basis of these results they argued against a single internal system of representation. There are also some famous cross-cultural studies that show differences in learning speed, flexibility of mental manipulations, and arithmetic tasks (Miura, Kim, Chang, & Okamoto, 1988, 1993). The better performance of Chinese children and adults is often attributed to the greater transparency of the language and to more drill, but Butterworth has recently raised the possibility that different teaching methods (e.g., where only half of a multiplication table together with a rule for commutativity is taught) might play a significant role in explaining these patterns of behavior based on different internal representations (Butterworth, 2005, 11).

Although this chapter dealt mainly with the role of Indo-Arabic numerals in mathematical cognition, I must emphasize that the main point—namely, the importance of understanding the structure of external representations, can be generalized. External representations need not be formal representations like numerals, but can also be artifacts, like an abacus representation. Indeed, investigations of the mathematical abilities and brain activity of children trained in abacus-based mental calculation suggest that mathematical processing is not abstract, but does depend on the format of external representations (Tang et al., 2006; Huang et al., 2015). Similarly, Frank and Barner (2012) argued that a mental abacus is represented in visual working memory and they showed that exact computations using this format are relatively insensitive to linguistic interference. In other words, also in the case of abacus users, the external form of the representation appears to have direct effects on the internal representations.

3.3.3 Internal Representations Revisited

The empirical findings discussed in the previous sections make it difficult to hold on to a picture in which the ANS grounds all numerical cognition. In fact, even a binary division between an approximate and an exact number system (as shown in Table 11.1) seems too crude to account for the various ways in which different external representations affect our arithmetical abilities. Dehaene (1997) suggested a triple-code representation of numbers based on an analog magnitude system (similar to the ANS discussed earlier), a visual representation of the Indo-Arabic numerals, and a verbal representation. However, in light of the earlier discussions, it does not seem enough to have a representation of the Indo-Arabic numerals "as identified strings of digits" (Dehaene & Cohen, 1998, 337), because this does not imply an understanding of the internal structure of the notation. Representational effects have led Zhang and Wang (2005) to argue for a distributed number representation on the basis of their investigations of multi-digit number processing, while Nuerk, Moeller, and Willmes (2015) have suggested that the representations and processes involved are (i) visual number form. (ii) Semantic representation of numerical magnitude. (iii) Verbal numerical representations. (iv) Spatial representation of numbers. (v) Strategic, conceptual, and procedural components. (vi) Structural representation of the symbolic number system (place-value representation). The last item clearly reflects the importance they put on the structure of the formal notation, but this is of course relative to the particular system that is being used. Presumably, people who perform their arithmetical computations on an abacus would have a different structural representation of the symbolic number system; that of people who use Roman numerals would be different again, and people who are conversant in various symbolic systems could have internal representations for each individual system.

4   Concluding Remarks

In the earlier discussion we saw that contemporary accounts of the development of mathematical cognition often do not consider the differences between verbal and formal representations of numbers and end with the abilities of naming numerals and of performing simple additions by counting-on. These, however, are not sufficient for an advanced conception of numbers—i.e., one that goes beyond counting and the direct comparison of numerical magnitudes—but underlies the development of more advanced arithmetical abilities, such as the addition of larger numbers. To achieve this level of mathematical understanding, I have argued that it is necessary to understand the internal structure of a system of numerals. Particular claims that were argued for are, in Section 3.1: (i) formal representations of numbers (i.e., numerals) are more conducive to conveying the ideas of potential infinity and of the structure of the natural numbers than verbal ones. In Section 3.2: (ii) being able to read and write Indo-Arabic numerals does not entail understanding their internal structure. (iii) Being able to count is not sufficient for performing arithmetic operations with larger numbers (e.g., greater than 100). (iv) Carrying out arithmetic algorithms improves the learning of the internal structure of a numeral system and thereby also the conceptual understanding of numbers. And, finally, in Section 3.3: (v) formal representations are not merely vehicles for expressing numbers and external representations are not merely a convenient way of expressing one's mathematical thoughts, but mental arithmetic bears traces of internalized operations with external representations (both in the case of Indo-Arabic numerals and abacus computations). In other words, the external representations affect our thinking itself. Therefore, the use of numerals and other formal systems of representations is constitutive for the development of an advanced conception of numbers, and the study of structural features of such systems should be taken seriously in cognitive science, mathematics education, and philosophy of mathematical practice.

Acknowledgments

Thanks to Sorin Bangu, Andrea Bender, David Gaber, Valeria Giardino, Matthew Inglis, Daniel Lovsted, and Mario Santos-Sousa for comments on an earlier draft. This research was supported by the Social Sciences and Humanities Research Council of Canada.

References

Ashlock, R. B. (1998). *Error patterns in computation: Using error patterns to improve instruction* (7th ed.). Upper Saddle River, NJ: Merrill.

Beller, S., & Bender, A. (2008). The limits of counting: Numerical cognition between evolution and culture. *Science, 319*, 213–215.

Bender, A., Schlimm, D., & Beller, S. (2015). The cognitive advantages of counting specifically: A representational analysis of verbal number systems. *Topics in Cognitive Science, 7*(4), 552–569.

Bergeron, J., & Herscovics, N. (1989). A model to describe the construction of mathematical concepts from an epistemological perspective. In L. Pereira-Mendoza & M. Quigley (Eds.), *Proceedings of the 1989 annual meeting, Brock University, St. Catharines, ON, May 27–31* (pp. 99–114). Memorial University of Newfoundland. Canadian Mathematics Education Study Group.

Butterworth, B. (1999). *The mathematical brain*. London: Papermac.

Butterworth, B. (2005). The development of arithmetical abilities. *Journal of Child Psychology and Psychiatry, 46*(1), 3–18.

Campbell, J. I. D. (1994). Architectures for numerical cognition. *Cognition, 531*–544.

Campbell, J. I. D. (2015). How abstract is arithmetic? In R. Cohen Kadosh & A. Dowker (Eds.), *The Oxford handbook of numerical cognition*, chapter 8 (pp. 140–157). Oxford: Oxford University Press.

Campbell-Jones, S. (1986). *Twice five plus the wings of a bird*. BBC.

Carey, S. (2009). *The origin of concepts*. Oxford: Oxford University Press.

Chrisomalis, S. (2010). *Numerical notation: A comparative history*. Cambridge: Cambridge University Press.

Chrisomalis, S. (2016). Umpteen reflections on indefinite hyperbolic numerals. *American Speech, 91*(1), 3–33.

Coolidge, F. L., & Overmann, K. A. (2012). Numerosity, abstraction, and the emergence of symbolic thinking. *Current Anthropology, 53*(2), 204–225.

Crollen, V., Seron, X., & Noel, M.-P. (2011). Is finger-counting necessary for the development of arithmetic abilities? *Frontiers in Psychology, 2*(242), 1–3.

De Cruz, H., & De Smedt, J. (2013). Mathematical symbols as epistemic actions. *Synthese, 190*(1), 3–19.

De Cruz, H., Neth, H., & Schlimm, D. (2010). The cognitive basis of arithmetic. In B. Lowe & T. Müller (Eds.), *Philosophy of mathematics: Sociological aspects and mathematical practice* (pp. 39–84). London: College Publications.

Dedekind, R. (1888). *Was sind und was sollen die Zahlen? Vieweg, Braunschweig.* Reprinted in Dedekind (1930–1932), III (pp. 335–391). English translation by Wooster W. Beman, revised by William Ewald, in Ewald (1996), (pp. 787–833).

Dedekind, R. (1930–1932). *Gesammelte mathematische Werke.* F. Vieweg & Sohn, Braunschweig (3 vols.), Edited by Robert Fricke, Emmy Noether, and Oystein Ore.

Dehaene, S. (1997). *The number sense: How the mind creates mathematics.* New York: Oxford University Press.

Dehaene, S., & Cohen, L. (1998). Levels of representation in number processing. In B. Stemmer & H. A. Whitaker (Eds.), *The handbook of neurolinguistics* (pp. 331–341). San Diego: Academic Press.

DeStefano, D., & LeFevre, J.-A. (2004). The role of working memory in mental arithmetic. *European Journal of Cognitive Psychology*, 16(3), 353–386.

Dowker, A., Bala, S., & Lloyd, D. (2008). Linguistic influences on mathematical development: How important is the transparency of the counting system? *Philosophical Psychology*, 21(4), 423–538.

Dutilh Novaes, C. (2012). *Formal languages in logic: A philosophical and cognitive analysis.* Cambridge, UK: Cambridge University Press.

Dutilh Novaes, C. (2013). Mathematical reasoning and external symbolic systems. *Logique et Analyse*, 56(221), 45–65.

Ewald, W. (1996). *From Kant to Hilbert: A source book in mathematics* (2 vols.). Oxford: Clarendon Press.

Fayol, M., & Seron, X. (2005). About numerical representations. In J. I. D. Campbell (Ed.), *Handbook of mathematical cognition*, chapter 1 (pp. 3–22). New York: Psychology Press.

Feigenson, L., Libertus, M. E., & Halberda, J. (2013). Links between the intuitive sense of number and formal mathematics ability. *Child Development Perspectives*, 7(2), 74–79.

Ferreirós, J. (2015). *Mathematical knowledge and the interplay of practices.* Princeton: Princeton University Press.

Frank, M. C., & Barner, D. (2012). Representing exact number visually using mental abacus. *Journal of Experimental Psychology: General*, 141(1), 134–139.

Fuson, K. C. (1988). *Children's counting and concepts of number.* New York: Springer-Verlag.

Gaber, D., & Schlimm, D. (2015). Basic mathematical cognition. *WIRE Cognitive Science*, 6(4), 355–369.

Gebuis, T., Cohen Kadosh, R., & Gevers, W. (2016). Sensory-integration system rather than approximate number system underlies numerosity processing: A critical review. *Acta Psychologica*, 171, 17–35.

Gelman, R., & Gallistel, C. R. (1986). *The child's understanding of number.* Cambridge, MA: Harvard University Press. With a new preface by the authors. 1st edition, 1978.

Giaquinto, M. (2001). Knowing numbers. *Journal of Philosophy*, 98(1), 5–18.

Göbel, S. M., Moeller, K., Pixner, S., Kaufmann, L., & Nuerk, H.-C. (2014). Language affects symbolic arithmetic in children: The case of number word inversion. *Journal of Experimental Child Psychology*, 119, 17–25.

Greenberg, J. H. (1978). Generalizations about numeral systems. In J. H. Greenberg, C. A. Ferguson, & E. A. Moravcsik (Eds.), *Universals of human language* (Vol. 3: Word Structure, pp. 249–295). Stanford, CA: Stanford University Press.

Greeno, J. G. (1991). Number sense as situated knowing in a conceptual domain. *Journal for Research in Mathematics Education*, 22(3), 170–218.

Harnad, S. (1990). The symbol grounding problem. *Physica D*, 42, 335–346.

Hiebert, J., & LeFevre, J.-A. (1986). Conceptual and procedural knowledge in mathematics: An introductory analysis. In J. Hiebert (Ed.), *Conceptual and procedural knowledge: The case of mathematics* (pp. 1–27). Hillsdale, NJ: Erlbaum.

Huang, J., Du, F.-L., Yao, Y., Wan, Q., Wang, X.-S., & Chen, F.-Y. (2015). Numerical magnitude processing in abacus-trained children with superior mathematical ability: An EEG study. *Journal of Zhejan University-Science B* (Biomedicine & Biotechnology), 16(8), 661–671.

Hurford, J. R. (1987). *Language and number: The emergence of a cognitive system.* Oxford: Blackwell.

Kasner, E., & Newman, J. (1949). *Mathematics and the imagination.* London: B. Bell and Sons.

Kilpatrick, J., Swafford, J., & Findell, B. (Eds.). (2001). *Adding it up: Helping children learn mathematics.* National Research Council. Mathematics Learning Study Committee, Center for Education, Division of Behavioral and Social Sciences and Education. Washington, DC: National Academy Press.

Klein, E., Bahnmueller, J., Mann, A., Pixner, S., Kaufmann, L., Nuerk, H.-C., & Moeller, K. (2013). Language influences on numerical development—Inversion effects on multi-digit number processing. *Frontiers in Psychology*, 4(480), 1–6.

Kolkman, M. E., Kroesbergen, E. H., & Leseman, P. P. M. (2013). Early numerical development and the role of non-symbolic and symbolic skills. *Learning and Instruction*, 25, 95–103.

Krajcsi, A., & Szabó, E. (2012). The role of number notation: Sign-value notation number processing is easier than place-value. *Frontiers in Psychology*, 3(463), 1–15.

Kramer, S. (2003). Writing, notational iconicity, calculus: On writing as a cultural technique. *Modern Language Notes—German Issue*, 118(3), 518–537. Johns Hopkins University Press.

Larkin, J. H., & Simon, H. A. (1987). Why a diagram is (sometimes) worth ten thousand words. *Cognitive Science*, 11, 65–99.

LeFevre, J.-A. (2016). Numerical cognition: Adding it up. *Canadian Journal of Experimental Psychology*, 70(1), 3–11.

Leibovich, T., & Ansari, D. (2016). The symbol-grounding problem in numerical cognition: A review of theory, evidence, and outstanding questions. *Canadian Journal of Experimental Psychology*, 70(1), 12–23.

Lengnink, K., & Schlimm, D. (2010). Learning and understanding numeral systems: Semantic aspects of number representations from an educational perspective. In B. Lowe & T. Müller (Eds.), *Philosophy of mathematics: Sociological aspects and mathematical practice* (pp. 235–264). London: College Publications.

Lyons, I. M., & Ansari, D. (2015). Foundations of children's numerical and mathematical skills: The roles of symbolic and non-symbolic representations of numerical magnitude. *Advances in Child Development and Behavior*, 48, 93–116.

Lyons, I. M., Ansari, D., & Beilock, S. L. (2012). Symbolic estrangement: Evidence against a strong association between numerical symbols and the quantities they represent. *Journal of Experimental Psychology: General*, 141(4), 635–641.

McCloskey, M. (1992). Cognitive mechanisms in numerical processing: Evidence from acquired dyscalculia. *Cognition*, 44, 107–157.

Miller, K. F., Kelly, M., & Zhou, X. (2005). Learning mathematics in china and the United States: Cross-cultural insights into the nature and course of preschool mathematical development. In J. I. D. Campbell (Ed.), *Handbook of mathematical cognition* (pp. 163–178). New York: Psychology Press.

Miura, I. T., Kim, C. C., Chang, C.-M., & Okamoto, Y. (1988). Effects of language characteristics on children's cognitive representation of number: Cross-national comparisons. *Child Development*, 59(6), 1445–1450.

Miura, I. T., Okamoto, Y., Kim, C. C., Steere, M., & Fayol, M. (1993). First grad-ers' cognitive representation of number and understanding of place value: Cross-national comparisons—France, Japan, Korea, Sweden, and the United States. *Journal of Experimental Psychology*, *85*(1), 24–30.

Moeller, K., Pixner, S., Zuber, J., Kaufmann, L., & Nuerk, H.-C. (2011). Early place-value understanding as a precursor for later arithmetic performance I a longitu-dinal study on numerical development. *Research in Developmental Disabilities*, *32*, 1837–1851.

Moyer, R. S., & Landauer, T. K. (1967). Time required for judgements of numerical inequality. *Nature*, *215*(21), 1519–1520.

Muntersbjorn, M. (2003). Representational innovation and mathematical ontology. *Synthese*, *134*, 159–180.

Neth, H. (2004). *Thinking by doing: Interactive problem solving with internal and external representations*. PhD thesis, School of Psychology, Cardiff University, Wales.

Nuerk, H.-C., Moeller, K., & Willmes, K. (2015). Multi-digit number process-ing: Overview, conceptual clarifications, and language influences. In R. Cohen Kadosh & A. Dowker (Eds.), *The Oxford handbook of numerical cognition*, chapter 7 (pp. 106–139). Oxford: Oxford University Press.

Nuerk, H.-C., Wegner, U., & Willmes, K. (2001). Decade breaks in the mental number line? Putting the tens and units back in different bins. *Cognition*, *82*(1), B25–B33.

Overmann, K. A. (2016). The role of materiality in numerical cognition. *Quaternary International*, *405*, 42–51.

Plunkett, S. (1979). Decomposition and all that rot. *Mathematics in School*, *8*(3), 2–7.

Rips, L. J., Asmuth, J., & Bloomffeld, A. (2006). Giving the boot to the bootstrap: How not to learn the natural numbers. *Cognition*, *101*(3), B51–B60.

Rittle-Johnson, B., & Sigler, R. S. (1998). The relation between conceptual and pro-cedural knowledge in learning mathematics: A review. In C. Donlan (Ed.), *The development of mathematical skills* (pp. 75–110). Hove: Psychology Press.

Rittle-Johnson, B., Sigler, R. S., & Alibali, M. W. (2001). Developing conceptual understanding and procedural skill in mathematics: An iterative process. *Journal of Experimental Psychology*, *93*(2), 346–362.

Sarnecka, B. W. (2008). SEVEN does not mean NATURAL NUMBER, and children know more than you think. *Behavioral and Brain Sciences*, *31*, 668–669.

Sarnecka, B. W. (2015). Learning to represent exact numbers. *Synthese*, 1–18. Online: doi: 10.1007/s11229-015-0854-6

Sasanguie, D., Gobel, S. M., Moll, K., Smets, K., & Reynvoet, B. (2013). Approxi-mate number sense, symbolic number processing, or number-space mappings: What underlies mathematics achievement? *Journal of Experimental Child Psy-chology*, *114*, 418–431.

Schlimm, D. (2018). Pasch's empiricist structuralism. In E. Reck & G. Schiemer (Eds.), *The pre-history of mathematical structuralism*. Oxford: Oxford Univer-sity Press. (To appear).

Schlimm, D., & Neth, H. (2008). Modeling ancient and modern arithmetic prac-tices: Addition and multiplication with Arabic and Roman numerals. In V. Slout-sky, B. Love, & K. McRae (Eds.), *Proceedings of the 30th annual meeting of the cognitive science society* (pp. 2007–2012). Austin, TX: Cognitive Science Society.

Seitz, K., & Schumann-Hengsteler, R. (2000). Mental multiplication and working memory. *European Journal of Cognitive Psychology*, *12*(4), 552–570.

Simon, H. A. (1978). On the forms of mental representation. In C. W. Savage (Ed.), *Perception and cognition: Issues in the foundations of psychology* (Vol. 9, pp. 3–18). Minneapolis: University of Minnesota Press.

Tang, Y., Zhang, W., Chen, K., Feng, S., Ji, Y., Shen, J., Reiman, E., & Liu, Y. (2006). Arithmetic processing in the brain shaped by cultures. *Proceedings of National Academy of Sciences, 103,* 10775–10780.

Verguts, T., & de Moor, W. (2005). Two-digit comparison: Decomposed, holistic, or hybrid? *Experimental Psychology, 52*(3), 195–200.

Widom, T. R., & Schlimm, D. (2012). Methodological reflections on typologies for numeral systems. *Science in Context, 25*(2), 155–195.

Zhang, J., & Norman, D. A. (1995). A representational analysis of numeration systems. *Cognition, 57,* 271–295.

Zhang, J., & Wang, H. (2005). The effect of external representations on numeric tasks. *The Quarterly Journal of Experimental Psychology, 58A*(5), 817–838.

12 Making Sense of Numbers Without a Number Sense

Karim Zahidi and Erik Myin

1. Introduction: Neorationalism

Traditional rationalism holds that individuals have in their minds a stock of knowledge, the acquisition of which is independent of learning or experience. Experience is at best a necessary condition to awaken this slumbering knowledge. Importantly, traditional rationalism—found in the Western philosophical tradition from Plato to Descartes to latter-day Cartesians such as Fodor—holds that such innate knowledge is conceptual. It is not easy to pinpoint what conceptual knowledge is. However, for our purposes, it will suffice to say that it implies knowledge of facts (e.g., knowledge that the earth is spherical) and that having such knowledge depends on having certain ideas or concepts in the mind (e.g., EARTH, SPHERE). In what follows, unless otherwise specified, we will take knowledge to be conceptual knowledge.

The problem for rationalism has always been to account for the provenance of the mind's innate ideas. Traditional answers have been that they derive from the communion between the soul and Platonic Forms, or that they are implanted by a benevolent God. However, due to their appeal to non-naturalistic entities and explanatory vacuity, such answers have lost favor among philosophers and cognitive psychologists. Faced with this problem, a new form of rationalism has come in vogue. It agrees with traditional rationalism that the individual mind contains ideas and knowledge prior to learning or experience. It departs from traditional rationalism, however, by relying on evolutionary theory to give what it claims to be a naturalistically plausible answer to the question of the origins of innate ideas and knowledge. This 'evolutionary nativism' (Carruthers, 1992) holds that evolution not only produces organisms whose bodies and behaviors are adapted to the environment prior to the individual's immersion in it, but is also able to produce individual minds with knowledge prior to learning. In neorationalism, 'species learning' takes over the role of individual learning in the individual's acquisition of (some) knowledge.

Given the fact that mathematical knowledge has been presented by rationalism as the prime showcase for the theory of innate knowledge, it is perhaps unsurprising that neorationalism is currently a prominent position in the field of mathematical cognition. In this chapter, we will critically examine neorationalistic theories in the domain of numerical cognition. Such theories are meant to account for the behavior of young infants, which can be interpreted as 'number-sensitive', in terms of innate numerical knowledge.[1] We will argue that, despite their initial plausibility, the neorationalistic theories fail to account for the observed behavior. In particular, we will argue that the key assumption—viz., that the behavior to be explained is a product of knowledge deployment—is not only unwarranted but also that the postulation of the required knowledge raises unsolved problems.

Although we focus on and are critical of neorationalistic explanations in the domain of numerical cognition, we do not therefore adhere to more empiricist theories of numerical cognition. The critique we level at neorationalistic numerical cognition is to a large extent applicable to various empiricist alternatives in so far as these share a basic assumption with neorationalism (e.g., Prinz's (2002) empiricist interpretation of the experimental findings). Note that what separates empiricist theories from rationalistic theories is the issue of whether the knowledge postulated to explain the behavior can be acquired via experience or not. What all rationalists and many empiricists agree on, however, is that to explain the behavior in question, one needs to postulate conceptual knowledge that can be acquired by isolated individuals.[2] Furthermore, they agree that the difference between the infant's postulated knowledge and that of adults is a difference in degree and not in kind. Hence sophisticated adult knowledge can in principle be acquired by isolated individuals.

But the dichotomy rationalist versus empiricist does not exhaust the theoretical options available to account for the acquisition of knowledge. There is a third alternative: the situated or socio-historic view, according to which

> knowing [. . .] is both an attribute of groups that carry out cooperative activities and an attribute of individuals who participate in the communities of which they are members. A group or individual with knowledge is attuned to the regularities of activities, which include the constraints and affordances of social practices and of the material and technological systems of environments.
>
> (Greeno et al., 1996, 17)

---

1 We use scare quotes because the description of the behavior of infants in terms of *number* sensitivity is misleading. A neutral way to describe the behavior is by saying that the behavior displays sensitivity to differences in stimuli that can be described in numerical terms.

2 Even empiricist accounts of mathematical knowledge that depart from the individualism of more classical empiricist theories in stressing the importance of social aspects in learning do retain a form of individualism when it comes to the acquisition of basic numerical knowledge. For example, Kitcher (1983, 108) does not deny that children's "learning is aided by teachers and parents", but he does seem to imply that the acquisition of numerical knowledge is in principle possible on an individualistic basis. For example, he remarks that the mere manipulation of objects in the environment could be sufficient for acquiring basic numerical knowledge. For more detailed criticism of Kitcher along these lines, see (Tymoczko, 1984).

Accordingly, to be an individual knower presupposes a community of knowers, and hence the idea that an isolated individual could be, or could become, a knower on his or her own is mistaken. In this chapter, we will use insights from recent work in the philosophy of mind to argue that the socio-historic view offers the best option to account for the acquisition of numerical knowledge. In particular, we will argue that both neorationalism and empiricism are based on a problematic unrestricted representational view of cognition— viz., that cognition is *always* the manipulation of mental representations.

While the unrestricted representational view of cognition is still dominant in the cognitive sciences, it has been severely criticized both from within the cognitive sciences (e.g., by ecological psychologists) and from adjacent philosophical disciplines, such as the philosophy of cognitive science and the philosophy of mind (Chemero, 2009, Hutto & Myin, 2013, 2017, Ramsey, 2007). Although such theorists differ in their analysis of what is the core problem facing the traditional picture of cognition, they nonetheless share a similar critical attitude to the central theoretical notion of mental representation. For example, Hutto and Myin (2013) have argued that mental representations lack the required naturalistic credentials to figure in a scientifically respectable theory of cognition. More precisely, they argue that thus far, the only plausible account of how representations can acquire semantic content is one that invokes the conditions under which truth-telling practices emerge in a sociocultural environment. If this is correct, then the idea that infants possess innate conceptual knowledge becomes suspect. Similarly, the empiricist's idea that individuals could bootstrap their way into conceptual knowledge on the basis of individual experience is equally in trouble.

The chapter is organized as follows. In Section 2, we outline the neorationalistic argument for the conclusion that some numerical knowledge is innate. Although we raise some concerns about the various steps in the argument, our main critique focuses on the first step—viz., that the behavior observed in 'numerical experiments' is knowledge driven. This critique is fully developed in Section 3. We start by pointing out that, according to a philosophical understanding of conceptual knowledge, there is no reason to assume that behavior is knowledge driven. While a psychological understanding of conceptual knowledge seems to offer more promising prospects, this reading also faces a number of objections drawn from the philosophy of mind. In Section 4, we offer an alternative view on the acquisition of numerical knowledge. This view emphasizes the constitutive role of the sociocultural process in the acquisition of knowledge.

To be absolutely clear about what we are arguing for and against in this chapter, we find nothing objectionable to the hypothesis that humans have innate discriminatory abilities—e.g., the ability to discriminate collections of objects in accordance with the cardinalities. Indeed, that humans have these abilities will be the starting point for our account of numerical knowledge acquisition. What we argue against, however, is the neorationalistic interpretation of these abilities according to which they are themselves driven by conceptual knowledge. This inflated interpretation has been popularized by Dehaene under the umbrella of the 'number-sense' hypothesis (cf. infra). What we show, then, is that we can make sense of numbers without possessing a number sense.

2. The Neorationalistic View of Numerical Cognition

Several theories that have been put forward as an explanation of our numerical abilities form a prime example of a neorationalistic account of cognition. These theories have two aims. First, they are offered as an explanation for the performance of young infants in numerous experiments. For example, Dehaene, Dehaene-Lambertz, and Cohen (1998) showed that infants distinguish between sets of objects in accordance to cardinality. Similarly, Jordan and Brannon (2006) found that infants match sets of equinumerous stimuli across modalities.[3] Second, neorationalistic theories try to give an account of the ontogenesis of our sophisticated numerical knowledge—i.e., they attempt to answer questions such as, "How can we learn numbers?", "How do we learn the basic arithmetic operations such as addition and subtraction?" By postulating innate numerical conceptual knowledge, neorationalists hope to provide an answer to both questions.[4]

The argument that leads to the postulation of innate knowledge has the following structure:

1. The observed behavior is knowledge driven, because it is intelligent;
2. The behavior is present in very young infants;
3. Hence the knowledge that drives the behavior must be innate.

As we will argue next, the first premise of the argument is highly problematic. Indeed, many proponents of the neorationalistic view on numerical cognition simply assume, without any argument, that the observed behavior is the product of knowledge. For example, in Stanislas Dehaene's defense of the number-sense hypothesis, the reformulation of the explanandum (i.e., the observed behavior of infants in the experiments) in epistemic or knowledge-laden terms is presented without any comment or argument:

> How can a 5-month-old baby *know* that 1 plus 1 equals 2? How is it possible for animals without language, such as chimpanzees, rats,

---

3 Although the idea that the experiments do show that infants are sensitive to numerical (and thus discrete) aspects of stimuli seems well entrenched, it is not uncontested. For example, Leibovich, Katzin, Harel, and Henik (2016) argue that the experimental findings are consistent with the idea that infants track continuous quantities (such as area or time) rather than discrete quantities (cardinalities).

4 Although theories differ somewhat with respect to the amount of numerical knowledge that is ascribed to neonates, typically, they assume that neonates innately have exact concepts for small numbers and approximate concepts for higher numbers.

and pigeons, to have some *knowledge of elementary arithmetic*? My hypothesis is that the answers to all these questions must be sought at a single source: the structure of our brain.

(Dehaene, 2011, xvii)

However, once this step has been taken, the natural subsequent question becomes how the knowledge that drives the behavior is acquired. In reply, two types of answers are offered: the empiricist's answer (individual experience and learning) and the rationalist's answer (innate). Neorationalists argue that in the light of the experimental evidence, the empiricist's answer cannot be sustained:

Indeed, it is hard to see how children could draw from the environment sufficient information to learn the numbers one, two, and three at such an early age. Even supposing that learning is possible before birth, or in the first few hours of life—during which visual stimulation is often close to nil—the problem remains, because it seems impossible for an organism that ignores everything about numbers to learn to recognize them.

(Dehaene, 2011, 50)

The argument for innateness is a poverty of stimulus argument. It is based on the idea that infants already possess knowledge at a stage in which past sensory stimulation was simply too impoverished to deliver that knowledge (for a criticism of this argument see, e.g., (Zahidi & Myin, 2016)).

The explanation for the presence of innate knowledge is then sought in evolutionary theory (see, e.g., Dehaene, 2011, 30). If innate numerical knowledge is the product of evolution, then species that are evolutionary close to us should display similar abilities as humans. And this is indeed the case. For example, the ability to distinguish between sets that differ in cardinality has been experimentally established for evolutionary close species such as chimpanzees (Boysen, Bernston, & Mukobi, 2001) and rhesus macaques (Hauser, Carey, & Hauser, 2000).[5] These experimental findings, when given an epistemic interpretation, are taken to lend support to the neorationalistic idea that some form of conceptual knowledge of number is part of our genetic endowment. In what follows however, we will question whether the conclusion—viz., that infants have innate knowledge of numbers—is sufficiently justified by the experimental observations. To be clear about the scope of our criticism, we do not deny that there are innate mechanisms pertaining to our discriminatory abilities—e.g., the ability to discriminate between sets of objects in accordance with their cardinality. We will, however, argue against the idea that these mechanisms are driven by conceptual knowledge about numbers.

5 These abilities are not restricted to phylogenetic close cousins but are also found in more distant relatives (e.g., chickens (Rugani, Regolin, & Vallortigara, 2008) and guppy fish (Lucon-Xiccato, Petrazzine, Agrillo, & Bisazza, 2015)).

In order to do so, we will focus on an idea that, although seldom explicitly stated, seems to motivate much of neorationalistic theorizing—viz., the idea that postulating innate knowledge offers the best explanation of the observed abilities in infants. If we are correct that there is insufficient justification for seeing these abilities as a display of knowledge, then the problem of giving an explanation for the origins of such conceptual knowledge becomes moot.

3. Neorationalism's Core Presupposition

As already noted, (neo)rationalism seems to be predicated on the idea that intelligent—i.e., flexible and adaptive, behavior is always the product of knowledge. For (neo)rationalism, then, any type of behavior falls between two extreme types of behavior. On the one hand, there is behavior that is automatic, reflex-like, and fixed. That kind of behavior can be fully explained without appealing to any knowledge. Then there is intelligent behavior, whose explanation does require the postulation of knowledge. All behavior, according to the neorationalist, can be classified along this epistemic dimension. But as Ryle pointed out it

> is a ruinous but popular mistake to suppose that intelligence operates only in the production and manipulation of propositions, i.e., that only in ratiocinating are we rational.
>
> (Ryle, 1946/2009, 228)

If Ryle is right, then behavior can be classified along two different dimensions: the intelligence dimension and the knowledge dimension. Unless provided with a substantial argument, we cannot assume that these two dimensions coincide. Rationalism, old or new, collapses two (prima facie) independent dimensions into one dimension, without offering an argument as to why we should do so.

The moral of this digression is that when trying to explain a form of intelligent behavior, such as the behavior displayed by infants in the aforementioned 'numerical' experiments, postulating knowledge as explanans requires a prior argument showing that, in the case at hand, the intelligent behavior is indeed knowledge driven. However, neorationalists usually omit this crucial step. A notable example is De Cruz and De Smedt (2010), who actually argue that the 'number-sensitivity' of infants results from knowledge deployment.

De Cruz and De Smedt focus on a line of experimental investigations started by Karen Wynn (1992). These experiments show that infants look longer at a scene when the occurring changes seem to be contrary to the rules of addition. This is considered indicative of a violation of the infants' expectations. Explaining the behavior thus requires explaining the infants' capacity to form expectations, which, according to Wynn, is best done by assuming that

> infants can compute the results of simple arithmetical operations. In sum, infants possess true numerical concepts—they have access to the

ordering of and numerical relations between small numbers, and can manipulate these concepts in numerically meaningful ways. [. . .] The existence of these arithmetical abilities so early in infancy suggests that humans innately possess the capacity to perform simple arithmetical calculations, which may provide the foundations for the development of further arithmetical knowledge.

(Wynn, 1992, 750)

While Wynn does not develop an extended argument for these conclusions, this task has been taken up by De Cruz and De Smedt (2010). They present an argument whose structure mirrors that of Dehaene's (as we expounded in Section 2 and which defends the idea that these infant abilities are driven by innate conceptual knowledge). In what follows, we will focus on the first step of the argumentation. This is the point at which they state that the performance can be best explained by the agent's 'conceptual knowledge of number'. However, it is important to note that De Cruz and De Smedt do not define the notion of 'conceptual knowledge'. This makes their claims, as we will see, ambiguous, for there are at least two possible readings of this claim: a philosophical and a psychological reading. These different readings turn on the different meanings of the notions of knowledge and concepts in philosophical and psychological theorizing about the mind (Machery, 2009).

While, we will provide more detail on these different meanings as we develop our argument, the crucial difference can be succinctly described as follows. For psychologists, knowledge is conceptual knowledge if it is knowledge that is stored in long-term memory. The concept of x just is the body of knowledge pertaining to x. For philosophers, on the other hand, concepts are those entities that allow us to have propositional attitudes (beliefs, desires, etc.) about features of the world. An individual has the concept of x if he can think about x. Furthermore, if knowledge is conceived of, as philosophers are prone to do, in terms of belief, truth, and justification, then having knowledge implies having concepts. What the precise relation is between the psychological and philosophical notions is a hotly debated issue, into which we need not enter here.[6] For our purposes, it will suffice to note that the ascription of conceptual knowledge in philosophical and psychological theories is based on different criteria.

With this distinction in mind, we can now turn to De Cruz and De Smedt's argument. In order to argue for the conclusion that the infants in Wynn's experiment exercise conceptual knowledge, De Cruz and De Smedt

---

6 This distinction does not imply that the philosophical notion is irrelevant for psychology or cognitive science. Arguably, cognitive science and psychology aim at describing the cognitive mechanisms that make conceptual knowledge (in the philosophical sense) acquisition possible. Invoking psychological conceptual knowledge to explain (philosophical) conceptual knowledge acquisition is but one option available to psychologists and cognitive scientists.

survey a number of alternative explanations based on lower cognitive (i.e., non-epistemic) faculties (such as long-term sensory persistence or familiarity preference) to explain the longer looking time of the agent. However, all of these rival explanations are rejected for experimental reasons. De Cruz and De Smedt argue that while these explanations might work to explain the results of the original experiments of Wynn, they have been falsified by subsequent refined versions of the original experiments. Hence the only hypothesis that accounts for all experimental findings, so they argue, is the hypothesis that in Wynn's experimental setup, infants exercise conceptual knowledge. Note that the argument is a negative argument: since no explanations involving only lower cognitive faculties can account for the behavior, the mechanism or faculty must be epistemic. But why should we accept this verdict since we have no idea how that mechanism is actually functioning as a *conceptual* mechanism? In other words, while we are given reasons as to why the mechanism cannot be a perceptual mechanism, we are given no positive reasons as to why this should be a conceptual mechanism. In short, what one needs is some set of criteria that allows us to ascribe conceptual knowledge to the agent.

First, on the philosophical understanding of concepts and conceptual knowledge, we may well admit that the intelligent behavior (i.e., behavior that is flexible and sensitive to changing features of the environment) as shown by the infants will figure among the criteria to ascribe numerical knowledge to them. There are good reasons, however, to assume that other criteria must be relevant as well. Ryle provides a strong case for the idea that "the propositional acknowledgment of rules, reasons or principles is not the parent of the intelligent application of them; it is a stepchild of that application" (Ryle, 1946/2009, 229). Ryle points to the fact that while behavior (e.g., that of the infants in Wynn's experiment) may look like an application of the conceptual knowledge that '1 + 1 = 2', the behavior in question is not sufficient for theoretical knowledge ascription. Put differently, on the philosophical reading of concepts, the fact that infants display certain behavior does not imply, pace Dehaene, that they have any propositional attitude toward, let alone knowledge of, the proposition '1 + 1 = 2'. Hence the experiments in no way show that they have conceptual knowledge about numbers (since the experiments do not show that they have the concept of the number one or the concept of addition).

What if we turn to the psychological notions of concepts and knowledge? Machery (2009) makes the following observation with respect to psychologists' use of the term knowledge:

> By "knowledge", psychologists mean any contentful state that can be used in cognitive processes. So defined, "knowledge" does not refer to states that are necessarily true and justified. Furthermore, "knowledge" does not refer to states that are necessarily explicit or propositional.

Rather, knowledge can be implicit or explicit; it can also be propositional, imagistic, or procedural.

(Machery, 2009, 8)

On this account, knowledge is constituted by contentful states that are
stored in memory, ready for recall by cognitive processes. Importantly, while
lower cognitive faculties access "their own proprietary memory stores",
higher cognitive faculties all have access to the same (multi-purpose) long-
term memory store. Concepts are those bodies of knowledge that are stored
in long-term memory for access by higher cognitive faculties.

On such a construal, then, the argument by De Cruz and De Smedt might
look more promising. Having the (psychological) concept of numbers '1'
and '2' or having conceptual knowledge concerning addition does not imply
that one is in a position to have beliefs concerning the proposition '1 + 1 = 2'.
Hence the criteria for knowledge ascription seem to be more relaxed than in
the philosophical case. The only criterion, apart from the behavior, is that
the knowledge is stored in long-term memory—i.e., in memory that is not
devoted to one type of cognitive function. Here the remark that no lower
cognitive mechanisms can account for the behavior (coupled with various
other experimental findings—e.g., that infants are able to match equinumerous series of stimuli across modalities) seems highly relevant. Nonetheless,
another worry arises. In contrast with the philosophical interpretation of
'conceptual knowledge', we could perhaps attribute conceptual knowledge
to the experimental agents on the basis of the experiments. What would still
be lacking, however, would be an explanation of how this knowledge results
in the behavior. In other words, if behavior is a criterion for the ascription
of knowledge, then we have still not begun to explain the behavior. For the
criteria are not explained by what they are criteria for.

The same worry can be phrased somewhat differently. Saying that up
until now reliance on lower cognitive faculties has not generated a good
explanation for the observed behavior does not imply that the ascription of
conceptual knowledge results in an explanation, let alone a good explanation. An explanation relying on conceptual knowledge would require identifying what kind of knowledge is coded in the concepts employed, what
format these concepts are supposed to have, how the concepts get applied,
how the agent knows what bodies of knowledge to apply in a given context,
etc. These and other related questions are not answered by Wynn or De
Cruz and De Smedt. Hence glossing the argument for conceptual knowledge
as an inference to the best explanation is incomplete since it leaves a number
of important questions unanswered.

However, even if an answer to these questions were given, a more fundamental problem remains. Recall that the psychological notion of knowledge
is defined in terms of contentful states, or equivalently in terms of mental representations. Hence the explanation of the performance of infants in terms of
conceptual knowledge stands or falls with the explanatory power of theories

in which contentful states play an important role. Ultimately, the explanatory power of contentful states turns on their existence as mental states with semantic content, which in turn depends on giving an answer to the hard problem of content—viz., the question of whether one can give a scientifically respectable account of how mental representations acquire content. Since no naturalistically plausible answer to this problem has been provided, the explanatory value of relying on contentful mental states remains unestablished.

4. The Ontogenesis of Numerical Knowledge

As we have seen the neorational picture of numerical cognition (and cognition more general) faces a number of serious obstacles. One way to avoid these obstacles is to abandon that picture and replace it with a less intellectual view of the mind. Recall that the central intuition behind this picture is that intelligent behavior is always the product of conceptual knowledge deployment.[7] Hence breaking the spell of the neorational view of cognition implies, inter alia, that we should allow for intelligent behavior that does not involve knowledge.

An example of such a view of cognition is provided by radical enactivism (REC) (Hutto & Myin, 2013, 2017). A central tenet of REC is that basic intelligent behavior can be explained without invoking contentful representations and hence without invoking knowledge. This is not to deny that contentful cognition exists or is possible, but is rather to insist that contentful cognition only arises through sociocultural practices. Content is then not a natural property of individual minds, but primarily a public affair. Since concepts are contentful representations, REC denies that basic minds operate with concepts and, a fortiori, negates the intuition that drives neorationalism—viz., all intelligent behavior is driven by conceptual knowledge. More can be said, of course, about how this view of cognition can be developed (Hutto & Myin, 2017). Here we will suffice with illustrating the REC picture of mind by giving a RECish view on the ontogenesis of numerical knowledge. In other words, we will give an account of how an individual can acquire numerical knowledge without relying on mental representations. In order to do so, we will assume that the individual is situated in an environment in which numerical knowledge is present—i.e., an environment in which there are already numerical knowledge users. Of course, in order for this to be plausible, one would also need to show that the phylogenetic origins of numerical knowledge can be explained in a way that is compatible with REC (see Zahidi & Myin, 2016) for a sketch of such a story).

---

7 In Section 3, we distinguished between a philosophical and a psychological notion of conceptual knowledge in order to be clear which notion of knowledge we are dealing with. We pointed out that psychological conceptual knowledge is routinely invoked to explain the acquisition of conceptual knowledge (in the philosophical sense). We argued that the psychological notion was problematic because of its reliance on contentful mental states. Hence we will present an alternative account for the acquisition of conceptual knowledge (in the philosophical sense).

It is necessary to make more explicit what we mean by the acquisition of numerical knowledge. On the basis of what criteria do we judge that someone has acquired knowledge? On the socio-historic view of knowledge, to have acquired knowledge in a certain domain is to be able to participate in the relevant sociocultural practices. The relevant practices for numerical knowledge are those practices that deal with, or target, numerical properties (size, cardinality, order) of the environment. Participation in such a practice requires more than merely imitating what the other participants do. It requires, for example, that one can autonomously initiate actions that are in accordance with the norms governing the practice. It means, for example, that one should be able to judge whether certain actions (e.g., utterances) are acceptable in that practice. Consider the criteria for determining whether one has acquired numerical knowledge. Having numerical knowledge requires that one has mastered the techniques of counting and calculating. That one has mastered the technique of counting is shown by being able to correctly answer questions about the numerosity of a given collection of objects, that when one is presented with a collection of objects one can correctly answer the question how many there are, or when told that there are 27 marbles in a bag, one can check whether the given number is correct, etc. Similarly, one has mastered the technique of calculating when one is able to do sums and subtractions one has never encountered before. Or, as Ryle (1967/2009) puts it,

> Nor does the pupil know that 7 + 7 = 14 while this is for him only a still undetachable bit of a memorized sing-song, but only when, for example, he can find fault with someone's assertion that 7 + 8 = 14, or can answer the new-type question How many 7s are there in 14? or the new-type question "If there are seven boys and seven girls in a room, how many children are in the room?" etc. Only then has he taken it in.
> (Ryle, 1967/2009, 470)

Since, on this account, having knowledge consists in being able to participate in practices, acquiring knowledge requires that one should learn how to adapt one's behavior in accordance to the rules and norms governing that practice.

To explain how an individual acquires numerical knowledge, we will start from the two crucial resources that are available to the infant: on the one hand, a set of innate behavioral dispositions and non-epistemic abilities and, on the other hand, the epistemically loaded sociocultural environment. Examples of innate behavioral dispositions include suckling behavior, crying when hungry, etc. Among those behavioral dispositions and abilities, we can include the non-epistemic abilities exemplified by the aforementioned 'numerical' experiments. It is important to note that these behaviors are abilities that are universal (i.e., they are available for any healthy infant). And to reiterate, these behaviors are not governed by semantic norms (and hence non-epistemic) but are part of a repertoire of (contentless) causal propensities.[8] Apart from those innate behaviors and causal propensities that are directly relevant for the acquisition of numerical knowledge, there are more general (biological) properties of the human behavioral repertoire, body, and brain that are important. For example, if learning is to take place, then both body and brain need to be plastic. This plasticity ensures that learning or training can have long-term effects on the structure of behavior, body, and brain.

So how does an organism with those abilities acquire numerical knowledge—i.e., becomes an autonomous participant in numerical practices? To answer this question, we will rely on Meredith Williams's Wittgensteinian account of language learning. Williams (2010) argues that the learning situation is a triadic relation between novice, master, and the world. For learning to start, it needs to be the case that both novice and master share a way of responding to the world "including perceptual saliences in the world" (Williams, 2010, 217). That shared way of responding to the world includes, inter alia, the differential responses to the world with respect to numerical properties of the stimulus. These shared responses form the basis for training. For example, the master can point to two puppets and put two fingers up or utter the word 'two' and prompt a child to imitate him. Or she can show three puppets and after removal exclaim, "One is missing". The aim of such games or interchanges is to prompt the novice to imitate the master's behavior, but this would be impossible of the novice did not make the same kind of distinctions that the master does. In this way, the novice is trained to display a patterned type of behavior (e.g., when being presented with four items, the child exclaims, "Four", or puts four fingers up). Of course, this does not make the novice an autonomous participant in the numerical practices, since although his behavior after training does conform to the norms of the practice, he will not perform these actions because of his knowledge of the norms. Or, as Williams puts it,

> Just as ostensive teaching exploits the causally based perceptual sensitivities [. . .] so training in practices exploits the malleability of our behavior and responsiveness to sanctioning. [. . .] Neither of these causal propensities, which we share with many other animals, is to be confused with normative structuring, governed as it is by standards of correctness, not laws of causal connection. These natural dispositions of acquiring regularities in our lives are exploited in initiating the novice into the normative regularities of custom or practices.
>
> (Williams, 2010, 362)

---

8 Bangu (2012) shares our skepticism about the neorational interpretation of these experiments. He uses the results of Wynn's experiment to develop a Wittgensteinian view on the nature of mathematical propositions. Although Bangu's project differs from ours, there is a connection with the Wittgensteinian account of learning that we develop next.

There is, however, something that the master brings to the training situation that is initially lacking in the novice and what can serve as a scaffold for the novice to go from an imitator to an autonomous practitioner. Indeed, the master has already acquired numerical knowledge and thus knows the norms of numerical practices. In particular, this means that it is only the presence of the master that allows for judging the correctness of the novice's actions. Absent a master there simply are no standards of correctness. For example, when a novice is reciting the number words while subsequently stretching fingers on the hand, that behavior, by itself, can neither be correct nor incorrect as counting behavior since the novice is not an autonomous practitioner of counting. However, the master treats the novice as if he or she were a competent practitioner of the counting practice and hence open to correction. It is by being informed as to what the correct patterns of behavior are, and by being trained to behave in the correct way in certain circumstances, that the novice can become an autonomous practitioner. Importantly, numerical training involves training in counting and calculation techniques. What that means is that a novice is not only trained to respond "12" to the question, "How much is 3x4?", but also trained in various techniques to work out multiplications he has never encountered before. To use slightly different terminology, training involves training in the application of various numerical algorithms. Such training does not require any numerical knowledge; the novice is simply trained to perform a certain set of actions in a set of similar circumstances.

However, it might be objected that this introduces a circularity. For we have argued that training in a practice does not presuppose any knowledge of norms, yet if training involves learning to perform a certain set of actions in similar circumstances, it seems that the novice should be able to judge whether a given situation is similar to a situation in which the set of actions should be applied. Consider, for example, a novice who has been trained in techniques or algorithms to perform addition and multiplication by way of many examples of applications of these techniques. The novice is then asked, "How much is 7 times 8?" In order to answer that question, the novice has to know that the actions he has to perform are those that go with the techniques of multiplication and not those that go with the technique of addition. It would seem that in order to decide, he has to know already what addition and multiplication are. In other words, it would seem that the novice has to know the relevant norms of similarity (this problem is akin to the problems encountered before—viz., which multiplication technique was appropriate) in order to be able to be trained. This would be the case if the novice were an isolated individual.

By contrast, however, in the learning situation, he is constantly open to sanctioning and approval by the teacher. Suppose that, in the earlier example, the novice starts performing actions that belong to the technique of addition. The master can interrupt the novice and point out that, in this situation, the technique for multiplication is required. She may further offer guidance, such as, "When the word 'times' is used in the question, you have to use the multiplication technique". This kind of guidance neither presupposes knowledge of relevant similarity norms nor knowledge of numerical operations. It merely presupposes that the novice can recognize certain sounds or words and can associate a certain behavioral pattern to that sound or word. This training might involve the manipulation of body parts or external artifacts that, in the given practice, come to serve as numerical representations. However, for the novice in the initial learning situation, these artifacts are at first merely objects to be manipulated in the way he is told. It is only after the trainee has become an autonomous practitioner that these objects acquire their meaning as numerical representations. After completing a successful training period, the novice is then able to participate fully in the numerical practices of his community. It is at this point that the novice has acquired numerical knowledge and has become himself a competent 'mathematician'.

5. Conclusion

In this chapter, we have critically examined the widespread idea that infants possess knowledge of numbers independent of learning. We found this idea wanting on two accounts. First, we have argued that on a philosophical account of knowledge, the experiments on which neorationalistic arguments have been based do not show that any conceptual knowledge is involved. Second, we have argued that, on the psychological understanding of conceptual knowledge, the proposed neorationalistic 'explanations' face a number of obstacles.

If our analysis is correct, then it would seem that the main problem of traditional rationalism is not only that it lacks a convincing story about the origins of innate knowledge (which neorationalism supposedly solves) but also that it has a view of the mind and cognition that equates intelligence with knowledge. Since neorationalism inherits this assumption, it is no advance on traditional rationalism in explaining the acquisition of knowledge. Our arguments should not give much joy to empiricists who share the basic assumption about the mind since their position is vulnerable to the same objections. Rather than taking sides in the debate between rationalism and empiricism, our arguments point to a way beyond it by recognizing that knowledge and knowledge acquisition constitutively involve the sociocultural world. This is not to deny the importance of certain innate behavioral dispositions or non-epistemic abilities. It is merely to point out that these do not constitute knowledge. Only making this distinction between capacities that don't and that do involve knowledge (and the related distinction between intelligence and knowledge) allows us to do justice to the inherently normative structure of knowledge fully.

Acknowledgments

The research of both authors is supported by the Research Foundation Flanders (FWO, project *Getting Real about Words and Numbers* [G0C7315N]). The authors would like to thank Victor Loughlin for helpful comments on a previous version.

232   *Karim Zahidi and Erik Myin*

References

Bangu, S. (2012). Wynn's experiments and later Wittgenstein's philosophy of mathematics. *The Jerusalem Philosophical Quarterly, 61*, 219–241.

Boysen, S. T., Bernston, G. G., & Mukobi, K. L. (2001). Size matters: Impact of item size and quantity on array of choice by a chimpanzee (*Pan troglodytes*). *The Journal of Comparative Psychology, 115*, 106–110.

Carruthers, P. (1992). *Human knowledge and human nature: An introduction to an ancient debate.* Oxford: Oxford University Press.

Chemero, A. (2009). *Radical Embodied Cognitive Science.* Cambridge, MA: MIT Press.

De Cruz, H., & De Smedt, J. (2010). The innateness hypothesis and mathematical concepts. *Topoi, 29*(1), 3–13.

Dehaene, S. (2011). *The number sense: How the mind creates mathematics* (Revised and Expanded ed.). Oxford: Oxford University Press.

Dehaene, S., Dehaene-Lambertz, G., & Cohen, L. (1998). Abstract representations of numbers in the animal and human brain. *Trends in Neuroscience, 21*, 355–361.

Greeno, J., Collins, A., & Resnick, L. B. (1996). Cognition and learning. In B. Berliner & R. Calfee (Eds.), *Handbook of educational psychology* (pp. 15–46). New York: Simon & Schuster MacMillan.

Hauser, M. D., Carey, S., & Hauser, L. (2000). Spontaneous number representation in semi-free-ranging rhesus monkeys. *Proceedings of the Royal Society London B, 267*, 82–833.

Hutto, D. D., & Myin, E. (2013). *Radicalizing enactivism: Basic minds without content.* Cambridge, MA: MIT Press.

Hutto, D. D., & Myin, E. (2017). *Evolving enactivism: Basic minds meet content.* Cambridge, MA: MIT Press.

Jordan, K. E., & Brannon, E. M. (2006). The multisensory representation of number in infancy. *Proceedings of the National Academy of Sciences of the United States of America, 103*(9), 3486–3489.

Kitcher, P. (1983). *The nature of mathematical knowledge.* Oxford: Oxford University Press.

Leibovich, T., Katzin, N., Harel, M., & Henik, A. (2016). From 'sense of number' to 'sense of magnitude'—The role of continuous magnitudes in numerical cognition. *Behavioral and Brain Sciences*, 1–62. doi: 10.1017/S0140525X16000960

Lucon-Xiccato, T., Petrazzine, M. E. M., Agrillo, C., & Bisazza, A. (2015). Guppies discriminate between two quantities of food items but prioritize item size over total amount. *Animal Behavior, 107*, 183–191.

Machery, E. (2009). *Doing without concepts.* Oxford: Oxford University Press.

Prinz, J. (2002). *Furnishing the mind: Concepts and their perceptual basis.* Cambridge, MA: MIT Press.

Ramsey, W. (2007). *Representation reconsidered.* Cambridge: Cambridge University Press.

Rugani, R., Regolin, L., & Vallortigara, G. (2008). Discrimination of small numerosities in young chicks. *Journal of Experimental Psychology: Animal Behavior Processes, 34*(3), 388–399.

Ryle, G. (1946/2009). Knowing how and knowing that. In *Collected essays: Volume 2* (pp. 222–235). London: Routledge.

Ryle, G. (1967/2009). Teaching and training. In *Collected essays: Volume 2* (pp. 464–478). London: Routledge.

Tymoczko, T. (1984). Gödel, Wittgenstein and the nature of mathematical knowledge. In *PSA: Proceedings of the biennial meeting of the philosophy of science association* (pp. 449–468). Chicago: Philosophy of Science Association.

Williams, M. (2010). Normative naturalism. *International Journal of Philosophical Studies, 18*(3), 355–375.

Wynn, K. (1992). Addition and subtraction by human infants. *Nature, 358,* 749–750.

Zahidi, K., & Myin, E. (2016). Radically enactive numerical cognition. In G. Etzelmüller & C. Tewes (Eds.), *Embodiment in evolution and culture* (pp. 57–71). Tübingen: Mohr Siebeck.

13 Beyond Peano
Looking Into the Unnaturalness of Natural Numbers

Josephine Relaford-Doyle and Rafael Núñez

Introduction

The counting numbers 1, 2, 3, . . . are ubiquitous in our daily lives. They are the first numbers we meet as young children and provide the building blocks for the arithmetic we learn later. Perhaps on account of their importance in our daily experience, these numbers are often assumed to be a 'natural domain of competence' (Antell & Keating, 1983), a 'natural perceptual category' (Nieder, 2016) about which we must certainly develop meaningful and usable concepts. Importantly, the counting numbers overlap with the formal mathematical set of natural numbers {1, 2, 3, 4, . . .}, and it is generally assumed that the number concepts we develop as children are in fact concepts of *natural number*. The importance of natural numbers in both practical and formal mathematics has led philosophers of mathematics and developmental psychologists to dedicate tremendous effort to the question, how do we develop concepts of natural number?

This deceptively simple question can in fact be broken down into two parts: (a) what is the nature of the natural number concept?, and (b) how do we come to develop that concept? These two questions have received an imbalance of empirical attention. Question (b), the question of *how*, forms the basis of a huge research program in developmental psychology and sees a large degree of disagreement. One line of research suggests that natural number concepts are innate and hardwired into every human's neural machinery (Gelman & Gallistel, 1978; Leslie, Gelman, & Gallistel, 2008), while others argue that developing concepts of natural numbers requires an inductive inference as the child learns to count (Carey, 2004; Sarnecka & Carey, 2008). Yet another group claims that knowledge of natural numbers must be learned as generalizations in a 'math schema' (Rips, Bloomfield, & Asmuth, 2008). While inconsistent, these views are all (in principle) empirical claims that can be subject to support and refutation from observable phenomena.

By contrast, question (a), the question of *what* the natural number concept is, shows a remarkable degree of consensus among developmental psychologists. In all major lines of research, the 'mature' (i.e., fully developed) natural number *concept* is assumed to mirror the *formal characterization* of natural numbers given in the Dedekind-Peano axioms (although see DeCock, 2008 for an alternate proposal based on Hume's Principle). Specifically, it is assumed that concepts of natural number are characterized by a unique starting value ('one') and knowledge of the *successor principle*: for any natural number n, the next natural number is given by $n + 1$. The assumption that the concepts are constrained by the formalization given in the Dedekind-Peano axioms is taken as unproblematic; in fact, some psychologists go so far as to thank Dedekind and his contemporaries for giving the field "a firm idea about the constituents of the natural number concept" (Rips et al., 2008, 640).

So while researchers conduct careful empirical work into *how* natural number concepts are developed, it is taken as given that the concept they are investigating is consistent with the formal mathematical characterization of natural numbers. Is this a valid assumption? It has been shown in higher level domains such as calculus that human mathematical conceptualizations can differ dramatically from mathematical formalizations (Núñez, 2006; Marghetis & Núñez, 2013). While it may be tempting to think of natural numbers as a basic concept, Peano didn't provide his axioms until 1889, two centuries *after* the invention of calculus. Further, while the natural numbers are ubiquitous in *our* lives, healthy adults in many isolated cultures around the world speak languages which have very limited number lexicons, often ending at 'six' or below (Núñez, 2017). Like calculus, the infinite set of natural numbers is *not* foundational to human experience, and so perhaps there is room for even this seemingly simple concept to vary from its formal characterization.

In this chapter, we challenge the assumption that the 'mature' natural number concept mirrors the formal characterization of natural numbers given by the Dedekind-Peano axioms. We first critically examine the empirical work focusing on children's understanding of the successor principle and find little evidence suggesting children ever come to understand its deeper implications. We then describe insights into the natural number concept that we gained via a new method using visual proofs to study informal mathematical induction in adults. Specifically, we present evidence suggesting that in adult mathematical practice, even highly educated people often seem to rely on conceptualizations of natural number that are different from, and even at odds with, the characterization given by the Dedekind-Peano axioms. We do not claim to be able to characterize natural number concepts on the basis of this evidence alone. Rather, we take it as an indication that those interested in the development of natural number concepts would be wise to open the question of *what* those concepts are to empirical investigation rather than relying on the mathematical formalization to provide the answer.

1.   What Do Children Know About the Successor Principle?

The successor principle, given in function notation as $S(n) = n + 1$, is a defining element of the Peano axioms and is assumed to be a key piece of our natural number concept. Researchers believe that understanding the successor principle allows children to develop exact representations of large numbers (Sarnecka & Carey, 2008), eventually understanding "the logic of natural number" (Cheung, Rubenson, & Barner, 2017) and recognizing the natural numbers as an unbounded and countably infinite set (Rips et al., 2008). Substantial empirical work has focused on children's understanding of the successor principle. However, because it is taken as given that this principle exists in the mature natural number concept, the question motivating this work is *when* (not whether) this principle is learned. Because of this, the extensive research regarding the successor principle reveals surprisingly little about what children actually know about it.

The successor principle itself seems simple enough: for any natural number n, the next natural number is given by $n + 1$ (and named by the next item in the count list). However, as in the axioms, this principle carries important and deep implications about natural numbers, including,

I.   The natural numbers are unbounded, since one can always be added to produce the next number.
II. Adding one is the *only* way to generate the next natural number.
III. The same relation holds between any natural number and its successor (e.g., the pairs 3 & 4, 8 & 9; 9,381,763 & 9,381,764; and 10^{23} & $10^{23} + 1$ are all governed by the same *+1* relation).
IV. No natural number is more 'natural' than any other.

These properties distinguish the natural numbers—the target of developmental accounts—from the counting numbers. The natural numbers extend infinitely, while the set of nameable counting numbers—the numbers a person could actually recite in a count sequence—is finite. In our experiences with counting numbers, not all numbers are alike. The smallest counting numbers 'one' through 'nine'—the *prototypical counting numbers* (PCNs)—are frequently encountered and highly familiar (Núñez & Marghetis, 2014). The PCNs can be written using single-digit numerals, and the associated number words are monomorphemic. Larger counting numbers (like, say, 197) are encountered far less frequently, require composite notation, and are lexically more complex. In the domain of *natural* number, however, these differences cease to be relevant. Any number that can be reached by successive additions of one, no matter how large, is a natural number; there are no 'degrees' of naturalness.

A theory of *natural number* concept development that includes the successor principle and does not specify or restrict its application commits itself to the deeper properties outlined above. Specifically, by any such theory, anyone who has learned the successor principle should demonstrate at least implicit understanding of (I)–(IV). In fact, there is very little evidence to suggest that children ever grasp these facts. Empirical work is limited by the assumption that the successor principle is a constituent of the mature natural number concept that all children eventually develop. This assumption has important implications for research:

- First, researchers often choose to examine only a *tiny* subset of natural numbers, assuming (falsely) that evidence for the successor principle anywhere is evidence for it everywhere.
- Second, due to convenience, researchers focus primarily on children's understanding of the PCNs—a highly peculiar subset of the natural numbers.
- Third, with presuppositions about the nature of the concept in mind, researchers seek out *any* (even very weak) evidence that it exists, while ignoring evidence that suggests children may in fact have other, perhaps conflicting conceptualizations.

Sarnecka and Carey (2008) examined children's understanding of the successor principle within the PCNs. In their Unit Task, the researcher began each trial by placing a number of items in a box and telling the child, for example, "I'm putting FOUR frogs here". The researcher then added either one or two more items to the box, and asked the child, "Now is it FIVE, or is it SIX?" Importantly, the highest number that a child was ever asked about was seven. If a child could answer these questions correctly, he or she was considered to have learned the successor principle—namely, that the numerals in the count sequence represent cardinalities increasing by exactly one unit. Certainly, though, a child who knows that adding one frog to a box with six frogs results in seven frogs has not demonstrated understanding of a full-fledged successor function. This issue would likely not bother Sarnecka and Carey, who have stated that they are interested only in how children generate specific natural number concepts (such as 'seven'), not concepts of the natural numbers themselves, including (I)–(IV) (e.g., Sarnecka, 2008). However, without further specification of the successor function, it is unclear why their account should explain one and not the other.

When researchers do look beyond the PCNs, they quickly find evidence that the story of the successor principle is not so straightforward. Two studies used versions of the Unit Task to assess children's knowledge of the successor function, but extended the questions to 'large numbers' (24 and 25; Davidson, Eng, & Barner, 2012) and 'very large numbers' (53, 57, 76, 77; Cheung et al., 2017). Both studies found that many children could identify successors only for the smallest numbers (i.e., the PCNs), even when they could comfortably count to the larger numbers and even higher. Only older children with considerable counting experience could successfully name the successors for all items in their count lists. These studies thus revealed that knowledge of the successor principle is not immediately applied to all number words, but initially restricted only to the smallest, most familiar numbers. For a time, at least, it would seem that not all natural numbers are equally natural.

Most studies in number cognition, including those described earlier, focus on children's understandings about particular familiar natural numbers and can thus tell us very little about whether children understand deeper facts such as (I)–(IV). However, some studies have conducted interviews with children to assess when they understand (I) that the successor principle implies the natural numbers are unbounded. For example, Cheung et al. (2017) supplemented their study of children's knowledge of the successor principle in "very large numbers" with an interview probing whether they understood that the natural numbers extend infinitely. The following transcript (Cheung et al., 2017, Appendix B) shows a conversation between an experimenter and a 5-year-old child. On the basis of this interview, the child was classified as understanding that every natural number has a successor and thus that the natural numbers are infinite.

(1) Experimenter: So if we thought of a really big number, could we always add to it and make it even bigger, or is there a number so big we couldn't add any more?
(2) Child: No, we can add to it.
(3) E: Why is that?
(4) C: Because there would be a million, and then you could keep adding numbers if you know them.
(5) E: So you said that the biggest number you know is a million. Is it possible to add one to a million, or is a million the biggest number possible?
(6) C: You can add one to it.
(7) E: Why?
(8) C: Because it would be like a million one, a million two.
(9) E: Could I keep adding one?
(10) C: [child nodded]
(11) E: Why?
(12) C: Because, um, because if you want to you can keep adding and adding and make it fun.

Suppose this child had also earlier demonstrated that they could consistently identify the successor for all the numbers in their count list, including 'very large numbers' like 77. In that case, the authors would conclude that this child understood "the successor function as defined in the Peano axioms—i.e., that every number n has a successor $n + 1$, such that numbers never end" (25). Does the evidence support this claim? Notice that it is the experimenter, not the child, who first mentions adding one as a way to make a number bigger (line 5)—that is, the experimenter explicitly mentions the successor function in their question. In five separate responses regarding how to make bigger numbers (lines 2, 4, 6, 10, 12), the child only explicitly mentions adding one once (6), immediately following the experimenter's mention of it in (5). In most of their responses, the child doesn't specify a number to be added, and the child's response in line 4 ("keep adding numbers") indicates that the child may in fact recognize that any number will do the job. Thus, by our reading, this child seems to understand two separate facts: first, that *counting* represents an increase by one each time ("a million one, a million two", line 8), and second, that you can always generate bigger numbers by adding. We see no evidence that this child's understanding of the unbounded nature of natural numbers relies *uniquely* on the successor principle. That is, this child may understand the logic of *counting*, but there is no evidence that the same logic governs the child's understanding of the infinite set of natural numbers.

Researchers' assumption that the successor function is a constituent of the mature concept infuses their interview protocol and narrows their lens on the data. This is especially clear in the original work using this interview method (Hartnett & Gelman, 1998). In this excerpt (354), a researcher interviews a 7-year-old who is subsequently classified as understanding the successor function:

(1) Experimenter: If I thought of a really big number, could I always add to it and get a bigger number? Or is there a number so big that I couldn't add anymore; I would have to stop?

(2) Child: You could always make it bigger and add numbers to it.

(3) E: If I count and count and count, will I ever get to the end of the numbers?

(4) C: Uh uh.

(5) E: Why not?

(6) C: Because there isn't one.

. . .

(7) E: How can there be no end to the numbers?

(8) C: Because you see people making up numbers. You can keep making them, and it would get higher and higher.

. . .

(9) E: You mean there's no last number?

(10) C: Uh uh. 'Cause when you get to a really high number, you can just keep on making up letters and adding one to it.

(11) E: Okay. Well, what if somebody came up to you and said that a googol was the biggest number there could ever be? . . . When you make a googol, you write a one and a hundred zeroes. It's a really big number. So if somebody came up to you and said, "I think a googol is the biggest number there could ever be", would you believe him?

(12) C: Uh uh. Because you could put one and a thousand zeroes on it.

This interview clearly reveals that this child's understanding of the unbounded nature of natural numbers does not rely exclusively on a successor principle.

In this brief exchange, the child offers *four* descriptions of how to generate bigger numbers: adding numbers (line 2), making them up (line 8, perhaps referring to children's use of nonsense words such as 'bajillion' and 'gazillion' to indicate bigger and bigger numbers), "making up letters and adding one to it" (line 10, possibly referring to a successor function, or possibly terms such as 'bajillion and *one*'), and adding *digits* to create a longer numeral string (line 12). This child certainly understands that there is no largest number and may also understand that adding one is *one way* to produce a larger number, but this interview does not give us any reason to believe that this child understands that every natural number has a unique successor that is produced by adding exactly one. In classifying this child as understanding the successor principle, Hartnett and Gelman ignore the extensive evidence, suggesting a much wider variety of conceptualizations. This is a missed opportunity, since it is exactly these conceptualizations that, if explored, could give us a deeper insight into the true nature of the child's natural number concept.

In sum, these studies reveal a pattern: researchers design their studies with the assumption that the successor principle is a constituent of the 'mature' natural number concept, seek out any evidence that it exists, and ignore data suggesting alternative conceptualizations. As a result, we know surprisingly little about the true role of the successor principle in the fully developed natural number concept. In particular, we have very little empirical evidence that can shed light on whether people—children or adults—understand deeper implications of the successor principle like (I)–(IV). How could we go about exploring this empirically, without making assumptions about the nature of the concept we're investigating? One potentially fruitful course would be to examine people's understanding of general mathematical principles that rely on the structure of the natural numbers rather than asking directly about the natural numbers themselves. By looking into people's understanding of principles that hold for *all* natural numbers, we may get some insight into how they conceptualize, not only particular natural numbers but also natural numbers as an infinite set (see Rips et al., 2008 for a discussion of this approach using the general principle of commutativity of addition). And perhaps the most effective general principles to study would be those that could be formally proven using *mathematical induction*, a formal proof method rooted directly in Peano's axioms and which explicitly relies on the successor principle.

2.   Mathematical Induction and Visual Proofs

A formal proof by mathematical induction demonstrates that a theorem is true of all (infinitely many) natural numbers in two steps. First it is shown by direct verification that the theorem is true of a base case, generally $n = 1$.

Next, in the inductive step, it is shown that if the theorem is true for some natural number n, then it must necessarily be true of its successor $n+1$. This formal proof strategy hinges on the definition of the natural numbers given in the Peano axioms and so provides a promising arena in which to explore the extent to which natural number concepts mirror the formal characterization.

A drawback, of course, is that mathematical induction is a highly technical, specialized proof-writing practice that requires explicit and extensive training. In fact, with its notational complexity and algebraic sophistication, mathematical induction is a notoriously difficult method for people, even educated adults, to learn (Avital & Libeskind, 1978; Fischbein & Engel, 1989; Movshovitz-Hadar, 1993). Thus if we seek to explore natural number concepts by looking at how people engage in mathematical induction, we restrict ourselves to the tiny subset of the adult population that is proficient in this proof method—namely, skilled mathematics students and professionals.

Visual proofs provide an exciting opportunity to study the intuitions of mathematical induction in a much wider population. Visual proofs are images, often quite simple and ideally notation- and text-free, which are designed to demonstrate that a mathematical theorem is true. Visual proofs exist for theorems in a variety of domains in mathematics (for a collection of examples see Nelsen, 1993), but of particular interest here are visual proofs for theorems that could be formally proven by mathematical induction. For instance, consider the theorem that, for any natural number n, the sum of the first n odd numbers is equal to n^2 (i.e., $1 + 3 + \ldots + (2n-1) = n^2$). A visual proof of this theorem is given in Figure 13.1 (after Brown, 1997).

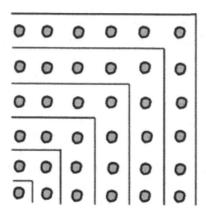

Figure 13.1 Visual proof that $1 + 3 + \ldots + (2n-1) = n^2$

Drawn after Brown (1997)

In the image, consecutive odd numbers of dots are arranged in layers, beginning with one in the lower left corner. When the dots in the first n layers are considered together, the resulting array forms an $n \times n$ square, and so the total number of dots in the array is given by n^2. While the image displays only the first six cases of the general theorem, a viewer might conclude that the pattern will continue to hold as more layers are added and therefore realize "by an intuitive version of mathematical induction" that the theorem holds for all natural numbers (Chihara, 2004). Importantly, the image is free of technical notation or computational complexity, and so we would expect it to be accessible to adults without any particular training in mathematics.

We conducted a study to investigate how people reason with 'visual proofs by induction' and the nature of the conclusions they draw from the image (for details, see Relaford-Doyle & Núñez, 2017). Forty-nine undergraduates participated in the study. We drew half of our participants from the general undergraduate population; these participants had no university-level training in mathematical induction (MI), and so we refer to them as 'MI-untrained'. We recruited the rest of our participants through the mathematics department. These students had all received high marks in 'Mathematical Reasoning', a university-level mathematics course on formal proof strategies including mathematical induction (we refer to these participants as 'MI-trained'). Importantly, all of our participants were highly educated adults, enrolled at a prestigious university, such that we would expect all of them to have developed meaningful and functional concepts regarding natural number.

We gave each participant the visual proof in Figure 13.1 and asked them to explain how the image represented the statement that "the sum of the first n numbers is equal to n^2". Each participant first filmed a 'tutorial video' in which they explained the image to an imagined third-party audience. Following this, we interviewed the participants to explore the conclusions they had drawn from the image, in particular whether after working with the visual proof they believed the theorem to be true of *all* natural numbers. We assessed this with two key questions. First, we asked the participants if they thought that the statement was true "in all cases", and what the sum of the first eight odd numbers would be. Importantly, simply answering "yes" and "64", respectively, cannot be taken as evidence that the participant believed the statement to be true of *all* natural numbers, in the mathematical sense. To assess whether the participant had truly generalized the statement to *all* natural numbers, we then raised the possibility that large-magnitude counterexamples—"very large numbers where the statement actually isn't true"—may exist and asked what they thought about that suggestion.

Our results (Relaford-Doyle & Núñez, 2017) revealed that all participants, regardless of their familiarity with MI, were willing to generalize the theorem to nearby cases not depicted in the image, such as $n = 8$ (Figure 13.2a). Crucially, however, only MI-trained participants showed significantly higher resistance to the suggestion that large-magnitude counterexamples may exist, while most MI-untrained participants showed little to no doubt

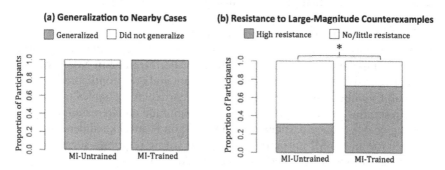

Figure 13.2 Participants in both groups believed the theorem would be true for nearby cases (a), but MI-trained participants were significantly more likely to show high resistance to the possibility of large-magnitude counterexamples (b)

Figures adapted from Relaford-Doyle and Núñez (2017)

that this could be the case (Figure 13.2b; see Relaford-Doyle & Núñez, 2017 for a discussion of these results, including more general consequences of being trained in proof-writing). As relates to concepts of natural number, this is surprising for two reasons. First, if we assume that all educated adults' natural number concept includes an unrestricted successor function, then it is unclear why MI-untrained participants should generalize the theorem only partially; if they are willing to generalize to nearby cases such as $n = 8$, what's keeping them from extending the theorem to *all* natural numbers, regardless of magnitude? MI-untrained participants' reasoning with the visual proof looks less like informal *mathematical* induction (which is based on a successor function) and more like standard *empirical* induction, where effects of similarity and typicality are well documented (e.g., Osherson, Smith, Wilkie, Lopez, & Shafir, 1990; Hampton & Cannon, 2004). Second, if an educated adult's concepts of natural number mirrors the Peano axioms, then learning MI should not fundamentally alter the nature of the concept. Why, then, the clear difference between our MI-trained and MI-untrained groups?

Although these results are not conclusive in regards to concepts of natural number, they do suggest that MI-trained and MI-untrained participants may have been conceptualizing this set in different ways. A closer look at participants' responses to the suggestion of large-magnitude counterexamples reveals that conceptualizations of natural number—either consistent or inconsistent with Peano-like characterizations—may well have contributed to each group's performance on the task.

3.   Examples

After working with the visual proof, we asked participants to respond to the suggestion that large-magnitude counterexamples to the theorem may exist. While this question was not specifically intended to assess natural number

244 Josephine Relaford-Doyle and Rafael Núñez

concepts, we found that at this moment in the interview participants often spontaneously produced utterances that were revealing of their conceptualization of natural number. Participants trained in MI were more likely to demonstrate understandings that were consistent with Peano's characterization of natural numbers. Participants who had not been trained in MI, however, often produced explanations suggesting different and even conflicting notions of natural number. Next, we examine some particular examples of participant's responses to the suggestion of large-magnitude counterexamples, first focusing on MI-trained participants who relied on Peano-like understandings and then turning to MI-untrained participants whose responses suggest other conceptualizations, inconsistent with Peano-like notions.

3.1 Consistent Notions of Natural Number

We will first examine two MI-trained participants who resisted the suggestion that large-magnitude counterexamples were possible. Although their explanations differ in level of technical sophistication, both demonstrate conceptualizations of natural number that are consistent with Peano-like characterizations. In particular, both indicate a conceptualization of natural number as a set that extends indefinitely in a uniform manner, such that all numbers, no matter their magnitude, should be governed by the same logic.

In the first part of the interview, participant JC explained that he thought the theorem was "true in all cases" because it could be proven using MI. He specifically referred to the steps of the proof, mentioning that you first demonstrate the statement for a base case, then in the inductive step "prove the process from ith to (i + 1)st case". When the suggestion of large-magnitude counterexamples is raised, he refers again to the inductive proof:

JC: I find it kind of hard to believe, um, for some arbitrary big number, because if this seems really consistent for, like, small n, and I know from induction that, hey, if you can do the inductive step and inductive reasoning, it should work for all n.

By invoking MI, JC is not only expressing a conceptualization that is *consistent* with the Peano axioms, but is in fact relying directly on the formal characterization of the natural numbers. JC explicitly mentions that the *inductive step*—in which we show that if the property holds for n, then it must also hold for its successor $n + 1$—implies that the statement should be true for all n.

While technical explanations involving MI are not surprising coming from math students, an explanation like JC's certainly seems unlikely to be generated by someone who has never learned the formal proof strategy. Importantly, however, participants did not need to mention mathematical formalizations in order to demonstrate Peano-like conceptualizations of natural number. JG, another MI-trained participant, explained her resistance to the suggestion of large-magnitude counterexamples using nontechnical vocabulary:

JG: I wouldn't be able to see it, right off the bat, how it doesn't work for the large numbers, I guess. Just because this picture seems to portray it on a smaller scale, and that shouldn't change for a larger number because you just keep *counting along this way*.

As JG refers to *"counting along this way"*, she produces a quick (600 millisecond) linear pointing gesture along the bottom of the image. She begins at the bottom left corner of the image, which she has labeled with '1' (Figure 13.3a). She then traces her finger along the bottom of the image, through sections that she has labeled '2' and '3', eventually extending beyond the image (Figure 13.3b). Her point then continues its linear trajectory, extending well off the page before coming to rest (Figure 13.3c).

Unlike the first participant, JG does not refer to technical notions such as the inductive step, but relies instead on the relatively simple idea of counting. JG's resistance is based in her knowledge that any natural number, no matter how large, can theoretically be reached by counting. The smoothness and linearity of her gesture reinforce the regularity of the natural numbers such that the pattern displayed here on a 'smaller scale' will be maintained even as the numbers become larger and larger. Her explanation suggests a conceptualization that is consistent with Peano's characterization of natural numbers; specifically, she seems aware that all natural numbers are 'equally natural', such that one would not expect large numbers to behave any differently than smaller numbers.

While we wouldn't expect MI-untrained participants to produce technical explanations like JC's, counting-based explanations like JG's seem like they should be relatively accessible to nonexperts. Surprisingly, however, not a single MI-untrained participant mentioned counting in their response to the suggestion of large-magnitude counterexamples. Instead, MI-untrained participants often indicated conceptualizations of natural number that were inconsistent with the formal characterization.

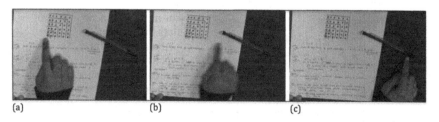

(a)   (b)   (c)

Figure 13.3 As she says "counting along this way", JG produces a smooth, linear pointing gesture starting at the bottom left corner of the image (a), extending horizontally (b), and eventually coming to rest well off the page (c).

3.2   *Inconsistent Notions of Natural Number*

MI-untrained participants were significantly more willing to accept the possibility that large-magnitude counterexamples existed, and in some cases, their reactions to this suggestion indicated conceptualizations of natural number that were inconsistent with Peano-like characterizations. MI-untrained participants often produced utterances that were at odds with the conceptualization of natural numbers as an infinite set extending indefinitely in a regular, uniform fashion, suggesting instead that, for these participants, some numbers are more natural than others.

When asked if she believed the statement to be "true in all cases", DA said yes, explaining, "It's a pattern, there's nothing that's going to change about it . . . the picture's going to stay the same, it's just the number that's changing". As part of this initial explanation, she adds the next layer—13 dots—to the image, but does so in a way that does not maintain the grid structure of the image (Figure 13.4). Specifically, the dots in her new layer do not line up with the dots in the preceding layers, and as a result, the outermost row and column no longer have equal numbers of dots.

The fact that DA adds the next layer in a way that is inconsistent with the structure of the preceding layers already hints that she may not be sensitive to the relational features of the diagram—that is, that each layer builds on the previous one in a consistent, uniform way. Her reaction to the possibility of large-magnitude counterexamples suggests that she may also be insensitive to the consistency of the relation between successive natural numbers, as characterized by the successor function. When the suggestion of large-magnitude counterexamples is raised, DA shows very little resistance:

DA: I guess that makes sense. Like the larger numbers could be, like, outliers, or something like that.

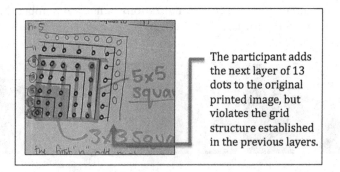

The participant adds the next layer of 13 dots to the original printed image, but violates the grid structure established in the previous layers.

Figure 13.4   MI-untrained participant adds the next layer of dots, violating the grid structure of the image.

Here DA may be using the word 'outlier' as it is defined in statistics, where an outlier is an observation that is distant or removed from other observations. Describing any natural numbers as 'outliers' in this sense is at odds with the system produced by the successor function, where all numbers are evenly spaced by exactly one unit. Of course, DA may have meant 'outlier' in its colloquial sense, meaning "different from all other members of a group", or simply "weird". This is still inconsistent with the formal characterization of natural numbers, in which all members of the set can be generated by successive additions of one and are thus 'equally natural'. In either interpretation, it is hard to imagine why, if DA's concept of natural number was characterized by knowledge of the successor principle, she would ever describe any natural numbers as 'outliers'. This, along with DA's violation of relational features in her drawing, makes it hard to believe that her natural number concept is best described by Peano's axioms.

KL's response to the suggestion of large-magnitude counterexamples is similarly inconsistent with the formal characterization. KL initially expresses some resistance to the possibility of counterexamples, but explains why they may in fact be possible:

KL: Based on my impression, just based on this observation, I think it would work, but when it gets to really high numbers, um, it's possible that, like *(pauses)*. I can see maybe it gets kind of fuzzy. Because at extremes things tend to not work as they do normally.

This response reveals a number of things about KL's natural number concept. First, KL conceives of the natural numbers as a set that has 'extremes'. This may refer to some sort of edge or ultimate ending point such that 'extremes' refer to 'outermost' or 'closest to the end' (as in 'the extreme northwest of California'). Or, rather than indicating proximity to an ending point, 'extremes' could refer to a number at which you are simply extremely far from the starting point; by this interpretation, KL is not necessarily implying that the natural numbers come to an end (a statement which is, of course, at odds with the formal characterization). However, KL also suggests that, once you reach the "extremes", things get "fuzzy" and no longer "work as they do normally". This statement—that very large numbers do not necessarily follow the same rules as smaller numbers—is clearly not consistent with the formal characterization of natural numbers. KL's response is reminiscent of Cheung, Rubenson, and Barner's findings (2017) in which children initially have knowledge of the successor function for just the smallest numbers and only later extend that knowledge to larger items in their count list. Developmental psychologists assume that the successor principle is eventually extended, not just to all items in the count list, but to "all possible numbers" (Cheung et al., 2017). KL's response here suggests that such an assumption may be unwarranted.

AT, another MI-untrained participant, echoes the idea that large numbers may be subject to different rules than smaller numbers. Throughout his interview, he refers to the visual proof as a "model"; when presented with the possibility of large-magnitude counterexamples, he explains,

AT: I guess this model proves to be true for, until, maybe like 99. I know it would be true. I don't know, I consider 99 a big number. I don't know how big this person . . . like maybe the model deconstructs at a thousand or a million, I don't know, but it's too hard to draw a million dots.

On the basis of the 'model', which shows the first six cases of the theorem, this participant concludes that the theorem will hold for numbers up to 99—'a big number'. However, past that, he is less confident. His response suggests that he is sensitive to the issue of *computability*—he is open to the possibility that the model 'deconstructs' at larger numbers like a million at least in part because it would simply be too hard to verify its truth by physically extending the image that far. The idea that small numbers and large numbers are different by nature of their computability was a common theme among MI-untrained participants and suggests that for these participants, there may be some qualitative difference between numbers that can be comfortably manipulated and larger numbers that take an unreasonable amount of time or effort to work with. In the formal characterization, of course, such differences are irrelevant—all natural numbers, no matter the magnitude, are equally natural.

Even MI-trained participants sometimes indicated a degree of tension around the idea of computability. In the following exchange, AG—a math and computer science major who was familiar with mathematical induction—explains her confusion about the possibility of large-magnitude counterexamples to the theorem:

AG: Yeah, large numbers have problems. In computer science too, even if it works on small, medium, most numbers, when you get to the large number, there's not enough space and it doesn't work anymore. So large numbers are weird cases.

Int: And so do you think that that issue of not having enough space that you're describing—

AG: Um, that wouldn't matter with this. I don't know why it wouldn't work on super large numbers though. Um . . . I'm not sure why. That's weird; it works on most, but not really big numbers.

This participant initially makes sense of the possibility of large-magnitude counterexamples by referring to computer science, where numbers can become so large that there is no longer enough "space" to perform computations. However, when probed by the interviewer it is clear that AG understands that issues of computer memory aren't relevant in the context of the mathematical theorem. Nonetheless, her beliefs about large numbers in the domain of computer science seem to have at least opened the door to the possibility that large numbers may "have problems", even in purely theoretical mathematics.

These examples demonstrate that, when engaged with a mathematical task that relies on the structure of the natural numbers, college-educated adults routinely made statements that are at odds with the formal characterization of natural numbers given in the Peano axioms. In contrast, adults who had been trained in mathematical induction—and thus explicitly taught about the successor function—demonstrated notions consistent with the formal characterization. Importantly, MI-trained participants did not rely exclusively on technical features of mathematical induction, but also produced explanations based on simple ideas, such as counting. The fact that MI-untrained participants did not produce such explanations suggests that—without explicit training—the logic that governs counting may not be spontaneously extended to produce the infinite set of natural numbers.

Concluding Remarks

The examples presented earlier provide compelling evidence that concepts of natural number—even among college-educated adults—are *not* best characterized by the Peano axioms. Educated adults demonstrated conceptualizations that were at odds with the formal characterization, first in their partial generalization of the theorem and, more clearly, in their explanations surrounding the possibility of large-magnitude counterexamples. Only participants with extra training—specifically, training in mathematical induction—consistently demonstrated understandings of natural number consistent with the Peano axioms. It would seem that the natural number concept does not come 'for free'; knowledge about the counting numbers— which, no doubt, all of our participants possessed—does not imply knowledge of the natural numbers.

Should this surprise us? In fact, there are clear differences between the collection of counting numbers and the set of natural numbers. There are only finitely many nameable counting numbers, yet the set of natural numbers is infinite. In our everyday experience with counting numbers, large numbers *are* different from smaller numbers: they are encountered less frequently, are lexically and notationally more complex, and are harder to manipulate in calculations. These issues are all irrelevant in the domain of *natural number*, where the successor function generates a system in which all natural numbers, regardless of magnitude, are governed by the same logic (including deep implications such as I–IV). Given the differences between counting numbers and natural numbers, we find it more surprising that one would assume concepts of natural number fall directly out of concepts of counting numbers. In our view, the assumption that people come to formally consistent understandings of natural number by virtue of their experience with familiar counting numbers is analogous to the claim that people come to scientifically valid understandings of the animal kingdom by virtue of their experience with their pet cats and dogs.

We began this chapter by noting that the extensively researched question of *how* we develop natural number concepts contains implicitly the question, what is the nature of those concepts? Currently it is widely assumed in developmental psychology that children eventually develop a concept of natural number that is consistent with Peano's formal characterization—our findings, however, suggest that researchers should reconsider this assumption. The question of the nature of our natural number concepts is an *empirical* question and should be informed by empirical work. This may also require updating the current theories of natural number concept development. If theorists are interested in deeper properties of natural number as governed by a successor principle, then they must consider the possibility that such understandings may not develop spontaneously and only arise with explicit technical training. If theorists are *not* interested in these properties—say, if they are interested in how we come to develop concepts of *counting numbers*—then they must restrict the scope of their theories by further specifying the role of the successor principle.

Finally, we hope we have convinced the reader that our method provides a promising first step by which to explore adult concepts of natural number. We were pleasantly surprised that our task, which was not specifically designed to investigate natural number concepts, was so effective at revealing a variety of adult conceptualizations. While future empirical work is needed to more conclusively characterize of the nature of our number concepts, one thing is clear: concepts of natural number may be surprisingly unnatural.

References

Antell, S. E., & Keating, D. P. (1983). Perception of numerical invariance in neonates. *Child Development, 54*, 695–701.

Avital, S., & Libeskind, S. (1978). Mathematical induction in the classroom: Didactical and mathematical issues. *Educational Studies in Mathematics, 9*, 429–438.

Brown, J. R. (1997). Proofs and pictures. *The British Journal for the Philosophy of Science, 48*(2), 161–180.

Carey, S. (2004). Bootstrapping & the origin of concepts. *Daedalus, 133*(1), 59–68.

Cheung, P., Rubenson, M., & Barner, D. (2017). To infinity and beyond: Children generalize the successor function to all possible numbers years after learning to count. *Cognitive Psychology, 92*, 22–36.

Chihara, C. S. (2004). *A structural account of mathematics.* New York, NY: Oxford University Press.

Davidson, K., Eng, K., & Barner, D. (2012). Does learning to count involve a semantic induction? *Cognition, 123*(1), 162–173.

DeCock, L. (2008). The conceptual basis of numerical abilities: One-to-one correspondence versus the successor relation. *Philosophical Psychology, 21*(4), 459–473.

Fischbein, E., & Engel, I. (1989). Psychological difficulties in understanding the principle of mathematical induction. In G Vergnaud, J. Rogalski, & M. Artigue (Eds.), *Proceedings of the 13th international conference for the psychology of mathematics education* (Vol. 1, pp. 276–282). Paris, France: CNRS.

Gelman, R., & Gallistel, C. R. (1978). *The child's understanding of number*. Cambridge, MA: Harvard University Press.

Hampton, J. A., & Cannon, I. (2004). Category-based induction: An effect of conclusion typicality. *Memory & Cognition, 32*(2), 235–243.

Hartnett, P., & Gelman, R. (1998). Early understandings of numbers: Paths or barriers to the construction of new understandings? *Learning and Instruction, 8*(4), 341–374.

Leslie, A. M., Gelman, R., & Gallistel, C. R. (2008). The generative basis of natural number concepts. *Trends in Cognitive Sciences, 12*(6), 213–218.

Marghetis, T., & Núñez, R. (2013). The motion behind the symbols: A vital role for dynamism in the conceptualization of limits and continuity in expert mathematics. *Topics in Cognitive Science, 5*(2), 299–316.

Movshovitz-Hadar, N. (1993). The false coin problem, mathematical induction and knowledge fragility. *Journal of Mathematical Behavior, 12*, 253–268.

Nelsen, R. B. (1993). *Proofs without words: Exercises in visual thinking (No. 1)*. Washington, DC: Mathematical Association of America.

Nieder, A. (2016). The neural code for number. *Nature Reviews of Neuroscience, 17*, 366–382.

Núñez, R. (2006). Do real numbers really move? Language, thought, and gesture: The embodied cognitive foundations of mathematics. In *18 unconventional essays on the nature of mathematics* (pp. 160–181). New York: Springer.

Núñez, R. (2017). Is there really an evolved capacity for number? *Trends in Cognitive Sciences, 21*(6), 409–424.

Núñez, R., & Marghetis, T. (2014). Cognitive linguistics and the concept(s) of number. In R. C. Kadosh & A. Dowker (Eds.), *The Oxford handbook of numerical cognition*. New York, NY: Oxford University Press.

Osherson, D. N., Smith, E. E., Wilkie, O., Lopez, A., & Shafir, E. (1990). Category-based induction. *Psychological Review, 97*(2), 185.

Relaford-Doyle, J., & Núñez, R. (2017). When does a 'visual proof by induction' serve a proof-like function in mathematics? In G. Gunzelmann, A. Howes, T. Tenbrink, & E. J. Davelaar (Eds.), *Proceedings of the 39th annual conference of the cognitive science society* (pp. 1004–1009). London, UK: Cognitive Science Society.

Rips, L. J., Bloomfield, A., & Asmuth, J. (2008). From numerical concepts to concepts of number. *Behavioral and Brain Sciences, 31*(6), 623–642.

Sarnecka, B. W. (2008). SEVEN does not mean NATURAL NUMBER, and children know more than you think. *Behavioral and Brain Sciences, 31*(6), 668–669.

Sarnecka, B. W., & Carey, S. (2008). How counting represents number: What children must learn and when they learn it. *Cognition, 108*(3), 662–674.

14 Beauty and Truth in Mathematics
Evidence From Cognitive Psychology

Rolf Reber

Beauty as an indicator for the truth in mathematics and science has been the object of both anecdotes and philosophical inquiry. While some of this work looked for characteristics of scientific theories that could be both beautiful and truth-conducive (e.g., McAllister, 1996), other parts included subjective elements in the form, for example, of the 'mere exposure' effect (Kuipers, 2002). The present chapter contributes an explanation for the idea that beauty is truth in mathematics and science by reviewing evidence that processing fluency (or, in short, fluency), which is the subjective ease with which a mental operation can be performed, is the experience underlying both beauty and truth.

After a quick attempt to define beauty and truth, I briefly introduce some anecdotes and philosophical viewpoints before I review empirical evidence for processing fluency as the link between beauty and truth. The chapter concludes with a discussion about the potential implications of this account on the emergence of mathematical thinking and theorizing. One limitation has to be acknowledged at the outset. As a psychologist, I am an expert on the empirical study of the mental processes underlying beauty and truth, but neither on the philosophical foundation of mathematics nor on the implications of this research for the history and the current practice of mathematics. The best a psychologist can do in this case is to embed the empirical research in philosophical thinking as he (in my case) knows it—philosophers can then judge and criticize the adequacy of the foundation—and to deliver options for implications for mathematics. Philosophers may judge which of these options seem most appropriate.

1. Defining Beauty and Truth

1.1 Defining Beauty

There are usually three different approaches to define beauty (see Reber, Schwarz, & Winkielman, 2004). One is objective in that scholars and artists try to find the features that render an aesthetic object beautiful. Such features include balance, symmetry, and roundness of form. Alternatively, beauty can be seen as subjective experience that is different from person to person. The synthesis of the objective and subjective approaches defines beauty as an interaction between features of the aesthetic object and the experience of the person. This interactive approach has been applied by Reber et al. (2004), whose naturalistic theory of beauty postulates that objective features interact with the processing dynamics of the human mind. Specifically, as we shall see later, the theory assumes that it is processing fluency that determines aesthetic pleasure. According to this account, an object is seen as beautiful if it can be processed with surprising ease.

However, is mathematical beauty really beauty, or could it be a related attribute, like elegance? In a hitherto unpublished study, Menninghaus, Wagner, Kegel, Knoop, and Schlotz (2017) examined the semantic differentials of the terms beauty, elegance, grace, and sexiness. Grace and elegance did not differ much. For our purpose, the difference between elegance and beauty is relevant (hopefully, I do not miss a point when assuming that mathematicians do not find sexiness in their formulas). When comparing elegance and beauty on individual items, the former is conceived as being significantly less *multi-colored*, more *refined*, less *natural*, more *rigorous*, more *elitist*, less *dreamy*, more *disciplined*, more *expensive*, more *exquisite*, more *cultured*, and more *skillful* than beauty. Although the authors were interested in a stereotype of the elegant versus beautiful person, it may be informative that elegance is seen as possessing attributes that would stereotypically better fit mathematical beauty, such as rigorous, elitist, disciplined, and skillful. However, it is too early to make generalizations from this study about person attributes to the same attributes as applied to mathematics.

Before beauty as an indication of truth in mathematics can be examined, we have to know that beauty in mathematics exists. Even if mathematicians and scientists talk about beauty, they may mean more formal criteria, such as simplicity. Zeki, Romaya, Benincasa, and Atiyah (2014) used brain imaging techniques to examine this question. They reasoned that if mathematicians appreciate the beauty of mathematical formulas, then the same brain areas should be activated as during the experience from other sources, such as art or music. Testing 15 mathematicians, they indeed found that formulas that were judged as being beautiful activated field A1 of the medial orbito-frontal cortex—a part of the emotional brain activated during the experience of beauty from other sources as well. This finding corroborates the subjective reports by mathematicians that mathematics gives rise to the experience of beauty and is in line with the notion that simple cognitive processes, for example Gestalt completion, have affective consequences (Erle, Topolinski, & Reber, 2017; Topolinski, Erle, & Reber, 2015). However, while such results and the subjective reports of scientists support the notion that mathematics or at least doing mathematics has an aesthetic component, it leaves open the question of whether this component is better captured by the term 'beauty' or 'elegance'.

1.2   *Defining Truth*

The term 'truth' has been used in different ways (see Kirkham, 1992, for an extensive review), and it is important to state in which sense we use the term. There are two uses that are relevant to the present chapter. First, truth is often defined as a belief that corresponds to an outer reality or systems of social constructions, such as norms or conventions. A derivative criterion for the truth of a system of propositions is internal coherence—that the propositions do not contradict each other. When mathematicians or scientists ask whether a solution is true, they ask for this kind of certainty about the epistemic status of a system of propositions. Second, a reliabilist notion of truth notes that a psychological process is truth-conducive if it yields the correct solution at above-chance accuracy (see Goldman, 1986). There are different versions of reliabilism, but for the present purposes, reliability of processes is most relevant. After reviewing psychological research, I shall argue that evidence accrued in the context of justification may achieve a high degree of correspondence to reality, while the sense of beauty in both the context of discovery and of justification may predict the truth of a proposition at above-chance level if the mental processes underlying the experience of beauty are well aligned to the truth of the proposition.

2.   Beauty Is Truth: Philosophical Theorizing

Anecdotes abound that mathematicians, scientists, artists, and poets saw beauty as a sign for truth. Toward the end of his career, the Nobel Prize-winning physicist Subrahmanyan Chandrasekhar (1987) published a book with the title *Truth and Beauty. Aesthetics and Motivations in Science.* He tells the story of Hermann Weyl who noted, "My work always tried to unite the true with the beautiful; but when I had to choose one or the other, I usually chose the beautiful" (Chandrasekhar, 1987, 65). Chandrasekhar continued,

> The example which Weyl gave was his gauge theory of gravitation, which he had worked out in his Raum-Zeit-Materie. Apparently, Weyl became convinced that this theory was not true as a theory of gravitation; but still it was so beautiful that he did not wish to abandon it and so he kept it alive for the sake of its beauty. But much later, it did turn out that Weyl's instinct was right after all, when the formalism of gauge invariance was incorporated into quantum electrodynamics.
>
> (Chandrasekhar, 1987, 65f)

That beauty could indicate truth has also been of interest in philosophy. Such discussions mostly start out from scientific anecdotes, like Breitenbach (2017), or from quotes of scientists, like Bangu (2006) or the chapter on beauty and truth in McAllister's (1996) book. Two main lines of arguments can be distinguished. The first is at its extreme attributed to Dirac (1939) who claimed that the

> trend of mathematics and physics towards unification provides the physicist with a powerful new method of research into the foundations of his subject, a method which has not yet been applied successfully, but which I feel confident will prove its value in the future. The method is to begin by choosing that branch of mathematics which one thinks will form the basis of the new theory. One should be influenced very much in this choice by considerations of mathematical beauty.
>
> (p. 125)

At another place in the same paper, he notes,

> The research worker, in his efforts to express the fundamental laws of Nature in mathematical form, should strive mainly for mathematical beauty. He should still take simplicity into consideration in a subordinate way to beauty. (For example Einstein, in choosing a law of gravitation, took the simplest one compatible with his space-time continuum, and was successful.). It often happens that the requirements of simplicity and of beauty are the same, but where they clash the latter must take precedence.
>
> (p. 124)

Like Weyl, Dirac preferred beauty to empirical evidence of truth of a theory by claiming, "It is more important to have beauty in one's equations than to have them fit experiment" (Dirac, 1963, 47).

Science-educated artist Roger Fry (1920), by contrast, drew a sharp distinction between aesthetic considerations and truth (see Kivy, 1991, for a summary and discussion). Fry noted that there certainly are complex systems of scholastic theology and modern metaphysics whose unity and coherence possess aesthetic value but they are not scientifically true.

The relationship of beauty and truth in scientific and mathematical reasoning has not only intrigued scientists but also philosophers. Kuipers (2002) concluded that aesthetic criteria may indicate truth but cannot be on equal terms with empirical criteria. He mentioned that aesthetic appeal might play a role in scientific revolutions where not enough evidence has been accrued to empirically support a new theory and that conservatives who deny aesthetics any role in scientific reasoning may impede scientific progress. Analyzing the history of the development of quantum mechanics, Bangu (2006) arrived at the similar conclusion that beauty may be useful in the context of theory discovery, but not in the context of theory verification.

Philosophical discussion accords some role to beauty for assessing the truth of a proposition, but only in the context in discovery, not in the context of justification. The reason is that in the context of justification, the truth of an idea is the most important criterion, and only evidence can adjudicate whether a proposition is true or not. It is essential to eliminate wrong ideas. In the context of discovery, by contrast, an idea is examined in accordance to its promise to offer a solution for a problem. It would be a pity to eliminate promising ideas. This approach does not clearly answer the question of why beauty is a valid cue in the context of discovery. Based on assumptions about the interaction of the rules of mathematics and the mathematician, I shall tentatively argue that beauty is a valid indicator of truth in the context of discovery because it predicts truth in the context of justification with above-chance accuracy. The novel contribution of this chapter is to provide a mechanism that plausibly explains why beauty could be a valid indicator for the correctness of a proposed solution (see Todd, 2017, for a related idea).

3. Beauty Is Truth: Empirical Evidence

For experimental psychologists, the challenge consists in bringing the essence of anecdotes and philosophical reasoning to the laboratory. As mathematical and scientific thought is very specialized so that it would be hardly possible to find one experimental paradigm that would capture beauty and truth in the thinking of all practitioners of mathematics and science, we relied on undergraduate students to find a preliminary answer to the question of whether beauty is taken as a sign for truth. In a speeded decision task, Reber, Brun, and Mitterndorfer (2008) conducted experiments in which they manipulated the symmetry of patterns in a geometrical addition task. Participants had to indicate whether the presented solution to the task was correct or not—a judgment related to true and false (see Figure 14.1). As symmetric patterns are judged to be more beautiful than asymmetric patterns (Makin, Pecchinenda, & Bertamini, 2012) these experiments

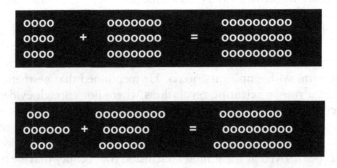

Figure 14.1 Examples of additions used in Reber et al. (2008); top: symmetric patterns; bottom: asymmetric patterns

examined whether tasks with beautiful patterns are also considered correct. Although symmetry of the patterns was unrelated to correct task solutions, participants were more likely to endorse solutions with symmetric patterns than solutions with asymmetric patterns. Although this is a situation remote from a mathematician having an intuitive hunch that a solution to a problem is true, the experiment shows that there is commonality between truth and symmetry—which is both easier to process and judged beautiful (see Makin et al., 2012). The study is therefore a first step at demonstrating the mechanisms behind the hitherto anecdotal evidence about mathematicians and scientists using the beauty of a theory as a guide to its truth.

In a further, yet unpublished study (Reber & Kogstad, 2011), we were able to replicate and extend these findings by using a different task (see Figure 14.2). Participants got equations with two pieces on the left of the equal sign and a rectangular piece on the right. The task of the participant was to determine whether the combined areas of the two pieces on the left were equal to the area of the pieces on the right. The crucial condition was the fit of the two pieces on the left that either matched like two jigsaw puzzle pieces or, apparently, mismatched because the left piece was reversed. Again, participants were more likely to believe that the piece on the right had the same area when the two pieces on the left matched than when they did not match, despite the fact that matching of pieces was unrelated to the correctness of the solution. Another group of participants had to rate the beauty of these configurations. It turned out that across items, the beauty of

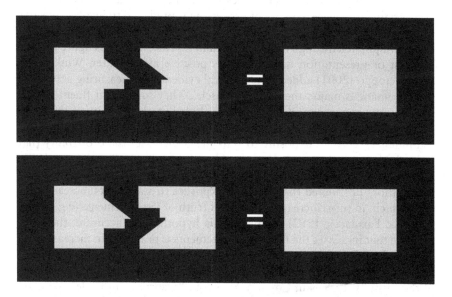

Figure 14.2 Examples of additions used in Reber and Kogstad (2011). Upper panel: the addends matched. Lower panel: the addends did not match

the configuration and the judged correctness of the equation correlated positively. These results obtained regardless of whether the task to determine the correctness of the solutions was sped up or not. This means that the results are robust and not limited to situations that are quite unnatural for problem solving. Again, like in the symmetry task used by Reber et al. (2008), beauty was used as an indicator of the truth of a proposition—here, of the correctness of equations. Moreover, we collected beauty judgments in the new task and found that beauty indeed correlated with judgments of truth.

4.   Processing Fluency, Beauty, and Truth

When both anecdotal evidence and laboratory experiments tell us that beauty equals truth in mathematics, the question arises whether there is a common variable underlying beauty and truth. One such variable is processing fluency, which plays a crucial role for aesthetic pleasure (see Reber et al., 2004).

First indications of parallel effects of fluency on beauty and truth came from research on effects of repeated exposure. Participants in experiments liked objects, such as line drawings and abstract shapes, more after they have been exposed to them than when they have never seen them (Zajonc, 1968; see Bornstein, 1989, for a review). One hypothesis has been that repeated exposure increases fluency of processing the stimulus, and fluency, in turn, increases liking (e.g., Bornstein & D'Agostino, 1994; Seamon, Brody, & Kauff, 1983; see Winkielman, Schwarz, Fazendeiro, & Reber, 2003, for a functional explanation). If the processing fluency account is accurate, then any variable that increases ease of processing is predicted to increase positive affect. Reber, Winkielman, and Schwarz (1998) found that processing fluency manipulated through means other than repetition, such as figure-ground contrast or presentation time, increases positive affect. Later, Winkielman and Cacioppo (2001) added physiological evidence by measuring activity of the zygomaticus major, the 'smiling muscle'. They found that fluent stimuli did not only increase ratings of positive affect but also activity of the zygomaticus major which is considered a reliable indicator of positive affect.

Similarly, repeated exposure to statements increases the probability that individuals judge them as being true (Hasher, Goldstein, & Toppino, 1977; see Dechêne, Stahl, Hansen, & Wänke, 2010, for a meta-analysis), and several authors postulated that repeated exposure to statements increases fluency which, in turn, increases the judged truth of the statements (e.g., Begg, Anas, & Farinacci, 1992). Again, if this hypothesis is accurate, then any variable that increases the fluency of statements is predicted to increase their perceived truth.

To test this assumption, Reber and Schwarz (1999) presented statements of the form "City X is in Country Y"—e.g., "Osorno is in Chile"—one by one at the center of a computer screen. Crucially, half of the statements were printed in a dark color on a white background, making them more fluent than the other half of the statements that were printed in a bright color. As predicted by a processing fluency theory, the authors found that statements presented in a fluent font were more likely to be judged true than statements presented in a non-fluent font. Later research replicated the basic effect (Unkelbach, 2007); other studies revealed that the effect is stronger when semantic instead of perceptual means to manipulate fluency were used (Parks & Toth, 2006) and when the experience of fluency was surprising (Hansen, Dechêne, & Wänke, 2008). In their experiment, they presented six statements with high contrast after six statements with low contrast. The authors reasoned that the fluency in high contrast statements that came at the beginning of the series after the low-contrast statements was more surprising than the fluency toward the end of the series of the high contrast statements. If it is not fluency per se that influences truth judgments but the discrepancy between the expected and the actual fluency, then participants should judge only high contrast statements at the beginning of a series as being more probably true than statements with low contrast. This is exactly what they found: participants judged only statements that were surprisingly fluent—because they followed statements with low fluency—as more probably true than non-fluent statements.

Note that the effect of fluency on truth generalizes to the quality of text in general. It is an old finding that bad handwriting leads to worse grades in essays (e.g., James, 1929; see Reber & Greifeneder, 2017)—an effect where the aesthetic quality of handwriting could be seen as a signal for essay quality. Greifeneder et al. (2010) examined this effect more closely and found that warnings about the disfluency eliminated the effect of bad handwriting on grades, presumably because disfluency deriving from bad handwriting was discounted as information for the judgment at hand. This finding, together with the studies on judged truth, suggests that fluency is a mechanism underlying a wide range of aesthetic and epistemic evaluations (see Reber, 2016).

5. The Epistemic Justification of Fluency for Judgments of Truth

From a rational point of view that values judgment accuracy, the impact of fluency on truth seems to lack epistemic justification and is therefore seen as a bias. Yet this view has been challenged. Reber and Unkelbach (2010) presented an analysis that revealed that fluency is a reliable cue to truth in contexts where the majority of statements that a person encounters are true. Mathematics education and research, especially in academic contexts, are such environments. The starting point was the aforementioned reliabilist approach to epistemic justification that assumes that reliable mental processes lead to above-chance accuracy of truth judgments (see Goldman, 1986). While verification or falsification of facts allow a person to hold or deny a belief with certainty, fluency is a probabilistic cue, and it depends on the ecology of information whether it reliably leads to truth. Reber and Unkelbach claimed that the judgment that a fluent statement is true is accurate at above-chance level if fluency stems from repetition and the repeated statement is true at above-chance level. However, fluency can be a misleading cue to truth if an information source is consistently wrong. The new phenomenon of fake news on social media is an example of biased information that makes fluency a reliable source to falsehood.[1]

Reber and Unkelbach (2010) argued that there is good reason to assume that, normally, people can trust that the information they read or hear is accurate. The few extant studies on the accuracy of encyclopedic entries reveal that there are few errors in the text, often less than 2% of all factual statements. Moreover, for many factual statements, such as the height of the Eiffel Tower, there is only one true statement (324 m) but various false statements (295 m; 313 m; 339 m, etc.). That is, even if people hear as many false statements about the height of the Eiffel Tower as true statements, they are more likely to encounter the figure for accurate height than each of the various erroneous figures. Finally, if we assume that a group of true statements represents a group of facts in the real world, then those statements are likely to build a coherent set of knowledge representations. False claims, by contrast, are more likely to lead to incoherent representations of the state of the world. However, conspiracy theories and ideologies might be examples of false statements that lead to coherent sets of representations.

This analysis may be transferred to mathematical thinking where practitioners may use fluency as a cue to truth. What is needed is an analysis of mathematical problems and their solutions in terms of ease of mental operation and truth. If such an analysis showed, in line with the analysis by Reber and Unkelbach (2010), that ease of operation indicates truth, then it would yield the interesting claim that beauty predicts truth in the context of justification. In line with current philosophical thinking, such a correspondence of beauty and truth is not perfect (therefore, the reliabilist approach); evidence is necessary to secure the truth status of a mathematical proposition. What has not been mentioned in philosophical analysis is that beauty would be a useful cue in the context of discovery, because it would be predictive of truth in the context of theory justification. This means that beauty is a useful guide in the context of theory discovery, because it predicts truth with above-chance probability for the later stage of justification. Provided that this probability is large enough, it may be adaptive and time-saving for a mathematician to follow beauty as an indicator for truth. If one knew the time used to empirically examine the justification of a theory and the

---

1 In an ingenious study, Unkelbach (2007) showed that learners usually use fluency as a cue to truth, but they do learn to reverse the fluency-truth link if they are consistently presented with fluent information that is false. This raises the question of whether the link between high fluency and truth comes with biological equipment or is acquired through learning.

probability that a theory is true given that it is beautiful, researchers could predict whether it is worthwhile, in terms of finding correct theories within a certain time, to follow the gut feeling produced by an aesthetic experience. For the same reason, we could calculate whether a mathematician or scientist should do the hard work of examining a theory step-by-step even if the gut feeling based on aesthetic considerations is negative and would indicate that a theory is false.

This analysis has to be taken with caution because there is to my knowledge no systematic analysis of the epistemic justification of beauty in mathematics. In analogy to the analysis of Reber and Unkelbach (2010), we would need a clear definition of beauty and a rigorous analysis of how the coherence of rules of mathematics facilitates mathematical problem solving.

6. The Nature of Insight

Mathematicians have not only been fascinated by beauty but also by the nature of insight. For example, the mathematician Henri Poincaré wrote,

> Just as I put my foot on the step, the idea came to me, though nothing in my former thoughts seemed to have prepared me for it, that the transformations I had used to define Fuchsian functions were identical with those of non-Euclidean geometry. . . . I made no verification . . . but I felt absolute certainty at once.
>
> (Poincaré, 1913/1996, 53)

Later, Poincaré again experienced an idea coming to him "with the same characteristics of conciseness, suddenness, and immediate certainty" (p. 54). For him, insight is "a real aesthetic feeling that all true mathematicians recognize, and this is truly sensibility", capable of eliciting "aesthetic emotion" (p. 59).

Again, the question for an experimental psychologist becomes how the experiences described in Poincaré's anecdote could be tested in the laboratory. Unlike the study by Reber et al. (2008), it was not possible to create one experimental setup that captured the multifaceted nature of an Aha-experience. Instead, Topolinski and Reber (2010a) built a model that depicts the interplay of four different attributes. This model was derived from empirical evidence that connected two of each of the attributes. The four attributes are suddenness, fluency, positive affect, and conviction that a solution is true. Suddenness denotes the fact that an insight comes as a surprise, often after a long incubation phase and at a moment when the problem solver is not thinking of the task. The cognitive basis of the phenomenal experience of suddenness is supposed to be a comprehensive, immediate restructuring of the representation of the problem, leading to discontinuity (Weisberg, 1995). The experience of immediacy that accompanies a sudden solution increases endorsement of the solution as true (Topolinski & Reber, 2010b). After this sudden restructuring, the work with the problem proceeds smoothly, which is the fluency component of the Aha-experience. Sudden fluency emerging from insight explains positive affect and perceived truth of a proposition. The pertinent experiments have been discussed earlier. Fluency yields positive affect (e.g., Winkielman & Cacioppo, 2001) and increases judged truth (e.g., Reber & Schwarz, 1999). Moreover, the effect of fluency on judged truth is especially pronounced when fluency is surprising (Hansen et al., 2008). This latter finding bolsters the notion that surprise after sudden restructuring of a problem plus fluency contribute in unison to the conviction that a statement is true. Finally, the fluency account of Aha-experiences would predict that beauty and truth are correlated, because they have a common underlying mechanism. Indeed, we have seen earlier that beautiful patterns were judged as more likely to be true (e.g., Reber et al. (2008).

Although experimental research provides support for individual links between the two variables, we have only recently begun to develop methods that allow the examination of the whole Aha-experience (Skaar & Reber, unpublished data). To this purpose, we developed surveys where participants first have to write down an Aha-experience in detail before they are asked questions about its phenomenological quality, such as suddenness of the experience; fluent processing of information related to a solution before, under, and after an Aha-experience; positive affect; and confidence in the truth of the solution as a consequence of the Aha-experience.[2] This survey study confirmed the findings of the experiments, further supporting the account by Topolinski and Reber (2010a). However, the Aha-experiences we assessed were not limited to mathematics and science, and some of the findings may not apply to insights in these fields. There is a debate, for example, whether Aha-experiences occur mostly when the problem solver is alone, as Csikszentmihalyi and Sawyer (1995) claimed, or when people interact with other people, as Liljedahl (2005) observed when undergraduate students had Aha-experiences during an introductory mathematics course. Our data reveal that roughly half of the Aha-experiences happen when the person does not interact with others—that is, either is alone or others are present but not relevant for the Aha-experience—and half of the

---

2 Note that we found a type of Aha-experience rarely discussed in the literature. While most of the pertinent observations found insights and Aha-experiences when students in laboratory experiments or eminent scientists and mathematicians in practice solve a problem, we found that some people experience Aha-experiences that were unrelated to any problem solving. An example is when a woman reads a book and suddenly realizes that it is similar to a book she already knows, or when a young man suddenly understands that the conflict with his parents does not make sense. Such Aha-experiences seem to "come out of the blue" (Skaar & Reber, unpublished data). However, it is an empirical question whether this type of spontaneous insight that is not bound to a problem-solving attempt is frequent in mathematical practice; I, therefore, discuss only problem-related insights.

experiences happen during interactions with other people. It will be interesting to see whether Aha-experiences of eminent mathematicians and scientists are alone during their Aha-experiences, as most accounts suggest where the circumstances are known (e.g., Poincaré, 1996/1913).

7. The Epistemic Justification of Insights for the Truth of a Claim

Similar to the analysis of epistemic justification of the experience of processing fluency, we could ask whether an Aha-experience is correlated with the truth of a solution. As we have seen, Aha-experiences increase both fluency and the confidence that a solution is true.

In the analysis on the role of fluency in judgments of truth, we have seen that fluency is a valid cue when a majority of propositions one has encountered is true—an analysis that assumes that fluency is increased by repetition. Reber and Unkelbach (2010) have mentioned that coherence plays a role in increasing both fluency and epistemic justification, but they did not offer a quantitative analysis of the impact of coherence on truth.

For Aha-experiences, repetition presumably plays a minimal role. Scientists or mathematicians most probably have not seen the solution before. However, the sudden reorganization of knowledge seems to transform formerly unrelated representations into a coherent whole. Coherence is relevant because mathematicians and scientists often look for patterns (see Simon, 2001). The argument for the epistemic justification of Aha-experiences may be similar as for the epistemic justification of fluency. If Aha-moments occur because coherence of representations increases, and this coherence maps onto the facts of the outside world, then Aha-experiences may indeed be predictive of the truth of a proposition.

8. Possible Implications for the Origin and History of Mathematics

While there is some agreement that theories in science reflect an outer reality, the nature of mathematics is debated. While platonic theories hold that there are mathematical objects that are independent of human or other intelligent agents and their thought, language, and practices (e.g., Frege, 1953; Hardy, 1967/1940; Linnebo, 2013), other philosophers, such as Wittgenstein (1978/1956; see Rodych, 2011), can be read as claiming that mathematics depends on inventions. Although this seems to be a radical constructivist view at first sight, there is reason to assume that Wittgenstein did not see mathematics as malleable to human will and practices—not anything goes in mathematics (see Bangu, 2012, for an extended discussion). For the possible implications of the findings reviewed in this chapter, it makes a difference whether one adheres to platonism or constructivism in mathematics.

From a platonist view of the emergence and development of mathematics (e.g., Frege, 1953, see Linnebo, 2013), the experiential elements in mathematical thinking reviewed in this chapter are epiphenomena, without further effects on how mathematics is practiced. Perceived beauty and elegance of mathematical proofs or Aha-experiences in mathematical problem solving are mere by-products that are negligible when the foundations of logic and mathematics are objectively given.

However, even if we assume that mathematical laws are givens, experiences may have changed the history of ideas, because mathematicians favored some lines of inquiry and had more difficulties to find others. This would mean that the historical development of mathematics could be understood as an interaction between the given mathematical objects and the processing dynamics of the human mind. Still, as discussed earlier, fluency and Aha-experiences might be valid predictors for the truth of a proposition.

A different picture may emerge when we endorse the assumption that much of logics and mathematics are human inventions and their operations defined by convention (Wittgenstein, 1978/1956; see Rodych, 2011 for a summary; see Bangu, 2012 for a discussion). Proponents of this view may respond like platonists and claim that these conventions cannot be explained by human experiences but by the reasoning capacities of the human mind. However, this is not the only answer constructivists could provide. Alternatively, they may assume that aesthetic considerations played a decisive role in the selection of the different options of logic and mathematics. In scientific reasoning, an infinite number of mathematical algorithms may explain a given observation. There is some plausibility in the claim that the principle of parsimony is driven by considerations of aesthetics and mental economy, which is related to fluency. However, to defend the principle of parsimony, recourse to aesthetics or mental economy is not necessary; it can be defended on purely cognitive grounds (see Simon, 2001, for discussion).

In a similar vein, Aha-experiences may be seen as a mere by-product of mathematical thinking, regardless of whether a theorist advocates platonism or constructivism in mathematics. However, constructivists may question the purely cognitive foundation of mathematical thinking and try to construct alternative developments of mathematics based on such experiences. One possibility is that mathematicians would have brought forward different conventions if they did not base them on their experiences. Alternatively, mathematicians might have chosen to address those problems that they found interesting, where this interest is based on Aha-experiences (see Liljedahl, 2005).

I do not advocate any of those views; they may well be wrong and my assumptions unfounded. The aim is to ask whether experiential components might have had an influence in the shaping of mathematics by the selection of problems, the finding of solutions, or the selection from alternative solutions (parsimony). This still leaves open the possibility that feelings in logical and mathematical reasoning are nothing else than by-products that may turn out to be a nuisance when leading to false solutions or a pleasure that motivates mathematicians to think without determining how they think. As psychologist, I now have to pass the torch to philosophers, historians, and practitioners of mathematics who might be in a better position to connect the findings reviewed in this chapter with the theory and history of mathematics.

References

Bangu, S. (2006). Pythagorean heuristic in physics. *Perspectives on Science, 14,* 387–416.

Bangu, S. (2012). Ludwig Wittgenstein: Later Philosophy of Mathematics. *Internet Encyclopedia of Philosophy.* Retrieved from www.iep.utm.edu/wittmath

Begg, I. M., Anas, A., & Farinacci, S. (1992). Dissociation of processes in belief: Source recollection, statement familiarity, and the illusion of truth. *Journal of Experimental Psychology: General, 121,* 446–458.

Bornstein, R. F. (1989). Exposure and affect: Overview and meta-analysis of research 1968–1987. *Psychological Bulletin, 106,* 265–289.

Bornstein, R. F., & D'Agostino, P. R. (1994). The attribution and discounting of perceptual fluency: Preliminary tests of a perceptual fluency/attributional model of the mere exposure effect. *Social Cognition, 12,* 103–128.

Breitenbach, A. (2017). The beauty of science without the science of beauty: Kant and the rationalists on the aesthetics of cognition. *Journal of the History of Philosophy,* in press.

Chandrasekhar, S. (1987). *Truth and beauty: Aesthetics and motivations in science.* Chicago: University of Chicago Press.

Csikszentmihalyi, M., & Sawyer, K. (1995). Creative insight: The social dimension of a solitary moment. In R. Sternberg & J. Davidson (Eds.), *The nature of insight* (pp. 329–361). Cambridge: MIT Press.

Dechêne, A., Stahl, C., Hansen, J., & Wänke, M. (2010). The truth about the truth: A meta analytic review of the truth effect. *Personality and Social Psychology Review, 14,* 219–235.

Dirac, P. A. M. (1939). Lecture delivered on presentation of the James Scott prize, February 6, 1939. *Proceedings of the Royal Society (Edinburgh), 59*(Part II), 122–129.

Dirac, P. A. M. (1963, May). The evolution of the physicists picture of nature. *Scientific American, 208,* 45–53.

Erle, T. M., Topolinski, S., & Reber, R. (2017). Affect from mere perception: Illusory contour perception feels good. *Emotion, 17,* 856–866.

Frege, G. (1953). *Foundations of arithmetic* (J. L. Austin, trans). Oxford: Blackwell.

Fry, R. (1920). *Vision and design.* London: Chatto & Windus.

Goldman, A. I. (1986). *Epistemology and cognition.* Cambridge: Harvard University Press.

Greifeneder, R., Alt, A., Bottenberg, K., Seele, T., Zelt, S., & Wagener, D. (2010). On writing legibly: Processing fluency systematically biases evaluations of handwritten material. *Social Psychological and Personality Science, 1,* 230–237.

Hansen, J., Dechêne, A., & Wänke, M. (2008). Discrepant fluency increases subjective truth. *Journal of Experimental Social Psychology, 44,* 687–691.

Hardy, G. H. (1967). *A mathematician's apology.* Cambridge: Cambridge University Press.

Hasher, L., Goldstein, D., & Toppino, T. (1977). Frequency and the conference of referential validity. *Journal of Verbal Learning and Verbal Behavior, 16,* 107–112.

James, H. W. (1929). The effect of handwriting upon grading. *The English Journal*, *16*, 180–185.

Kirkham, R. L. (1992). *Theories of truth*. Cambridge, MA: MIT Press.

Kivy, P. (1991). Science and aesthetic appreciation. *Midwest Studies in Philosophy*, *16*, 180–195.

Kuipers, T. A. F. (2002). Beauty, a road to truth. *Synthese*, *131*, 291–328.

Liljedahl, P. G. (2005). Mathematical discovery and affect: The effect of AHA! Experiences on undergraduate mathematics students. *International Journal of Mathematical Education in Science and Technology*, *36*, 219–234.

Linnebo, Ø. (2013). Platonism in the philosophy of mathematics. In E. N. Zalta (Ed.), *The Stanford encyclopedia of philosophy*. Retrieved from https://plato.stanford.edu/archives/win2013/entries/platonism-mathematics

Makin, A. D. J., Pecchinenda, A., & Bertamini, M. (2012). Implicit affective evaluation of visual symmetry. *Emotion*, *12*, 1021–1030.

McAllister, J. W. (1996). *Beauty and revolution in science*. Ithaca: Cornell University Press.

Menninghaus, W., Wagner, V., Kegel, V., Knoop, C. A., & Schlotz, W. (2017). *Elegance: Defining the sister of grace and beauty*. Unpublished manuscript.

Parks, C. M., & Toth, J. P. (2006). Fluency, familiarity, aging, and the illusion of truth. *Aging, Neuropsychology, and Cognition*, *13*, 225–253.

Poincaré, H. (1996/1913). *Science and method*. London: Routledge/Thoemess Press.

Reber, R. (2016). *Critical feeling: How to use feelings strategically*. Cambridge: Cambridge University Press.

Reber, R., Brun, M., & Mitterndorfer, K. (2008). The use of heuristics in intuitive mathematical judgment. *Psychonomic Bulletin & Review*, *15*, 1174–1178.

Reber, R., & Greifeneder, R. (2017). Processing fluency in education: How metacognitive feelings shape learning, belief formation, and affect. *Educational Psychologist*, *52*, 84–103.

Reber, R., & Kogstad, K. (2011). *Beauty as truth in mathematical intuition: Effects of simplicity and coherence on speeded judgments*. Unpublished manuscript.

Reber, R., & Schwarz, N. (1999). Effects of perceptual fluency on judgments of truth. *Consciousness and Cognition*, *8*, 338–342.

Reber, R., Schwarz, N., & Winkielman, P. (2004). Processing fluency and aesthetic pleasure: Is beauty in the perceiver's processing experience? *Personality and Social Psychology Review*, *8*, 364–382.

Reber, R., & Unkelbach, C. (2010). The epistemic status of processing fluency as source for judgments of truth. *Review of Philosophy and Psychology*, *1*, 563–581.

Reber, R., Winkielman, P., & Schwarz, N. (1998). Effects of perceptual fluency on affective judgments. *Psychological Science*, *9*, 45–48.

Rodych, V. (2011). Wittgenstein's philosophy of mathematics. In E. N. Zalta (Ed.), *The Stanford encyclopedia of philosophy*. Retrieved from https://plato.stanford.edu/entries/wittgenstein-mathematics

Seamon, J. G., Brody, N., & Kauff, D. M. (1983). Affective discrimination of stimuli that are not recognized: Effects of shadowing, masking, and central laterality. *Journal of Experimental Psychology: Learning, Memory and Cognition*, *9*, 544–555.

Simon, H. A. (2001). Science seeks parsimony, not simplicity: Searching for patterns in phenomena. In A. Zellner, H. A. Keuzenkamp, & M. McAleer (Eds.), *Inference and modelling: Keeping it sophisticatedly simple* (pp. 32–72). Cambridge: Cambridge University Press.

Todd, C. (2017). Fitting feelings and elegant proofs: On the psychology of aesthetic evaluation in mathematics. *Philosophia Mathematica*, in press.

Topolinski, S., Erle, T. M., & Reber, R. (2015). Necker's smile: Immediate affective consequences of early perceptual processes. *Cognition*, *140*, 1–13.

Topolinski, S., & Reber, R. (2010a). Immediate truth: Temporal contiguity between a cognitive problem and its solution determines experienced veracity of the solution. *Cognition, 114*, 117–22.

Topolinski, S., & Reber, R. (2010b). Gaining insight into the "Aha"-experience. *Current Directions in Psychological Science, 19*, 402–405.

Unkelbach, C. (2007). Reversing the truth effect: Learning the interpretation of processing fluency in judgments of truth. *Journal of Experimental Psychology-Learning Memory and Cognition, 33*, 219–230.

Weisberg, R. W. (1995). Prolegomena to theory of insight. In R. J. Sternberg & J. E. Davidson (Eds.), *The nature of insights* (pp. 157–196). Cambridge, MA: MIT Press.

Winkielman, P., & Cacioppo, J. T. (2001). Mind at ease puts a smile on the face: Psychophysiological evidence that processing facilitation elicits positive affect. *Journal of Personality and Social Psychology, 81*, 989–1000.

Winkielman, P., Schwarz, N., Fazendeiro, T. A., & Reber, R. (2003). The hedonic marking of processing fluency: Implications for evaluative judgment. In J. Musch & K. C. Klauer (Eds.), *The psychology of evaluation: Affective processes in cognition and emotion* (pp. 189–217). Mahwah, NJ: Lawrence Erlbaum Associates, Publishers.

Wittgenstein, L. (1978/1956). *Remarks on the foundations of mathematics* (Revised ed., G. E. M. Anscombe, trans). Oxford: Basil Blackwell.

Zajonc, R. B. (1968). Attitudinal effects of mere exposure. *Journal of Personality and Social Psychology, Monograph Supplement, 9*, 1–27.

Zeki, S., Romaya, J., Benincasa, D., & Atiyah, M. (2014). The experience of mathematical beauty and its neural correlates. *Frontiers in Human Neuroscience, 8*.

15 Mathematical Knowledge, the Analytic Method, and Naturalism

Fabio Sterpetti

1. The Problem of Mathematical Knowledge

It is often argued that mathematics is "the paradigm of certain and final knowledge" (Feferman, 1998, 77). The degree of certainty that mathematics is able to provide is considered one of its qualifying features by many authors. For example, Byers states that the certainty of mathematics is "different from the certainty one finds in other fields [. . .]. Mathematical truth has [. . .] [the] quality of inexorability. This is its essence" (Byers, 2007, 328).

It is also often claimed that mathematics is objective, in the sense that it is mind-independent, and so that it is independent from our biological constitution. For example, George and Velleman state that understanding the nature of mathematics does not require asking "such questions as 'What brain, or neural activity, or cognitive architecture makes mathematical thought possible?'", because "such studies focus on phenomena that are really extraneous to the nature of mathematical thought itself" (George & Velleman, 2002, 2).

Mathematics proved tremendously useful in dealing with the world. Indeed, current natural science is "mathematical through and through: it is impossible to do physics, chemistry, molecular biology, and so forth without a very thorough and quite extensive knowledge of modern mathematics" (Weir, 2005, 461). But despite its being so pervasive in scientific knowledge, we do not have yet an uncontroversial and science oriented account of what mathematics is. So "in a reality [. . .] understood by the methods of science", we are unable to answer to the following question, "Where does mathematical certainty come from?", even because most mathematicians and scientists "do not take seriously the problem of reconciling" the certainty of mathematical knowledge "with a scientific world-view" (Deutsch, 1997, 240).

Moreover, many authors are skeptical about the very possibility of developing a naturalist perspective on mathematics. They think that "mathematics is an enormous Trojan Horse sitting firmly in the center of the citadel of naturalism", because even if "natural science is mathematical through and through", mathematics seems to "provide a counterexample both to methodological and to ontological naturalism". Indeed, mathematics ultimately rests on axioms, which are "traditionally held to be known a priori, in some accounts by virtue of a form of intuitive awareness". The epistemic role of the axioms in mathematics seems "uncomfortably close to that played by the insights of a mystic. When we turn to ontology, matters are, if anything, worse: mathematical entities, as traditionally construed, do not even exist in time, never mind space" (Weir, 2005, 461–462). In fact, the majority of mathematicians and philosophers of mathematics argues for some form of mathematical realism (Balaguer, 2009).

Thus it is very difficult even to envisage how naturalistically accounting for what mathematics is and how we acquire mathematical knowledge. For example, Brown admits that he has "no idea how the mind is able to 'grasp' or 'perceive' mathematical objects and mathematical facts", but he nevertheless is certain that it "is not by means of some efficient cause" (Brown, 2012, 12). The difficulty of accommodating mathematical knowledge within a coherent scientific worldview is what Mary Leng called 'the problem of mathematical knowledge'. According to her, "the most obvious answers to the two questions 'What is a human?' and 'What is mathematics?' together seem to conspire to make human mathematical knowledge impossible" (Leng, 2007, 1).[1]

A clarification is in order. There is a huge number of works in cognitive science devoted to study numerical capacities in human and nonhuman animals (see e.g., Cohen Kadosh & Dowker, 2015; Dehaene & Brannon, 2011; Dehaene, Duhamel, Hauser, & Rizzolatti, 2005), but we will not be primarily concerned with these works here. Indeed, these researches do shed light on how to conceive of mathematics naturalistically (De Cruz, 2006). But they have so far investigated the origin and functioning of some very basic numerical abilities. These basic capacities are thought to have evolved because they allow us to approximately deal with numerosities sufficiently well to ensure the survival. This seems insufficient to justify the claim that mathematical knowledge is certain. However, no adequate scientific account of how we develop advanced mathematics starting from those basic numerical abilities has been provided yet (see e.g., Spelke, 2011). Thus, even if *prima facie* the study of such basic cognitive abilities does not support the traditional view of mathematics, it seems at the moment unable to definitely confute that view. Indeed, according to many authors that support the traditional view, throwing light on the evolutionary roots of these numerical

---

1 On the difficulty of making the traditional view of mathematics compatible with a naturalist stance, cf. Núñez (2009, 69): "Lakoff and I called this view the *romance of mathematics*, a kind of mythology in which mathematics has a truly objective existence, providing structure to this universe and any possible universe, independent of and transcending the existence of human beings or any beings at all. However, despite its immediate intuitiveness and despite being supported by many outstanding physicists and mathematicians, the romance of mathematics is *scientifically* untenable".

capacities is, in and of itself, insufficient to naturalistically explain two things: first, the degree of certainty of mathematics and, second, the effectiveness in dealing with the world that our advanced mathematics displays. For example, Polkinghorne states that it

> seems clear enough that some very modest degree of elementary mathematical understanding [. . .] would have provided our ancestors with valuable evolutionary advantage. But whence has come the human capacity [. . .] to attain the ability to conjecture and eventually prove Fermat's Last Theorem, or to discover non-commutative geometry? Not only do these powers appear to convey no direct survival advantage, but they also seem vastly to exceed anything that might plausibly be considered a fortunate spin-off from such mundane necessity.
>
> (Polkinghorne, 2011, 31–32)

Since we are dealing here with the issue of whether the traditional view is compatible with a naturalist stance, as it will become more clear in what follows, we will not dwell on those attempts that try to naturalize mathematics by focusing on discoveries related to our basic numerical abilities, without addressing the issue of whether the traditional view of mathematics should be maintained or not in the light of our scientific understanding of those abilities.

This chapter aims to suggest that a promising step toward the elaboration of an adequate naturalist account of mathematics and mathematical knowledge may be to take the method of mathematics to be the analytic method rather than the axiomatic method. Indeed, it seems impossible to naturalize mathematics without challenging at least some crucial aspects of the traditional view of mathematics, according to which mathematical knowledge is certain and the method of mathematics is the axiomatic method. Nor does it seem possible to keep maintaining that the method of mathematics is the axiomatic method and mathematical knowledge is certain, if we dismiss that view. The analytic view of the method of mathematics, which has been mainly advocated by Carlo Cellucci in recent years (Cellucci, *forthcoming*, 2017, 2013), will be illustrated in some detail; then I will argue that this view could contribute to develop a naturalist account of mathematics and mathematical knowledge. I will also take Cellucci's insight further and point out that the analytic view of method can do that at a cost: it forces us to rethink the 'traditional image' of mathematics. Indeed, if we take the method of mathematics to be the analytic method, then mathematical knowledge cannot be said to be certain, and the only kind of mathematical knowledge that we can have is *plausible* knowledge.

2. The Method of Mathematics

The certainty of mathematical knowledge is usually supposed to be due to the method of mathematics, which is commonly taken to be the axiomatic method.[2] In this view, the method of mathematics differs from the method of investigation in the natural science: whereas "the latter acquire general knowledge using inductive methods, mathematical knowledge appears to be acquired [. . .] by deduction from basic principles" (Horsten, 2015). According to Frege, when we do mathematics we form chains of deductive "inferences starting from known theorems, axioms, postulates or definitions and terminating with the theorem in question" (Frege, 1984, 204). In the same vein, Gowers states that what mathematicians do is that they "start by writing down some axioms and deduce from them a theorem" (Gowers, 2006, 183). So it is the deductive character of mathematical demonstrations that confers its characteristic certainty to mathematical knowledge since demonstrative "reasoning is safe, beyond controversy, and final" (Pólya, 1954, I, v), precisely because it is deductive in character. In this view, "deductive proof is almost the defining feature of mathematics" (Auslander, 2008, 62).

If the method of mathematics is the axiomatic method, mathematics mainly consists in deductive chains from given axioms. So, in order to claim that mathematical knowledge is certain, we have to know that those axioms are 'true', where 'true' is usually intended as 'consistent with each other'.

As well as the consistency of axioms, the problem of justifying our reliability about mathematics is also related to the problem of justifying our reliability about logic. Indeed, if we think that the method of mathematics is the axiomatic method, proving the reliability of deductive inferences is essential for claiming for the certainty of mathematical knowledge. For example, Franks states that in mathematics "deductive logic is the only arbiter of truth" (Franks, 1989, 68).

Thus there are two statements that we should be able to prove in order to safely claim that mathematical knowledge is certain: 1) axioms are consistent and 2) deduction is truth-preserving. Indeed, a deductive proof "yields categorical knowledge [i.e., knowledge independent of any particular assumption] only if it proceeds from a secure starting point and if the rules of inference are truth-preserving" (Baker, 2016, sec. 2.2).

---

2 Cf. Baker (2016, sec. 1): "It seems fair to say that there is a philosophically established received view of the basic methodology of mathematics. Roughly, it is that mathematicians aim to prove mathematical claims [. . .], and that proof consists of the logical derivation of a given claim from axioms".

Now, while whether it is possible to deductively prove 2) is at least controversial (see e.g., Haack, 1976; Cellucci, 2006), it is uncontroversial that it is generally impossible to mathematically prove 1)—i.e., that axioms are consistent—because of Gödel's results.[3] Indeed, by Gödel's second incompleteness theorem, for any consistent, sufficiently strong deductive theory T, the sentence expressing the consistency of T is undemonstrable in T. Usually, those authors that despite this result maintain that mathematical knowledge is certain, make reference to a sort of faculty that we are supposed to possess, and that would allow us to 'see' that axioms are consistent. For example, Brown states that we "can intuit mathematical objects and grasp mathematical truths. Mathematical entities can be 'perceived' or 'grasped' with the mind's eye" (Brown, 2012, 45).

This view has been advocated by many great mathematicians and philosophers. Detlefsen describes the two main claims of this view as follows: 1) "mathematicians are commonly convinced that their reasoning is part of a process of discovery, and not mere invention" and 2) "mathematical entities exist in a noetic realm to which the human mind has access", (Detlefsen, 2011, 73). With respect to the ability of grasping mathematical truths—i.e., accessing the mathematical realm—this view traditionally assumes "a type of apprehension, *noēsis*, which is characterized by its distinctly 'intellectual' nature. This has generally been contrasted to forms of *aisthēsis*, which is broadly sensuous or 'experiential' cognition" (Ibidem, 73). For example, Gödel states, "Despite their remoteness from sense experience we do have something like a perception also of the objects of set theory, as is seen from the fact that the axioms force themselves upon us as being true" (Gödel, 1947, 1990, 268).

The problem is that this view is commonly supported by authors that are anti-naturalists.[4] Hence they do not take care of articulating a scientifically plausible account of how such 'intuition' may work or may have evolved. The following question then arises: is it possible to naturalize the human ability to grasp mathematics and logic, and maintain the traditional view of mathematical knowledge—i.e., that mathematical knowledge is certain and the method of mathematics is the axiomatic method? In other words, how can we account for the reliability of mathematics and logic if we accept the idea that they are both produced by humans and humans are evolved organisms? (Schechter, 2013; Smith, 2012).

---

3 Cf. Baker (2016, sec. 2.3): "Although these results apply only to mathematical theories strong enough to embed arithmetic, the centrality of the natural numbers (and their extensions into the rationals, reals, complexes, etc.) as a focus of mathematical activity means that the implications are widespread".

4 On the anti-naturalism of many of the supporters of this view, cf. e.g. Gödel (1951, 1995, 323): "There exists [. . .] an entire world consisting of the totality of mathematical truths, which is accessible to us only through our intelligence, just as there exists the world of physical realities; each one is independent of us, both of them divinely created".

There is no clear answer to this question. Some authors have tried to naturalize mathematics and logic relying on evolutionism (see De Cruz, 2006; Krebs, 2011; Woleński, 2012). The main difficulties afflicting these approaches derive from the fact that they try to naturalize mathematics in a way that allow them to avoid the risk of being excessively revisionary on what we take mathematical knowledge to be. In other words, they try to show that mathematics rests on some evolved cognitive abilities and that this *evolutionary* ground confers a degree of epistemic justification to the mathematics we actually do, which is able to secure our convictions on what mathematical knowledge is.[5]

The fact is that it is not easy to defend the claim that evolution may provide the degree of justification needed to maintain the traditional view of mathematics as the paradigm of certain knowledge. Briefly, in order to claim that natural selection gave us the ability at attaining the truth with regard to mathematics, we should demonstrate that natural selection is an aimed-at-truth process. For example, Wilkins and Griffiths state that to "defeat evolutionary skepticism, true belief must be linked to evolutionary success in such a way that selection will favour organisms which have true beliefs" (Wilkins & Griffiths, 2013, 134). The problem is exactly *how* to justify such a link, and the issue is at least controversial (Vlerick & Broadbent, 2015; Sage, 2004).

Consider our confidence in the fact that deduction is truth-preserving. Kyburg states, "Our justification of deductive rules must ultimately rest, in part, on an element of deductive intuition: we see that M[odus] P[onens] is truth-preserving – this is simply the same as to reflect on it and fail to see how it can lead us astray" (Kyburg, 1965, 276). The problem is that our failing to conceive an alternative can justify the reliability of the deductive rules only if our ability to conceive alternatives could be shown to reliably exhaust the space of all the possible alternatives. But such a demonstration of the reliability of our ability to conceive alternatives doesn't exist, and we can rely only on our 'intuition' (i.e., on our failing in finding any counterexample) to convince ourselves that MP cannot lead us astray and on the fact that such 'intuition' appears self-evident to us.[6] The mere fact that some statements

---

5 Other proposals devoted to naturalize mathematics by considering our evolved cognitive abilities have been put forward by Núñez (2008, 2009) and Ye (2011). These interesting proposals cannot be discussed here for reason of space. What is nevertheless worth underlining here is that even if both these proposals, despite their full-blooded naturalism, seem to leave untouched the idea that the method of mathematics is the axiomatic method, they both explicitly claim that Mathematical Platonism is incompatible with a naturalist stance, because it is scientifically untenable. So they cannot be regarded as naturalized versions of the traditional view.

6 Sklar (1981) argues that in fact *we are unable* to explore the space of possible alternatives exhaustively, given that in the past, as history of science teaches us, we routinely failed in exploring the space of the conceivable alternatives to a given scientific hypothesis. More recently, Stanford (2006) has further developed Sklar's insight to defend his instrumentalist view of science.

appear to us as self-evidently true it is not by itself a guarantee of their truth, if our ability in evaluating the self-evident truth of a statement is an evolved capacity. Our 'sense' of the self-evident may be not only oriented toward contingent connections that were useful in the past and that do not reflect necessary and eternal truths, but given that we are not able to demonstrate that only true beliefs may lead us to successfully dealing with the world,[7] we cannot even eliminate the possibility that an ability in perceiving some falsities as self-evident has been selected, because perceiving such falsities as self-evident truths was adaptively useful (Nozick, 2001; see also Vaidya, 2016).

It is worth noticing that, if we want to maintain that mathematical knowledge is *certain*, and we want to naturalize mathematical knowledge, the evolved cognitive ability to grasp whether axioms are consistent, cannot be supposed to be *fallible*.

Indeed, if this faculty is fallible, and we are not able to determine correctly whether axioms are true in *all* the cases we examine, then we will be generally unable to claim that our mathematical knowledge is certain in any particular case. Indeed, as we have seen, a mathematical result is true and certain if the axioms from which it is derived are 'true' (in the sense of consistent), and deduction is truth-preserving. If naturalizing mathematics implies that our evolved ability in assessing the truth of the axioms is fallible, and we have no other way to verify our verdict, we find ourselves in a situation in which we may have erred and we are unable to detect *whether* we erred. Thus we would be unable to claim that we judged correctly, and so that our mathematical knowledge *is* certain. So if the justification of our mathematical knowledge rests on some fallible faculty, then the attempt to naturalize mathematics cannot maintain the traditional view of mathematics.[8]

7 Cf. McKay and Dennett (2009, 507): "In many cases (perhaps most), beliefs will be adaptive by virtue of their veridicality. The adaptiveness of such beliefs is not independent of their truth or falsity. On the other hand, the adaptiveness [. . .] of *some* beliefs is quite independent of their truth or falsity".

8 For an opposite view, see McEvoy (2004), who argues for the compatibility of reliabilism and mathematical realism. According to him, mathematical intuition may be at the same time an *a priori*, reliable, and fallible faculty of reason. In a similar vein, Brown (2012) maintains that platonism and fallibilism can be combined. But even if we concede that fallibilism in epistemology is compatible with platonism in ontology, this view seems not compatible with a *naturalist* stance, since it is not able to give a naturalist account of how intuition can provide mathematical knowledge that is certain. This view has to face the same difficulties discussed earlier with regard to the justification of the claim that deduction is truth-preserving: when evolution enters the picture, it is not easy to justify the claim that we are able to assess correctly what is possible or impossible through reasoning alone. This impinges on the possibility of claiming that our mathematical beliefs are certainly true. So any kind of evolutionary reliabilism seems not really able to provide a naturalist way of supporting the traditional view of mathematics, since it is not able to secure the certainty of our mathematical knowledge (Sage, 2004).

3. Mathematics and Naturalism

As to how to understand 'naturalism', we are not concerned here with any specific view of naturalism, nor we will survey the many criticisms that have been moved to this (family of) view(s) (for a survey, see Clark, 2016; Papineau, 2016; Rosenberg, 1996). For the purpose of this chapter, 'naturalism' can just be understood as the claim that we should refuse accounts or explanations that appeal to non-natural entities, faculties or events, where 'non-natural' has to be understood as indicating that those entities, faculties or events cannot *in principle* be investigated, tested, and accounted for in the way we usually do in science. In other words, non-natural entities, faculties, or events are those that are characterized and defined precisely by their inaccessibility, by the impossibility of being assessed, empirically confirmed, or even made compatible with what we consider genuine scientific knowledge in the same or in some close domain of investigation. In all those cases, we have to face a problem of access[9] and a claim of exceptionality that usually lacks sufficiently strong reasons to be conceded.

So in this chapter, we will exclusively be concerned with those strategies aimed at naturalizing a domain D, which has traditionally been considered to be affected by an 'access problem' (e.g., mathematics, morality, modality, etc.), by providing a plausible evolutionary account of some cognitive abilities that would make our knowledge of some D-truths a natural fact. As an example, consider Timothy Williamson's approach to modality. He first reduces the problem of explaining modal knowledge to the problem of explaining our capacity to perform counterfactual reasoning correctly.

---

9 See Benacerraf (1973). Benacerraf's famous epistemological challenge to Mathematical Platonism has been criticized because it assumes the causal theory of knowledge, which nowadays is discredited among epistemologists. But Benacerraf's argument may be raised against platonism without assuming the causal theory of knowledge, as Field maintains (Field, 1989). On this issue, cf. Baron (2015, 152): "Field's version of the access problem focuses on mathematicians' mathematical beliefs. The mathematical propositions that mathematicians believe tend to be true. If platonism is correct, however, then these propositions are about mathematical objects. So, the mathematical beliefs held by mathematicians [. . .] are reliably correlated with facts about such objects. The challenge facing the platonist, then, is to provide an account of this reliable correlation". It may be objected that this formulation implicitly assumes the correspondence view of truth, and that this view of truth has not necessarily been held by platonists. But, even if accepting that view of truth is not strictly mandatory for a realist, the correspondence view is in fact the view of truth usually adopted by realists of all stripes. And according to many authors, one of the major arguments *"for* mathematical realism appeals to a desire for a uniform semantics for *all* discourse: mathematical and non-mathematical alike [. . .]. Mathematical realism, of course, meets this challenge easily, since it explains the truth of mathematical statements in exactly the same way as in other domains" (Colyvan, 2015, sec. 5)—i.e., by assuming that there is a *correspondence* between the realm of mathematical objects and our mathematical knowledge. So if platonists try to avoid Benacerraf's challenge by rejecting the correspondence view of truth, they risk dismissing one of the most convincing reasons for adopting platonism in the first place.

Then he gives some reasons to think that an evolutionary account of the emergence of this capacity may be plausible (Williamson, 2000).

In this context, a naturalistic account of mathematics has to assume the hypothesis that the human mathematician is "a thoroughly natural being situated in the physical universe" and that therefore "any faculty that the knower has and can invoke in pursuit of knowledge must involve only natural processes amenable to ordinary scientific scrutiny" (Shapiro, 1997, 110).

Recently, Helen De Cruz argued that an evolutionary account of mathematics may well be compatible with a realist view of mathematics. According to her, "animals make representations of magnitude in the way they do because they are tracking structural (or other realist) properties of numbers" (De Cruz, 2016, 7). In this view, "realism about numbers could be true, given what we know about evolved numerical cognition" (Ibidem, 2). Indeed, "it seems plausible that numerical cognition has an evolved, adaptive function", and it has been demonstrated that numerical cognition "plays a critical role in our ability to engage in formal arithmetic" (De Cruz, 2016, 4).

The main problem with this line of reasoning is the following: if we try to naturalize mathematics and maintain the traditional view—i.e., the view according to which the method of mathematics is the axiomatic method and mathematical knowledge is certain—then our naturalized account of mathematics risks to become incompatible with Gödel's results. Indeed, in this view, as we have already noted, in order to justify mathematical knowledge, at least two requirements have to be fulfilled: axioms have to be shown to be consistent, and deduction has to be shown to be truth-preserving.

Let's focus on the first requirement. If we maintain that evolution is able to justify the traditional view of mathematics, this amounts to claim that evolution, in some way, gave us an ability to know with certainty whether axioms are true, at least in the sense that they are consistent. Let's name the 'result' that we obtain thanks to such evolved ability—i.e., "the axioms of the axiomatic system we are considering are consistent"—T.

The problem is that, by Gödel's second incompleteness theorem, it is impossible to demonstrate in any sufficiently powerful axiomatic system that the axioms of such system are consistent. Let's name this result G.

Now, if the method of mathematics *really* is the axiomatic method, how could we accept that T holds? Or, to put it slightly differently, should we consider T be part of our *mathematical* knowledge?

If T is part of our mathematical knowledge, then the axiomatic method is not really the unique method of mathematics, since a crucial result as T is not obtained by this method, and so the method by which T has been obtained has to be added to the list of the legitimate methods of mathematics. This would render G almost irrelevant. Indeed, if the axiomatic method is not the only acceptable method in mathematics, and if we can know that a set of axioms is consistent thanks to some evolved faculty, then that in some axiomatic systems we cannot prove whether a set of axioms is consistent is irrelevant to us. The fact that the axioms are consistent could be taken to be established by our evolved faculty.

But the majority of mathematicians, even of platonist mathematicians, would be unwilling to consider Gödel's contributions as irrelevant and the consistency of axioms establishable by merely relying on an evolved sort of intuition. Precisely because they do believe the axiomatic method to be the method of mathematics, they tend to confer a higher degree of certainty to Gödel's results, which have been established according to such method, than to the intuitions of an evolved faculty—the reliability of which cannot be proved by the same method. Indeed, in a naturalist framework, our evolved intuitions may be shown to be reliable only through some inductive method, which is peculiar of natural science. If we concede that the method of mathematics is distinct from the method of natural science, as the traditional view holds, and that the method of mathematics is the axiomatic method, then we will be unable to sufficiently justify the belief that our evolved intuition is reliable up to a degree which is comparable with the confidence that the axiomatic method is supposed to confer to mathematical results. Thus, even if our evolved intuition were in fact reliable and infallible, we would be unable to demonstrate its infallibility with the same degree of certainty with which Gödel's results can be proven, given that they are mathematical results.

If, on the other hand, we take T not to be part of our mathematical knowledge, and protest that T is not really a 'mathematical result', we will nevertheless find ourselves in an uncomfortable position: we should maintain that we possess some *knowledge* about some mathematical issue that is not part of our *mathematical* knowledge. It is not easy to accommodate this claim by the usual epistemological standards. Since knowledge requires (at least) truth and justification, if we take T to be knowledge, T is true and justified. If T expresses something true about some mathematical issue, then we can affirm that T expresses a mathematical truth. But if we refuse to consider T as a part of our mathematical knowledge, and we are not able to express the same mathematical truth that T expresses by the means of what we take to constitute our current mathematical knowledge, then T would be able to express a mathematical truth that our mathematical knowledge would be unable to express.

It may be objected that T expresses a mathematical truth that cannot be expressed on the basis of our mathematical knowledge, because the justification requirement that a true belief needs to fulfill in order to become *mathematical* knowledge is stricter than the justification requirement that has to be fulfilled in other domains. Let's concede, for argument's sake at least, such claim on the justification requirement for mathematics.[10] If this is the case, T could well be able to express a truth about some mathematical issue, but this truth may nevertheless be insufficiently justified in order to become part of

10 See, e.g., Kitcher (1988, 297): "The obvious way to distinguish mathematical knowledge from mere true belief is to suggest that a person only knows a mathematical statement when that person has evidence for the truth of the statement—typically, though not invariably, what mathematicians count as a proof".

our mathematical knowledge. And this would explain the fact that *T* is able to express a truth on a mathematical issue, and that this truth does not figure among our *known* mathematical truths. But this would amount saying that our mathematical knowledge is a kind of knowledge with a higher degree of certainty than *T*, since the true beliefs that constitute our mathematical knowledge are supposed to display a higher degree of justification.

But if we try to naturalize mathematics in the way here we are dealing with, things should go the other way round. Since *T*, in order to be able to justify the traditional view of mathematics, has to be certain, *T* has to be knowledge with the highest degree of certainty. Thus the degree of certainty that our mathematical knowledge may display is in some sense subordinated to *T*, since its certainty is dependent on the certainty of *T*. Mathematical knowledge would be in this way a kind of knowledge with a less high degree of certainty than *T*. Thus it cannot be the case that the mathematical truth expressed by *T* is not an instance of mathematical knowledge because it is insufficiently justified. So this objection is inadequate.

Since in both cases—i.e., either we take *T* to be part of mathematical knowledge or not—we end with implausible scenarios, the supporters of the traditional view seem unable to really find an adequate way of naturalistically justifying the claim that mathematics is the paradigm of certain knowledge.

4.   The Analytic Method

We will now present the analytic view of method in some detail and try to underline the reasons why it may be of use to those interested in the attempt of naturalizing mathematics. There are three main claims that characterize the analytic view of method: 1) the method of mathematics is the analytic method, and not the axiomatic method; 2) the concept of truth proved to be inadequate to account for the method of mathematics, natural sciences, and philosophy, and thus has to be replaced with the concept of 'plausibility'; and 3) since mathematics, natural sciences, and philosophy all aim at acquiring new knowledge, they all share the very same method—i.e., the analytic method, which is the method through which knowledge can be ampliated.

Let's start with 1), the claim that the analytic method is the method of mathematics. The analytic method may be defined as the method in which

> to solve a problem, one looks for some hypothesis that is a sufficient condition for solving it. The hypothesis is obtained from the problem, and possibly other data already available, by some non-deductive rule, and must be plausible [. . .]. But the hypothesis is in its turn a problem that must be solved, and is solved in the same way [. . .]. And so on, *ad infinitum*. Thus solving a problem is a potentially infinite process.
>
> (Cellucci, 2013, 55)

The origin of the analytic method may be traced back to the works of the mathematician Hippocrates of Chios and the physician Hippocrates of Cos, and was first explicitly formulated by Plato in *Meno, Phaedo*, and the *Republic*. As an example of the analytic method, consider the solution to the problem of the quadrature of certain lunules provided by Hippocrates of Chios:

> Show that, if *PQR* is a right isosceles triangle and *PRQ*, *PTR* are semicircles on *PQ*, *PR*, respectively, then the lunule *PTRU* is equal to the right isosceles triangle *PRS*.

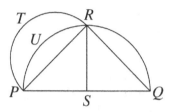

Figure 15.1 Drawn after Cellucci (2013, 61)

To solve this problem, Hippocrates of Chios states the following hypothesis:

(B) Circles are as the squares on their diameters.

Hypothesis (B) is a sufficient condition for solving the problem. For, by the Pythagorean theorem, the square on *PQ* is twice the square on *PR*. Then, by (B), the semicircle on *PQ*, that is, *PRQ*, is twice the semicircle on *PR*, that is, *PTR*, and hence the quarter of circle *PRS* is equal to the semicircle *PTR*. Subtracting the same circular segment, *PUR*, from both the quarter of circle *PRS* and the semicircle *PTR*, we obtain the lunule *PTRU* and the triangle *PRS*, respectively. Therefore, "the lunule" *PTRU* "is equal to the triangle". [Simplicius, *In Aristotelis Physicorum libros Commentaria*, A 2, 61]. This solves the problem. But hypothesis (B) is in its turn a problem that must be solved.

(Cellucci, 2013, 61)

And in fact, this new problem was solved, presumably by Eudoxus, proposing the following hypothesis: (B') Similar regular polygons inscribed in circles are as the squares on their diameters. Hypothesis (B') is a sufficient condition for solving the problem—i.e., hypothesis (B), but hypothesis (B') is in its turn a problem that must be solved. And so on, *ad infinitum*.

The analytic method is the method that from antiquity has been used to advance mathematical knowledge, to acquire *new* knowledge. On the contrary, the axiomatic method has been developed, and was probably already used by mathematicians at Plato's time, mainly as a *didactic* tool in order to teach *already acquired* knowledge:

> Aristotle states that, in order to present, justify, and teach an already acquired proposition, we must start from the principles proper to the subject matter of the proposition, and deduce the proposition from them, since "didactic arguments are those that deduce" propositions "from the principles proper to each subject matter" (Aristotle, *De Sophisticis Elenchis*, 2, 165 b 1–2).
>
> (Cellucci *forthcoming*, sec. 12.9)

According to Aristotle, who first explicitly formulated the axiomatic method in *Posterior Analytics*, the "mathematical sciences are learned in this way, and so is each of the other arts" (Aristotle, *Analytica Posteriora*, A 1, 71 a 3–4). The merely *didactic* nature of the axiomatic method clearly emerges from the fact that Aristotle distinguishes between the analytic-synthetic method, which according to him is the method of research, and the axiomatic method, which is the method to present results already acquired.

In the last century, the dominance of a foundationalist perspective on scientific and mathematical knowledge, the influence of Hilbert's thought, and the diffusion of the idea that a logic of discovery cannot exist, led to the widespread conviction that the method of mathematics and science is (or should be) the axiomatic method,[11] according to which, to demonstrate a statement, one starts from some given premises, which are supposed to be true, and deduces the statement from them. Analysis has been overlooked or neglected (Schickore, 2014; Cellucci, 2013).

For example, the hypothetico-deductive model of science, which is still dominant although it has been severely criticized,[12] is based on the axiomatic method. According to the hypothetico-deductive view of science, building "a scientific theory is a matter of choosing certain hypotheses, deducing consequences from them, and comparing consequences with the observation data" (Cellucci, *forthcoming*, sec. 13.5). In this perspective, the process of knowledge ampliation is accounted for in terms of deductive reasoning. The problem is that the "propositions deduced from the hypotheses contain

---

11 For a historical survey of the main conceptions of scientific and mathematical method that have been put forward so far, see Cellucci (2016, 2013). On the analytic method, see also Hintikka, Remes (1974) and Lakatos (1978), vol. 2, chap. 5. On Plato's formulation of the analytic method, see Menn (2002). For a historical survey of the axiomatic method, and the relevance of Hilbert's view, see Rodin (2014), part I.

12 See Schickore (2014) for a survey.

nothing essentially new with respect to the hypotheses", because deduction is non-ampliative, "it simply makes explicit what is implicitly contained in the hypotheses" (Ibidem, sec. 13.6).[13]

Thus the axiomatic method may illustrate how to derive a result from certain premises, but it does not explain *how* we got there. This is the reason why, according to the analytic view of method, the axiomatic method is inadequate to support a naturalist perspective on mathematics and science. Indeed, the axiomatic method does not account for the process of *hypotheses production*, and so it is not able to show the path that mathematicians and scientists follow to reach a result and solve a problem. Thus, the axiomatic method cannot improve our understanding of how we produce knowledge. From a naturalist perspective, it is unsatisfying that the alleged *method* of mathematics is unable to say anything relevant on *how* mathematical knowledge is acquired. If knowledge ampliation remains a mystery, naturalists cannot adequately counter those anti-naturalists who maintain that it is necessary to take into consideration some non-natural element in order to explain our ability to produce mathematical knowledge.

On the contrary, in the analytic view the path that has been followed to reach a result and solve a problem is not occulted, since in this view the context of discovery is not divorced from the context of justification. Indeed, an analytic demonstration consists in a non-deductive derivation of a hypothesis from a problem and possibly other data, where the hypothesis is a sufficient condition for the solution of the problem and is plausible (Ibidem, chap. 21).

It is important to underline that the analytic method involves both deductive and non-deductive reasoning. Indeed, to find a hypothesis we proceed from the problem in an ampliative (i.e., non-deductive) way, performing ampliative inferences, and then in order to assess the plausibility of such hypothesis, we deduce conclusions from it. But the role that deduction plays in the analytic view is not the crucial role that deduction is supposed to play in the 'received view'—i.e., the axiomatic method, which can be considered as a variant of the analytic-synthetic method. According to the analytic view, axioms are not the source of mathematical knowledge, and we shouldn't overestimate their role, which is limited to giving us the possibility of presenting, for didactic or rhetorical purposes,[14] some body of already acquired knowledge in deductive form. Axioms do not enjoy any temporal or conceptual priority

---

13 The characterization of deduction as non-ampliative has been questioned—e.g., by Haack (1996). For a rejoinder, see Cellucci (2006).

14 See Cellucci, *forthcoming*, sec. 22.7: "The purpose of axiomatic demonstration is to justify and teach an already acquired proposition. This serves to convince the audience—readers of research papers or textbooks, conference audiences or students in the classroom—that the proposition should be accepted. Therefore, several people have asserted that axiomatic demonstration has a rhetorical role".

in the development of mathematical knowledge, nor do they play any special epistemological role. As Hamming states, if "the Pythagorean theorem were found to not follow from postulates, we would again search for a way to alter the postulates until it was true. Euclid's postulates came from the Pythagorean theorem, not the other way" (Hamming, 1980, 87).

In order to avoid misunderstanding, it has to be underlined that the analytic method has not to be confused with the analytic-synthetic method. Indeed, it may be objected that when we stop, since it seems that we have to, the place where we stop can be regarded as an 'axiom', and so we are back in the axiomatic method. But it is not so. This objection derives precisely from confusing the analytic method with the analytic-synthetic method. According to the analytic-synthetic method, as stated by Aristotle, the search for a solution to a problem is a *finite* process, "so the ascending sequence of the premises must terminate", and once the prime premises have been found, "the only role which remains for analysis is to find deductions of given conclusions from prime premises", therefore the analytic-synthetic method "is primarily a method [. . .] for finding demonstrations in given axiomatic systems" (Cellucci, 2013, 75). In this view, once the process of discovery ends, there is no more a relevant role for the analytic method.

On the contrary, according to the analytic method, the process of discovery is a potentially *infinite* process. So, since the process of discovery never really ends, there is no relevant role for the axiomatic method in this view. Indeed, even if we provisionally stop, we know that we will have to go further in order solve the problem of justifying the hypothesis where we stopped.[15]

4.1 Plausibility in Place of Truth

Let's examine the second qualifying claim of the analytic view, which states that we should replace the concept of truth with the concept of 'plausibility' (see Section 4, this chapter). The ('Kantian') reasons for adopting plausibility in place of truth in order to account for human knowledge, may be briefly summarized as follows: since the main definitions of truth that have been

---

15 It may be objected that, if solving a problem is a potentially infinite process, we could be unable to assess whether we were *really* able to solve a given problem or not. This objection seems to rest on the idea that a genuine solution to a given problem has to be a *definite* solution—i.e., a solution that will never be altered, or overcome by a better solution. In this view, once a problem is solved, we should not keep trying to solve it. But there is no compelling reason why a solution to a problem should not be thought to be a genuine solution, unless it can be proved that it is a definite solution. Think of mathematics. In mathematics, solutions are routinely searched also for those problems for which a solution has already been found. And solutions that once were regarded as adequate, have often successively been regarded as inadequate. As Poincaré states, in mathematics there are not "solved problems and others which are not; there are only problems *more or less* solved", where "it often happens however that an imperfect solution guides us toward a better one" (Poincaré, 2015, 377–378).

proposed so far proved unable to give us a workable criterion of truth—i.e., a nonalgorithmic means to decide whether a given statement is true—they cannot avoid the skeptical argument of the criterion (see Cellucci, 2014a).[16] In other words, if we maintain that knowledge requires truth, and adopt one of the traditional definitions of truth, since our conception of truth cannot face the skeptical challenge of the criterion, we would never be able to assess whether we *really* reached some truth and so whether our knowledge is genuine knowledge. So if knowledge requires truth, knowledge is epistemically unattainable by humans. But if knowledge is epistemically unattainable, we will never be able to account for our having knowledge, let alone to account for our having knowledge *naturalistically*. Since truth is such an *unrealistic* aim for our human epistemic activities, we should instead take plausibility to be the central concept of epistemology:[17]

> The goal of science is plausibility. Scientific theories do not deal with the essence of natural substances, but only with some of their phenomenal properties, and deal with them on the basis of plausible hypotheses. Then a scientific theory is not a set of truths but rather a set of plausible hypotheses. Thus the goal of science is plausibility rather than truth.
>
> (Cellucci, 2013, 154)

But how should we conceive of plausibility? The plausibility of a hypothesis is assessed by a careful examination of the arguments for and against it. According to Cellucci, in order to judge over the plausibility of a hypothesis, the following 'plausibility test procedure' has to be performed:

(1) Deduce conclusions from the hypothesis.
(2) Compare the conclusions with each other in order to see that the hypothesis does not lead to contradictions.
(3) Compare the conclusions with other hypotheses already known to be plausible, and with results of observations or experiments, in order to see that the arguments for the hypothesis are stronger than those against it on the basis of experience.

(Ibidem, 56)

---

16 The problem of the criterion of truth is the ancient skeptical paradox of the wheel: "In order to know any proposition we must first know a criterion, but in order to know a criterion we must already know some proposition" (Cling, 1997, 109). For a survey, see Amico (1993).

17 This argument can be called 'Kantian' since it is analogous to the 'Kantian' argument against metaphysical realism. David summarizes this argument as follows: "We cannot step outside our own minds to *compare* our thoughts with mind-independent reality. Yet, on the realist correspondence view of truth, this is what we would have to do to gain knowledge of the world. We would have to *access* reality as it is *in itself*, to determine whether our thoughts correspond to it. Since all our access to the world is mediated by our cognition, this is impossible. Hence, on realism, knowledge of the world would be impossible. Since knowledge of the world is possible, realism must be wrong" (David, 2016, 173).

If a hypothesis passes the plausibility test procedure, it can be temporarily accepted. If, on the contrary, a hypothesis does not pass the plausibility test, it is put on a 'waiting list' since new data may emerge and a discarded hypothesis may successively be reevaluated.

Thus, according to the analytic view of method, what in the ultimate analysis we really do, and are able to do, in the process of knowledge ampliation, is non-deductively producing hypotheses, assessing the arguments for and the arguments against each hypothesis, and provisionally accept or refuse such hypotheses.[18]

4.2   The Analytic Method and the Biological
Role of Knowledge

To better understand the third qualifying claim of the analytic view listed earlier (Section 4), according to which mathematics, natural sciences, and philosophy share the same method—i.e., the analytic method—we have to focus on the biological role of knowledge. In this view, the main consequence of considering the biological role of knowledge is that knowledge cannot be related to the concept of truth.[19] Indeed, if knowledge is related to our biological makeup, and the latter is related to evolution, then in order to account for what knowledge is, it is better to abandon the concept of truth for at least three reasons, which can be briefly described as follows: 1) evolution is not an aimed-at-truth process, or, at least, the claim that natural selection leads to truth is so debated and controversial that it cannot represent a firm ground on which one can develop a satisfying account of knowledge;[20] 2) if knowledge is something that biological entities *produce* in order to deal with their environment, then it cannot be independent from the subject's biological make-up. Thus if truth has to be understood, as many realists say, as a non-epistemic concept—i.e., as unrelated to the subject who knows, then human knowledge cannot

---

18 This conception of 'plausible hypotheses' is similar, but not identical, to Aristotle's definition of *endoxa*. Cf. Cellucci *forthcoming*, sec. 9.5: "*Endoxa* are things which seem acceptable to everyone, or to the great majority, or to the wise, etc., on the basis of an examination of the arguments for and against them, from which the former turn out to be stronger than the latter. [. . .]. However, that plausibility is to a certain extent related to Aristotle's *endoxa* does not mean that plausibility and *endoxa* are identical. For Aristotle, *endoxa* are continuous with truth. Indeed, he states that 'an ability to aim at *endoxa* is a characteristic of one who also has a similar ability in regard to the truth'. (Aristotle, *Rhetorica*, 1355 a 17–18). Conversely, plausibility [. . .] is an alternative to truth. Therefore, plausibility and *endoxa* are not identical".

19 This is one of the main differences between the analytic view and the attempts aimed at naturalizing knowledge made by Kornblith (2002) and Plotkin (1997), since both Kornblith and Plotkin maintain that knowledge is related to truth.

20 On the difficulty of supporting the claim that evolution leads to truth, see Sage (2004).

be related to truth.[21] 3) If knowledge is related to truth, and truth is not attainable by humans, because its epistemic standard is so elevated that cannot be met by humans, then knowledge would be impossible. Since knowledge is not only possible, but it is necessary for the survival, knowledge cannot be related to truth (Cellucci, 2013). For those reasons, the analytic view considers 'truth' an unuseful concept in order to naturalistically account for knowledge—i.e., to see knowledge as a human activity that is made possible by our evolved abilities and which is indispensable for the survival.

A brief clarification is in order. Since the reference to the concept of truth is usually seen as a characterizing feature of realism, someone may deem the analytic view an 'antirealist' position. Obviously, it is undeniable that the analytic view gives up any reference to the notion of 'truth'. But the analytic view must not be confused with other views, which can be regarded, broadly speaking, as 'antirealists'. For instance, the analytic view has not to be confused with idealism. Indeed, the analytic view does not deny the existence, nor the independence of the external reality. It is not a form of *metaphysical* antirealism. Nor should the analytic view be confused with skepticism. Indeed, the analytic view does not deny the *possibility* of knowledge. It claims that knowledge is possible and that we do have knowledge, but it denies that knowledge needs to be true, simply because it denies that the concept of truth is necessary to define knowledge.

Turning to the issue at stake, as already noted, the analytic view considers knowledge to be necessary for the survival. Now, since "in order to survive, all organisms must acquire knowledge about the environment", this means that "knowledge is a natural phenomenon that occurs in all organisms" (Cellucci, 2013, 250). Since knowledge is necessary for the survival, and the analytic view sees knowledge as the result of solving problems by the analytic method, this means that knowledge is essentially a problem-solving process that is homogeneous throughout the biological realm.

Moving along this line, the analytic view supports the naturalist claim that mathematics, philosophy, and science are continuous. Indeed, mathematics, philosophy, and science are human attempts to acquire knowledge; knowledge is essentially problem solving, and problems are solved throughout the biological realm by the analytic method. So mathematics, philosophy, and science are continuous because their method is the very same method.[22] Moreover, in this perspective, mathematics, philosophy,

---

21 Cf. Sankey (2008, 112): "The realist conception of truth is a non-epistemic conception of truth, which enforces a sharp divide between truth and rational justification".

22 On the difference between science and philosophy, see Cellucci (2014b). On the method of science and the 'models of science', see also Cellucci (2016).

and science may be accounted for in naturalistic terms, because they solve human problems in the same way in which problems are solved in the rest of the biological realm—i.e., through the analytic method.

4.3   The Analytic Method and Fallibilism

As we have seen, the analytic view of method takes knowledge acquisition to be a potentially infinite process and knowledge to be always hypothetical and provisional:

> Solving problems by the analytic method, when successful, produces knowledge, though knowledge that, being based on hypotheses that are only plausible, is not absolutely certain. [. . .] there is a strict connection between knowledge and the analytic method: knowledge arises from solving problems by the analytic method.
>
> (Cellucci, 2015, 224–225)

Since the analytic method is the method of mathematics, this means that according to this view, mathematical knowledge cannot be said to be certain. And that the only mathematical knowledge we can have is knowledge that is plausible. This view may certainly appear *prima facie* extremely revisionary with respect to traditional epistemology and philosophy of mathematics. But it nevertheless displays some vantages, the most important of which is that it is compatible with both our current *mathematical* and *scientific* knowledge.

Indeed, the claim that mathematical knowledge is plausible is compatible with Gödel's results. A proposition cannot be more justified than the axioms from which it is deduced and, by Gödel's second incompleteness theorem, no absolutely reliable justification for the axioms is generally possible. Thus if we claim that our mathematical knowledge is certain, while the axioms from which this mathematical knowledge is derived cannot be justified up to certainty, this claim would be incompatible with an established mathematical result. While if we claim that, following the analytic view, our mathematical knowledge is plausible, this claim is not in contrast with any established mathematical result, since axioms as well as the propositions that we deduce from them may safely be deemed to be plausible.

Moreover, that mathematical knowledge is plausible is also compatible with our current scientific knowledge, according to which there is no evidence that we humans possess such a special cognitive faculty as the 'intuition' that we should assume to possess in order to guarantee that the axioms from which we derive our mathematical knowledge are true. Thus if we claim that mathematical knowledge is certain, and there is no mathematical way to justify up to certainty the axioms from which we derive our mathematical theories, then we have to subscribe to the view according to which we can say that our axioms are true because our 'intuition' is in some way able to know that they are true with certainty. Now, there are no cues that such 'intuition' may exist, nor any scientifically plausible account of how exactly it may work has been provided, nor any plausible account of how it could have evolved has been provided. On the opposite, cognitive science and evolutionary biology give us good reasons to think that such a faculty does not exist, and that perhaps it cannot even exist. So if we claim that mathematical knowledge is certain, this claim is incompatible with our current scientific knowledge. While if we claim that, following the analytic view, our mathematical knowledge is plausible, this claim is not incompatible with our current scientific knowledge, since in order to claim for the plausibility of our mathematical knowledge, there is no need to postulate the existence of any extravagant faculty.

From what we have seen, it is clear that the analytic view may be deemed, broadly speaking, a fallibilist epistemological approach. Therefore, it may be worth briefly stressing some of the differences that exist between the analytic view and other fallibilist perspectives—namely, 'Popperian' and 'Lakatosian' approaches—which certainly have been very influential in the development of the analytic view.

First of all, the analytic view shares with Popper the idea that knowledge is essentially problem solving, but, unlike Popper, it does not separate the context of discovery from that of justification, nor it denies the relevance of the context of discovery. And being especially focused on the context of discovery makes the analytic view well suited to figure as a component of a naturalist account of the process of knowledge ampliation, because it does not make of this crucial aspect of knowledge acquisition a mystery.[23]

Second, according to the analytic view, in order to solve problems, hypotheses are produced by non-deductive inferences, so logic is essentially a logic of discovery. Thus, in this perspective, contrary to Popper's view, logic has not to be regarded as a merely deductive enterprise, and the non-deductive ampliative rules are considered to be legitimate parts of logic as well as the deductive and non-ampliative ones.[24] Indeed, the analytic method "is a logical method" and from the fact that "knowledge is the result of solving problems by the analytic method, it follows that logic provides means to acquire knowledge" (Cellucci, 2013, 284).

It is worth noticing that claiming that the analytic method is a *logic* of discovery does not mean to deny the relevance of unconscious processes in scientific discovery. Indeed, "in the analytic method, some non-deductive inferences by which hypotheses are obtained may be unconscious"

---

23 See, e.g., Lakatos (1978, vol. 1, 140): "For Popper the logic of discovery [. . .] consists merely of a set of [. . .] rules for the appraisal of ready articulated theories".

24 On the fact that, since there is no non-circular justification of the validity of deductive inferences rules, nor is there an adequate justification of the acceptability of circular justifications, non-deductive rules and deductive rules are on a par with respect to the issue of the justification of their validity, see Cellucci (2006). See also Haack (1976) and Carroll (1895).

(Ibidem, 235). So, according to the analytic view, unconsciously formulated hypotheses should not be regarded as irrational. Moreover, the analytic view denies that a logic of discovery cannot exist because the processes involved in the discovery are purely subjective and psychological. Indeed, in this view, the non-deductive rules by which the analytic method "is implemented [. . .] are as objective as the deductive rules of a logic of justification" (Ibidem, 289).

Coming to 'Lakatosian' approaches, although these approaches and the analytic view both stress the role of heuristic in the ampliation of knowledge and strongly criticize the axiomatic view,[25] they are distinct positions. The main difference between those positions can probably be identified by analyzing their views on the issue of truth: while Lakatos maintains that the aim of science is truth, and truth is correspondence,[26] the analytic view replaces truth with plausibility for the reasons outlined earlier.[27]

4.4 The Analytic Method and Infinitism

The analytic view aims at avoiding both a) skepticism, since, as we have already stressed earlier, it takes knowledge not only to be possible, but even necessary to the survival, and b) foundationalism, given that it takes knowledge to be a potentially infinite process. To better clarify this point, it may be useful to point out some of the differences between the analytic view and infinitism—an epistemological position that is usually credited by its supporters to be similarly able to avoid both skepticism and foundationalism (Turri & Klein, 2014).

The analytic view and infinitism agree that the fact that knowledge acquisition may be a potential infinite process does not prevent us to consider genuine knowledge the portion of knowledge we reached so far. Nevertheless, they are distinct positions. Infinitism usually retains the relation between the concept of knowledge and the concept of truth, and thus has to face several difficulties. For example, Cling (2004) underlines that if knowledge has to be related to truth, it is not sufficient to consider infinite patterns of reasons as acceptable in order to claim that we have knowledge, we should also be able to distinguish those infinite patterns that allow us to reach the truth from those which don't. But to do that, we should be able to identify some properties which characterize true reasons. This would amount to already having a criterion of truth and would undermine the infinitist attempt to avoid the challenge of the criterion of truth exactly by allowing the infinite

---

25 Lakatos is considered to be the initiator of the 'maverick' tradition in the philosophy of mathematics (Kitcher & Aspray, 1988, 17). Relying on the work of Pólya, he strongly criticized the occultation of the heuristic elements that proved crucial in developing mathematics. See Lakatos (1976).

26 See Lakatos (1978).

27 There are several other differences between Lakatos's and Cellucci's views of heuristic, which cannot be discussed here for reason of space. See Cellucci (forthcoming, 2013).

regress of reasons. Thus infinitism *per se* seems unable to solve the skeptical problem of the criterion and the related problem of the infinite regress.

On the contrary, the analytic view can safely maintain that knowledge acquisition is a potentially infinite process and that the portion of knowledge we have produced so far is genuine knowledge, since the analytic view conceives of knowledge as plausible and provisional:

> If the series of the premisses is infinite, there will be no immediately justified premisses, so no knowledge will be definitive. But this does not mean that there will be no knowledge. There would be no knowledge only if the premisses, or hypotheses, occurring in the infinite series were arbitrary. But they need not be arbitrary [. . .] they must be plausible, namely the arguments for them must be stronger than the arguments against them, on the basis of the existing knowledge. If the hypotheses are plausible, then there will be knowledge, albeit provisional knowledge [. . .] since new data may always emerge.
>
> (Cellucci *forthcoming*, sec. 3.2)

It is worth underlining here that 'plausibility' has not to be confused with 'probability'. Indeed, as Kant points out, "plausibility is concerned with whether, in the cognition, there are more grounds for the thing than against it" (Kant, 1992, 331), while probability measures the relation between the winning cases and possible cases. Plausibility involves a comparison between the arguments for and the arguments against, so it is not a mathematical concept. Conversely, probability is a mathematical concept.

5 Conclusion

This chapter tried to suggest that, since it seems that the traditional view of mathematics cannot be naturalized, if we wish to maintain a naturalist stance, a promising way to account for mathematics and mathematical knowledge may be to take the method of mathematics to be the analytic method.

It has been argued that it is impossible to naturalize mathematics without challenging at least some crucial aspects of the traditional view. Nor does it seem possible to keep maintaining that the method of mathematics is the axiomatic method and mathematical knowledge is certain, if we dismiss that view.

The analytic view has been illustrated in some detail. To make the proposal of taking the method of mathematics, natural science, and philosophy to be the analytic method more intelligible, the differences between the analytic view and some related epistemological positions have been pointed out.

Certainly, the analytic view comes with a cost: it forces us to rethink the 'traditional image' of mathematics. Indeed, if we take the method of mathematics to be the analytic method, mathematical knowledge cannot be said to be certain, and the only kind of mathematical knowledge that we can have is knowledge, which is plausible. But also the alternative option of maintaining the traditional view comes with a cost for the naturalist: she is unable to scientifically account for mathematics, while she maintains the primacy of a thorough mathematized science in her worldview. We think that the analytic view may represent a promising route to take for the naturalist, and we tried to show that it is worthy of further investigation.

References

Amico, R. P. (1993). *The problem of the criterion*. Lanham, MD: Rowman & Littlefield.

Auslander, J. (2008). On the roles of proof in mathematics. In B. Gold & R. Simons (Eds.), *Proof and other dilemmas: Mathematics and philosophy* (pp. 62–77). Washington: The Mathematical Association of America.

Baker, A. (2016). Non-deductive methods in mathematics. In E. N. Zalta (Ed.), *The Stanford encyclopedia of philosophy*. Retrieved from https://plato.stanford.edu/archives/win2016/entries/mathematics-nondeductive/

Balaguer, M. (2009). Realism and anti-realism in mathematics. In D. Gabbay, P. Thagard, & J. Woods (Eds.), *Handbook of the philosophy of science* (Vol. 4, pp. 117–151). Amsterdam: Elsevier.

Baron, S. (2015). Mathematical explanation and epistemology: Please mind the gap. *Ratio, 29*, 14–167.

Benacerraf, P. (1973). Mathematical truth. *The Journal of Philosophy, 70*, 661–679.

Brown, J. R. (2012). *Platonism, naturalism, and mathematical knowledge*. New York: Routledge.

Byers, W. (2007). *How mathematicians think*. Princeton: Princeton University Press.

Carroll, L. (1895). What the Tortoise said to Achilles. *Mind, 4*, 278–280.

Cellucci, C. (2006). The question Hume didn't ask: Why should we accept deductive inferences? In C. Cellucci & P. Pecere (Eds.), *Demonstrative and nondemonstrative reasoning* (pp. 207–235). Cassino: Edizioni dell'Università degli Studi di Cassino.

Cellucci, C. (2013). *Rethinking logic: Logic in relation to mathematics, evolution, and method*. Dordrecht: Springer.

Cellucci, C. (2014a). Knowledge, truth and plausibility. *Axiomathes, 24*, 517–532.

Cellucci, C. (2014b). Rethinking philosophy. *Philosophia, 42*, 271–288.

Cellucci, C. (2015). Rethinking knowledge. *Metaphilosophy, 46*, 213–234.

Cellucci, C. (2016). Models of science and models in science. In E. Ippoliti, F. Sterpetti, & T. Nickles (Eds.), *Models and inferences in science* (pp. 95–122). Cham: Springer.

Cellucci, C. (2017). Is mathematics problem solving or theorem proving? *Foundations of Science, 22*, 183–199.

Cellucci, C. (forthcoming). *Rethinking knowledge: The heuristic view*. Dordrecht: Springer.

Clark, K. J. (2016). Naturalism and its discontents. In K. J. Clark (Ed.), *The blackwell companion to naturalism* (pp. 1–15). Oxford: Blackwell.

Cling, A. D. (1997). Epistemic levels and the problem of the criterion. *Philosophical Studies, 88*, 109–140.

Cling, A. D. (2004). The trouble with infinitism. *Synthese, 138*, 101–123.

Cohen Kadosh, R., & Dowker, A. (Eds.). (2015). *The Oxford handbook of numerical cognition*. Oxford: Oxford University Press.

Colyvan, M. (2015). Indispensability arguments in the philosophy of mathematics. In E. N. Zalta (Ed.), *The Stanford encyclopedia of philosophy*. Retrieved from https://plato.stanford.edu/archives/spr2015/entries/mathphil-indis/

David, M. (2016). Anti-realism. *Disputatio, 8*, 173–185.

De Cruz, H. (2006). Towards a darwinian approach to mathematics. *Foundations of Science, 11*, 157–196.

De Cruz, H. (2016). Numerical cognition and mathematical realism. *Philosophers' Imprint, 16*, 1–13.

Dehaene, S., & Brannon, E. M. (Eds.). (2011). *Space, time and number in the brain*. Amsterdam: Elsevier.

Dehaene, S., Duhamel, J.-R., Hauser, M. D., & Rizzolatti, G. (Eds.). (2005). *From monkey brain to human brain*. Cambridge, MA: MIT Press.

Detlefsen, M. (2011). Discovery, invention and realism: Gödel and others on the reality of concepts. In J. Polkinghorne (Ed.), *Meaning in mathematics* (pp. 73–94). Oxford: Oxford University Press.

Deutsch, D. (1997). *The fabric of reality*. New York: Penguin Books.

Feferman, S. (1998). *In the light of logic*. Oxford: Oxford University Press.

Field, H. (1989). *Realism, mathematics and modality*. Oxford: Blackwell.

Franks, J. (1989). Review of J. Gleick, Chaos: Making a new science. *The Mathematical Intelligencer, 11*, 65–69.

Frege, G. (1984). *Collected papers on mathematics, logic, and philosophy*. Oxford: Blackwell.

George, A., & Velleman, D. J. (2002). *Philosophies of mathematics*. Malden: Blackwell.

Gödel, K. (1947). What is Cantor's continuum problem? In K. Gödel (1990), *Kurt Gödel: Collected works* (Vol. 2, pp. 176–187). Oxford: Oxford University Press.

Gödel, K. (1951). Some basic theorems on the foundations of mathematics and their implications. In: K. Gödel (1995), *Kurt Gödel: Collected works* (Vol. 3, pp. 304–323). Oxford: Oxford University Press.

Gowers, T. (2006). Does mathematics need a philosophy? In R. Hersh (Ed.), *18 unconventional essays on the nature of mathematics* (pp. 182–200). New York: Springer.

Haack, S. (1976). The justification of deduction. *Mind, 85*, 112–119.

Haack, S. (1996). *Deviant logic, fuzzy logic*. Chicago: The University of Chicago Press.

Hamming, R. W. (1980). The unreasonable effectiveness of mathematics. *The American Mathematical Monthly, 87*, 81–90.

Hintikka, J., & Remes, U. (1974). *The method of analysis*. Dordrecht: Reidel.

Horsten, L. (2015). Philosophy of mathematics. In E. N. Zalta (Ed.), *The Stanford encyclopedia of philosophy*, Retrieved from http://plato.stanford.edu/archives/spr2015/entries/philosophy-mathematics/

Kant, I. (1992). *Lectures on logic*. Cambridge: Cambridge University Press.

Kitcher, P. (1988). Mathematical naturalism. In: W. Aspray & P. Kitcher (Eds.), *Minnesota studies in the philosophy of science* (Vol. 11, pp. 293–325). Minneapolis: University of Minnesota Press.

Kitcher, P., & Aspray, W. (1988). An opinionated introduction. In W. Aspray & P. Kitcher (Eds.), *Minnesota studies in the philosophy of science* (Vol. 11, pp. 3–57). Minneapolis: University of Minnesota Press.

Kornblith, H. (2002). *Knowledge and its place in nature*. Oxford: Oxford University Press.

Krebs, N. (2011). Our best shot at truth: Why humans evolved mathematical abilities. In U. J. Frey, C. Störmer, & K. P. Willführ (Eds.), *Essential building blocks of human nature* (pp. 123–141). Dordrecht: Springer.

Kyburg, H. (1965). Comments on Salmon's 'inductive evidence'. *American Philosophical Quarterly*, 2, 274–276.

Lakatos, I. (1976). *Proofs and refutations*. Cambridge: Cambridge University Press.

Lakatos, I. (1978). *Philosophical papers* (2 vol.). Cambridge: Cambridge University Press.

Leng, M. (2007). Introduction. In M. Leng, A. Paseau, & M. Potter (Eds.), *Mathematical knowledge* (pp. 1–15). Oxford: Oxford University Press.

McEvoy, M. (2004). Is reliabilism compatible with mathematical knowledge? *The Philosophical Forum*, 35, 423–437.

McKay, R. T., & Dennett, D. C. (2009). The evolution of misbelief. *Behavioral and Brain Sciences*, 32, 493–510.

Menn, S. (2002). Plato and the method of analysis. *Phronesis*, 47, 193–223.

Nozick, R. (2001). *Invariances*. Cambridge, MA: Harvard University Press.

Núñez, R. (2008). Mathematics, the ultimate challenge to embodiment: Truth and the grounding of axiomatic systems. In P. Calvo & T. Gomila (Eds.), *Handbook of cognitive science: An embodied approach* (pp. 333–353). Amsterdam: Elsevier.

Núñez, R. (2009). Numbers and arithmetic: Neither hardwired nor out there. *Biological Theory*, 4, 68–83.

Papineau, D. (2016). Naturalism. In E. N. Zalta (Ed.), *The Stanford encyclopedia of philosophy*, Retrieved from https://plato.stanford.edu/archives/win2016/entries/naturalism/

Plotkin, H. (1997). *Darwin machines and the nature of knowledge*. Cambridge, MA: Harvard University Press.

Poincaré, H. (2015). *The foundations of science*. Cambridge: Cambridge University Press.

Polkinghorne, J. (2011). Mathematical reality. In J. Polkinghorne (Ed.), *Meaning in mathematics* (pp. 27–34). Oxford: Oxford University Press.

Pólya, G. (1954). *Mathematics and plausible reasoning*. Princeton: Princeton University Press.

Rodin, A. (2014). *Axiomatic method and category theory*. Berlin: Springer.

Rosenberg, A. (1996). A field guide to recent species of naturalism. *The British Journal for the Philosophy of Science*, 47, 1–29.

Sage, J. (2004). Truth-reliability and the evolution of human cognitive faculties. *Philosophical Studies*, 117, 95–106.

Sankey, H. (2008). *Scientific realism and the rationality of science*. Aldershot: Ashgate.

Schechter, J. (2013). Could evolution explain our reliability about logic? In T. S. Gendler & J. Hawthorne (Eds.), *Oxford studies in epistemology* (Vol. 4, pp. 214–239). Oxford: Oxford University Press.

Schickore, J. (2014). Scientific discovery. In E. N. Zalta (Ed.), *The Stanford encyclopedia of philosophy*. Retrieved from http://plato.stanford.edu/archives/spr2014/entries/scientific-discovery/

Shapiro, S. (1997). *Philosophy of mathematics: Structure and ontology*. Oxford: Oxford University Press.

Sklar, L. (1981). Do unborn hypotheses have rights? *Pacific Philosophical Quarterly*, 62, 17–29.

Smith, J. M. (2012). Evolution and logic. In P. Dybjer, S. Lindström, E. Palmgren, & G. Sundholm (Eds.), *Epistemology versus ontology* (pp. 129–138). Dordrecht: Springer.

Spelke, E. S. (2011). Natural number and natural geometry. In S. Dehaene, E. M. Brannon (Eds.), *Space, time and number in the brain* (pp. 287–317). Amsterdam: Elsevier.

Stanford, P. K. (2006). *Exceeding our grasp*. New York: Oxford University Press.

Turri, J., & Klein, P. (2014). Introduction. In J. Turri & P. Klein (Eds.), *Ad infinitum: New essays on epistemological infinitism* (pp. 1–17). Oxford: Oxford University Press.

Vaidya, A. (2016). The epistemology of modality. In E. N. Zalta (Ed.), *The Stanford encyclopedia of philosophy*. Retrieved from https://plato.stanford.edu/archives/win2016/entries/modality-epistemology/

Vlerick, M., & Broadbent, A. (2015). Evolution and epistemic justification. *Dialectica, 69*, 185–203.

Weir, A. (2005). Naturalism reconsidered. In S. Shapiro (Ed.), *The Oxford handbook of philosophy of mathematics and logic* (pp. 460–482). Oxford: Oxford University Press.

Wilkins, J. S., & Griffiths, P. E. (2013). Evolutionary debunking arguments in three domains: Fact, value, and religion. In J. Maclaurin, & G. Dawes (Eds.), *A new science of religion* (pp. 133–146). New York: Routledge.

Williamson, T. (2000). *Knowledge and its limits*. Oxford: Oxford University Press.

Woleński, J. (2012). Naturalism and genesis of logic. *Studies in Logic, Grammar and Rhetoric, 27*, 223–240.

Ye, F. (2011). *Strict finitism and the logic of mathematical applications*. Springer: Dordrecht.

Contributors

Sorin Bangu is Professor in the Department of Philosophy, University of Bergen, Norway, and previously held appointments at University of Illinois at Urbana-Champaign and University of Cambridge HPS. He works in philosophy of mathematics and philosophy of physics, having longstanding interests in epistemology and the history of analytic philosophy. He is the author of *The Applicability of Mathematics in Science: Indispensability and Ontology* (Palgrave Macmillan, 2012).

Helen De Cruz is Senior Lecturer in philosophy at Oxford Brookes University. She investigates how we as human beings—embodied, socially and materially embedded, cognitively limited—acquire beliefs about subjects that seem far removed from our everyday experience, such as in mathematics, science, and theology. She has co-authored (with Johan De Smedt) *A Natural History of Natural Theology* (MIT Press, 2015) and co-edited (with Ryan Nichols) *Advances in Religion, Cognitive Science, and Experimental Philosophy* (Bloomsbury Academic, 2016). Currently, she is writing a book on religious disagreement with Cambridge University Press.

Mark Fedyk is an Associate Professor in the Department of Philosophy at Mount Allison University and a Research Associate with the Ottawa Hospital Research Institute. His research intersects with topics in cognitive science, value theory, and social sciences. His recent book, *The Social Turn in Moral Psychology* (MIT Press), combines findings from psychology and different social sciences to propose new ways of conducting research in both ethics and scientific moral psychology.

Max Jones is a Postdoctoral Research Fellow in philosophy at the University of Leeds. His research primarily focuses on the implications of recent developments in the cognitive sciences for traditional philosophical issues in metaphysics, philosophy of mathematics, philosophy of perception, and epistemology. His PhD research project, conducted at the University of Bristol, examined the impact of contemporary evidence about numerical cognition on the debate about the so-called access problem in the philosophy of mathematics.

Penelope Maddy is Distinguished Professor of Logic and Philosophy of Science at the University of California, Irvine. Her writings include *Realism in Mathematics* (1990), *Naturalism in Mathematics* (1997) (winner of the 2002 Lakatos award), *Second Philosophy* (2007), *Defending the Axioms* (2011), *The Logical Must. Wittgenstein on Logic* (2014) and *What Do Philosophers Do? Skepticism and the Practice of Philosophy* (2017).

Erik Myin is Professor of Philosophy at the University of Antwerp and director of the Centre for Philosophical Psychology. He has published papers on mind and cognition, sometimes in collaboration with scientists or with other philosophers. With Dan Hutto, he wrote *Radicalizing Enactivism: Basic Minds without Content* (2013) and *Evolving Enactivism: Basic Minds Meet Content* (2017), both published by MIT Press.

Rafael Núñez is a Professor of Cognitive Science at the University of California, San Diego. He grew up in Chile, obtained his doctoral degree in Switzerland, and completed his postdoctoral work at Stanford and UC Berkeley. He investigates abstraction and conceptual systems from the perspective of the embodied mind. His multidisciplinary approach uses methods drawn from experimental psychology, cognitive linguistics, gesture studies, brain imaging, human history, and fieldwork with isolated indigenous groups. His book *Where Mathematics Comes From* (with George Lakoff) presents a new theoretical framework for understanding the human nature of mathematics and its foundations.

Jean-Charles Pelland holds degrees in philosophy from McGill University and Université du Québec à Montréal. He was a Researcher for the Canada Research Chair in the Philosophy of Logic and Mathematics and for the *Existence et connaissance des objets mathématiques* interuniversity research group. He also held a visiting PhD fellowship at the Institute of Philosophy (London). His research interests include cognitive science, philosophy of mind, and philosophy of mathematics. His current work as a postdoctoral research fellow at New College of the Humanities (London) focuses on numerical cognition and its implications for psychologism, antirealism, and intuitionism.

Rolf Reber is Professor of Cognitive Psychology at the University of Oslo, Norway. The overarching theme of his research is feeling. With his colleagues, he developed the processing fluency theory of aesthetic judgments and extended it to shared tastes in art, rituals, mathematical intuition, and

Aha-experiences. A second line of research has explored interventions to increase interest in mathematics at school. The two strands of research culminated in his recent book *Critical Feeling*, published by Cambridge University Press.

Josephine Relaford-Doyle is a PhD student in the department of Cognitive Science at the University of California, San Diego. She received her BA in Cognitive Science from UC Berkeley, and prior to beginning her research career she spent five years teaching high school mathematics as a Math for America Los Angeles fellow. Josephine is interested in mathematical certainty and justification, and more generally in the extent to which conceptualizations of mathematical objects match their formal characterizations. Her work is often inspired by questions in the philosophy of mathematics, which she seeks to inform using empirical methods.

Paul D. Robinson is a PhD student in philosophy, with an interdisciplinary specialization in cognitive and brain sciences, at the Ohio State University. He is currently completing his dissertation on the "social turn" in theories about the nature and function of reasoning.

Richard Samuels is Professor of Philosophy and Cognitive Science at Ohio State University, and previously held appointments at King's College London and the University of Pennsylvania. He has published numerous papers on topics in the philosophy of psychology and foundations of cognitive science.

Dirk Schlimm is Associate Professor in the Department of Philosophy at McGill University. His research interests fall into the areas of history and philosophy of mathematics and science, epistemology, and cognitive science. In particular, he is interested in the developments in the 19th and early 20th century that led to the emergence of modern mathematics and logic, and in systematic investigations regarding axiomatics, empiricism, and analogical reasoning, as well as the use of external representations such as numerals and notations. He is also involved in editorial projects of the works of Pasch, Hilbert, Bernays, and Carnap.

Fabio Sterpetti is currently a Postdoctoral Research Fellow at Campus Bio-Medico University of Rome and contract professor in philosophy of science at the Department of Philosophy, Sapienza University of Rome. His main research areas are general philosophy of science, philosophy of biology, and philosophy of mathematics. He is also interested in epistemology and metaphilosophy. From 2013 to 2015, he was a postdoctoral research fellow at Sapienza University of Rome, where he also received his PhD in philosophy in 2012. He recently co-edited, with Emiliano Ippoliti and Thomas Nickles, *Models and Inferences in Science* (Springer, 2016).

Kristy vanMarle received her PhD in developmental psychology in 2004 from Yale University. After being a postdoctoral associate at Rutgers Center for Cognitive Science, she joined the faculty at the University of Missouri–Columbia in 2007, where she is currently an Associate Professor of Psychological Sciences. The central aim of her research is to understand the origins and development of quantitative concepts, such as number, space, and time, and how early emerging, nonverbal quantity representations are related to the acquisition of formal, symbolic mathematical skills in early childhood (e.g., learning the verbal count list, abstract concepts such as zero).

Karen Wynn is Professor of Psychology and Cognitive Science at Yale University. She obtained her BA at McGill University and a PhD in cognitive science at MIT. Throughout her scientific career, she has been investigating the fundamental nature of the human mind by studying infants, who provide a unique glimpse into the mind as it exists before the imprintings of language, culture, and formal education. Her current research explores the nature of human morality by examining babies' moral judgments and also their prejudices. Previous work has included pioneering research revealing numerical understanding and competence in human infants. She is a fellow of the Association for Psychological Science and has received numerous awards in recognition of her research, including a Troland Research Award from the National Academy of Sciences and a Distinguished Scientific Award for Early Career Contribution to Psychology from the American Psychological Association.

Byeong-uk Yi is Professor of Philosophy at the University of Toronto and Kyung Hee International Scholar at Kyung Hee University. He studied at Seoul National University, the University of Pittsburgh, and UCLA, where he received a PhD in philosophy, and held academic appointments at University of Alberta, University of Queensland, University of Glasgow, and University of Minnesota. He published a book, *Understanding the Many*, and articles in many areas in logic and philosophy, including logic, metaphysics, philosophy of language, philosophy of mathematics, and semantics of classifier languages.

Karim Zahidi is Lecturer and Postdoctoral Fellow at the Centre of Philosophical Psychology at the University of Antwerp. He has published papers in the philosophy of mind, logic, and philosophy of mathematics.

Index

Page numbers in italic indicate figures and in bold indicate tables on the corresponding pages.

Printed in the United States
by Baker & Taylor Publisher Services